Energetics of the Climate System

SCIENCES

Climate and Atmosphere, Field Director – Denis-Didier Rousseau

Climate Systems, Subject Head – Denis-Didier Rousseau

Energetics of the Climate System

Hervé Le Treut
Mounia Mostefaoui

iSTE

WILEY

First published 2025 in Great Britain and the United States by ISTE Ltd and John Wiley & Sons, Inc.

ISTE Ltd
27-37 St George's Road
London SW19 4EU
UK

www.iste.co.uk

John Wiley & Sons, Inc.
111 River Street
Hoboken, NJ 07030
USA

www.wiley.com

Library of Congress Control Number: 2025937569

British Library Cataloguing-in-Publication Data
A CIP record for this book is available from the British Library
ISBN 978-1-78945-230-3

ERC code:
PE10 Earth System Science
PE10_1 Atmospheric chemistry, atmospheric composition, air pollution
PE10_3 Climatology and climate change
SH7 Human Mobility, Environment, and Space
SH7_6 Environmental and climate change, societal impact and policy

The manufacturer's authorized representative according to the EU General Product Safety Regulation is Wiley-VCH GmbH, Boschstr. 12, 69469 Weinheim, Germany, e-mail: Product_Safety@wiley.com.

Contents

Preface

This book is based on a course that Hervé Le Treut has been giving for decades to students in the second year of a master's degree, from the École Normale Supérieure, the École Polytechnique and the Sorbonne University, all of which are part of the Institut Pierre Simon Laplace, as well as to students from many other French and international institutions. It delivers a solid grasp of the complexity related to the energetics of the Earth climate system. Our aim was also to try to inject the voice and pedagogical talent of Hervé Le Treut in clarifying such complexity, and to deliver a tribute to his unique storytelling style. It is also a tribute to this world-known savant and humanist who has trained generations of climatologists, including the co-author, as well as engineers and economists. The style of this book deliberately maintains some oral-style formula in order to let his voice be heard.

This pedagogical opus is organized in five chapters quite independent but intimately complementary, with an appendix with exercises.

The aim of the first three, science-oriented chapters is to deliver a presentation of the natural energetic phenomena governing the Earth. The main constraints regulating the Earth climate coming from the laws of energy conservation and the rotation of the Earth at different time and space scales are described as introductory material in Chapter 1. The vertical dimension of the energy processes impacting the climate system is then presented through the description of the radiative and radiative–convective phenomena at play in Chapter 2. Furthermore, the horizontal motion of the atmosphere and of the ocean is discussed in Chapter 3 as a key phenomenon to balance the very differentiated distribution of solar energy flux between low latitudes, where it enters in the Earth system, and high latitudes, where it leaves it.

Energetics of the Climate System,
by Hervé LE TREUT and Mounia MOSTEFAOUI. © ISTE Ltd 2025.

The last two chapters of this pedagogical book (Chapters 4 and 5) are focused on the anthropogenic forcing upon the Earth's energetic processes (as detailed in the first chapters) and its socio-political impacts. Chapter 4 discusses the transition from quasi-equilibrium to forcing from anthropogenic sources. Chapter 4 also provides a clear cartography of predictable phenomena, uncertainties and feedbacks related to the climate response to the anthropogenic disturbance.

Finally, a valuable overview of the role of scientific diagnostics for decision-making is offered in Chapter 5 by discussing the place of uncertainties on assessing global warming, accounting for natural variability and estimating climate impacts. The shift from global to local climate features is also highlighted as a key element in the interaction between science and the decision-making process after the Paris Agreement.

An appendix with past exams and element solutions related to the chapters of this book serve as a tool to evaluate the understanding of some main notions of the energetics of the climate system, especially for master-level students.

The two authors warmly thank Dr. Denis-Didier Rousseau for the invitation on this collection and for his very valuable support, as well as all of the ISTE team for their precious help.

June 2025

1

A Description of the Main Constraints Regulating the Earth Climate

1.1. Generalities about the atmosphere and the ocean

The general objective of this introductory material is to provide a description of the main constraints regulating the Earth climate, which come from the laws of energy conservation and from the rotation of the Earth, by moving from the global picture of the Earth to more detailed processes.

The word "climate" comes from the ancient Greek "Klima" which literally means "inclination of the Earth towards the pole". During the third century BCE, the polymath and former library of Alexandria's director, Eratosthenes of Cyrene, calculated with a negligible error the obliquity of the ecliptic, i.e. the tilt of the Earth axis comparing to its axis of rotation around the Sun. Eratosthenes' eclectic work, going from science to philosophy and poetry, is still partially available.

Therefore, the Ancient Greeks already knew that the Earth was a sphere and that it would be warmer near the equator and colder at the poles. From this, they also deduced the existence of another livable "unknown" world at mid-latitudes on the other side of the equator.

For a color version of all of the figures in this chapter, see www.iste.co.uk/letreut/energetics.zip.

Energetics of the Climate System,
by Hervé LE TREUT and Mounia MOSTEFAOUI.

Figure 1.1. *Eratosthenes teaching in Alexandria (Source: Bernardo Strozzi (1635))*

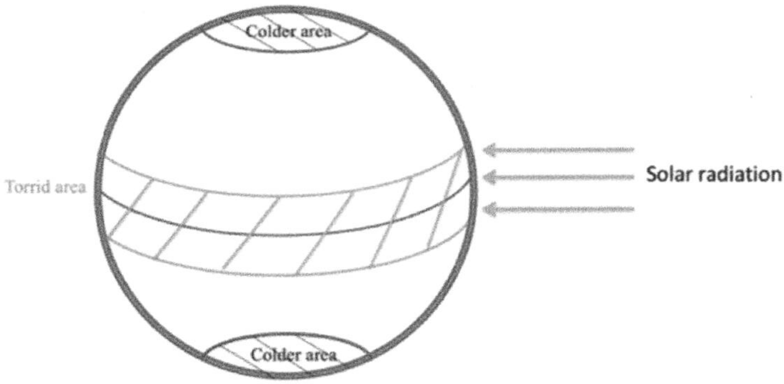

Figure 1.2. *A simplified scheme of the conception of the Earth by Eratosthenes of Cyrene*

Later, by 150 CE, Claudius Ptolemy wrote a summary of the state of knowledge in geography in his opus named *Geographia*.

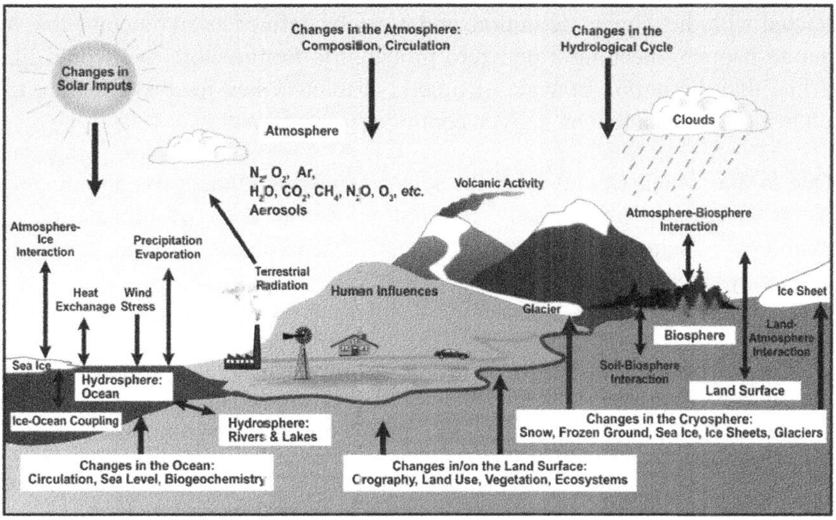

Figure 1.3. *Schematic view of the components of the climate system, their processes and interactions (Le Treut et al. 2007)*

Figure 1.3 represents a modern scheme of the ancient idea of a climate system organized in a large number of subcomponents and different "spheres" linked by different processes, among which the energy exchanges play an essential role.

"Atmo-sphere" stands for the sphere of the Earth, "hydro-sphere" stands for the sphere of water, "cryo-sphere" stands for the sphere of ice and "geo-sphere" stands for the sphere of the land beyond the surface that can be compressed. As for the word "biosphere", it was invented in the 20th century by the Ukrainian–Russian scientist Vladimir Vernadsky at Sorbonne University and stands for the sphere of life.

Figure 1.3 also illustrates the Earth complex system characterized by exchanges of energy, which represent the guiding thread of this book. No other types of exchanges such as water or momentum will be discussed in detail in this opus.

Beyond its etymological significance, we can define the word "climate" as the ensemble of statistics that describe the state of the climate system in a given location and over a given period of time (some decades). It is important to note that "statistical" does not only mean average. Moreover, the concept of risk is intimately associated with the climate definition, and it can be defined as an outcome that has a chance to happen, meaning a non-zero probability. Furthermore, we would like to underline that the notion of scale of time is absolutely key for this complex topic, which is highly dependent on a given contingency's context.

Due to the complexity at stake, the science studying climate has no one defined name. It is usually alternatively referred to as "physics of climate change", "climatology", "sciences of the environment", "geosciences", "earth, atmospheric and planetary sciences", etc.

1.1.1. *The atmosphere*

The Earth's atmosphere is 100 km thick, but 99% of its mass lies in its first 21 kilometers.

Figure 1.4. *A side view of the atmosphere from a balloon (Source: François Danis – LMD 2000)*

In this chapter, we will focus on the lower part of the stratosphere, below 20 km, where the atmosphere has a mass. In Figure 1.4, we can also distinguish the ionosphere layer.

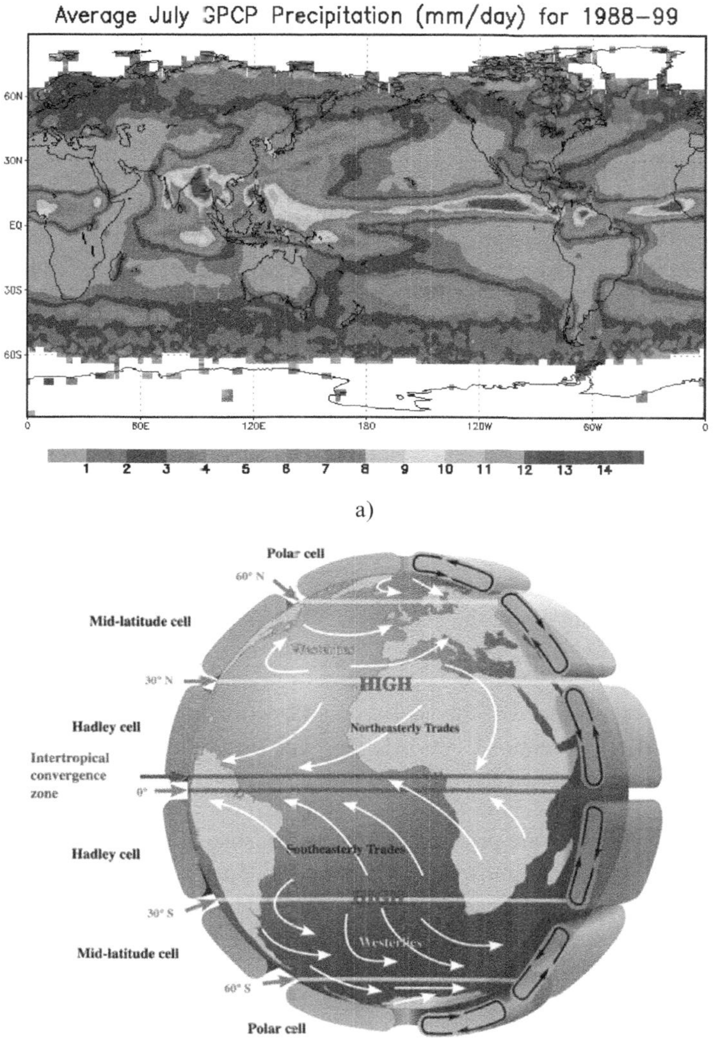

a)

b)

Figure 1.5. *a) Map of average July GCP precipitation for 1988–1999 (Source: World Climate Research Program); b) Scheme of Hadley cells*

In Figure 1.5, which shows a map of average July GCP precipitation in Bonn for 1988–1999, monthly means show an extremely organized atmospheric circulation, with a moving system from West to East. This is very different from the situation on other planets, with a complex organization. The scheme on the right in Figure 1.5 also shows Hadley cells that are defined below, with warm air climbing at the equator where they lose their water and descending cooler air, as well as a desert belt over continents.

The Hadley cells correspond to a relatively slow movement of the atmosphere in a mean meridian plane: with ascendance near the equator and descending air near the subtropical regions. On an annual mean, this circulation is fed by a convergence of air in the low layers around the equatorial region and a divergence higher in the atmosphere from this same area.

The fuel of this circulation comes from the more intense heating of equatorial regions compared to the other regions, getting more solar energy. This more intense heating creates a dilatation of the atmosphere at low latitudes, whose accumulated effect induces a gradient of pressure which is accentuated in the high layers of the atmosphere.

If we consider the atmosphere in its meridian plane, we can also diagnose this situation as a generator of vorticity because the gradients of pressure and density will not be aligned anymore: the air density varies not only with altitude, but also in latitude.

Despite the weak speed of this meridional flow (the winds are much more violent in the West–East direction), the Hadley cells play an essential role in the climatology of intertropical regions.

Ascending air zones are precipitating zones because of the very rapid decrease of the level of saturation of atmospheric water vapor upper in a troposphere, which gets colder with altitude. The seasonal movements of Hadley cells come together with the interplay between dry seasons or wet monsoons in the equatorial regions.

The Hadley cells play a main role in the climate system, as they transport the surplus of energy from equatorial regions to the Poles. The latitudes which limit the domain of Hadley cells in the North and in the South are therefore main indicators of the climate zones on the Earth surface.

Figure 1.6 is a representation of Hadley cells. On the y-axis on the left, the height is given in decibars; in the center, the lines are a representation of the mass of the troposphere; and the y-axis on the right, the vision is given in latitudes for three different scales of time: annual mean, winter (December/January/February) and summer (June/July/August). The horizontal lines indicate the latitudes.

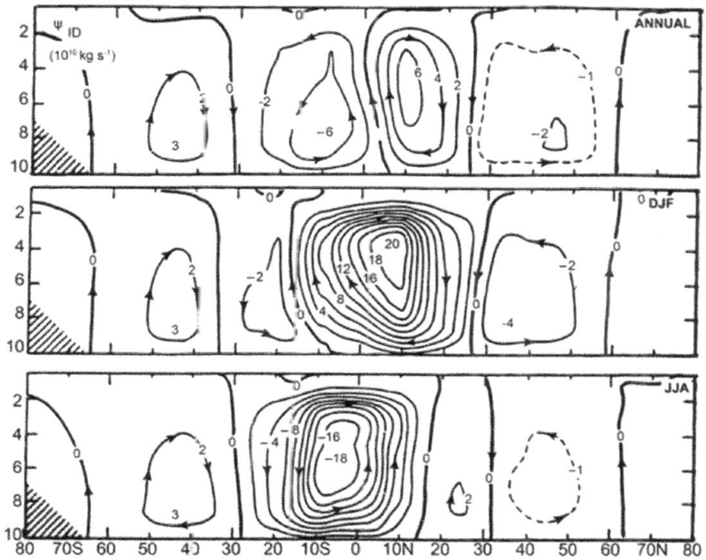

Figure 1.6. *A representation of Hadley cells (Source: Peixoto and Oort (1992))*

The stream function of meridian flow (flow averaged over longitudes and the time) is represented on a diagram of altitude/latitude. The abscissas are expressed in degrees of latitude. The lines of stream "iso-function" are by definition parallel to the flow. In the case of a stationary mean flow, we can consider that the flow is between the isolines. The values of stream functions are expressed in mass flow rate (the unity is 1010 kg/s): between two separated isolines of 1010 kg/s and 1010 kg of air every second. The diagram shows from the top to bottom in three situations: (1) the yearly mean of the flow, characterized by an approximate symmetry relatively to the equator; (2) the mean for boreal winter (December/January/February), where the main ascend is displaced from the equator toward the southern hemisphere; and (3) the mean for the boreal summer (June/July/August), where the main ascent is displaced from the equator toward the north hemisphere.

Figure 1.7 represents the atmospheric circulation viewed by the satellite radar TRMM, providing an instantaneous view of the Earth (synthesis image for demonstration by NASA). The radar echoes provoked by the presence of condensed water show ascending air. We can note, even on this instantaneous image corresponding to the month of July, an opposition between intertropical and extratropical regions. The intertropical zones are characterized by zones of ascending air essentially near the equator, and zones of air ascent more frequent approximately 30°N or 30°S. The extratropical regions are characterized by disturbances where the air ascends progressively all along “spirals” structures of several thousands of kilometers.

Figure 1.7. *Atmospheric circulation viewed by the satellite radar TRMM (Source: NASA Project TRMM)*

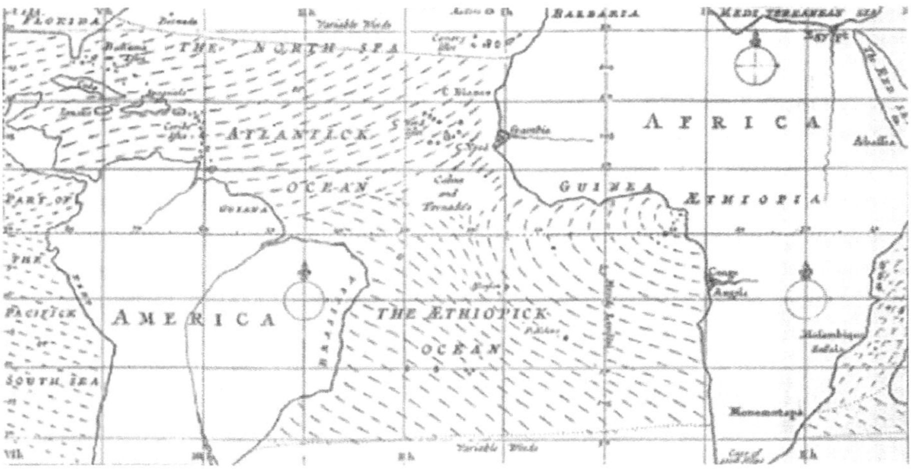

Figure 1.8. *Perhaps the world’s first thematic map: Edmund Halley’s map of trade winds (Source: Halley, Philosophical Transactions (1686))*

To mention some elements of history, part of this organization was already known by Halley (17th century) and Hadley (18th century). The map presented in Figure 1.8 was the result of important efforts at that time. Halley was the captain of a military boat; he was also a specialist in Earth Sciences and an astronomer well-known for comets. After his death, a comet came at the exact moment that he foresaw through his calculations.

1.1.2. *A brief comparison between some oceans and the atmosphere's main orders of magnitude*

Atmosphere	Oceans
Height $\approx$ 20 km	Depth $\approx$ 3.8 km
Average wind velocity $\approx$ 10 ms^{-1}	Average surface velocity $\approx$ 0.01 $m.s^{-1}$
Heat capacity: C_p = 1,004 J kg^{-1} K^{-1}	C = 4,180 J kg^{-1} K^{-1}
r = P /RT $\approx$ 1 kg m^{-3}	r = 10^3 kg m^{-3}
m (column) = 10^4 kg m^{-2}	m (column) $\approx$ 3.8 10^6 kg m^{-2}

Table 1.1. *Atmosphere and oceans: some figures*

For the atmosphere, the height of 20 km is a very small cover around the Earth. The heat capacity is linked to the mass; it represents the capacity of energy to absorb per kilogram to change the temperature by 1°C. We also note that the density of the atmosphere is not particularly light for a fluid: 1 $kg.m^{-3}$ is indeed not negligible.

The ocean is the timekeeper of the climate occurring now and stores about 90% of the greenhouse gases. Its depth (3.8 km) is much less than the radius of the Earth (6,371 km).

The atmosphere is responsible for the geography of the Earth. It transports very fast (with an average wind velocity of 10 $m.s^{-1}$) a lower amount of energy than the ocean, which, on the contrary, transports slowly, with an average surface velocity of 0.01 $m.s^{-1}$, a larger amount of energy.

1.2. A global view of radiative processes

Figure 1.9 shows the sunspots that characterize a situation where the emissions of the Sun are cooler on average, as well as the convective motion of the magma. Every day, a scientist scrutinizes the Sun to look for sunspots, and it has been the case for centuries from the Paris Observatory. The observation of the sunspots/changes in the solar radiation started under the impulsion of the French Academy of Sciences during the reign of Louis XIV, who was at the origins of the creation of the Paris Observatory in 1667.

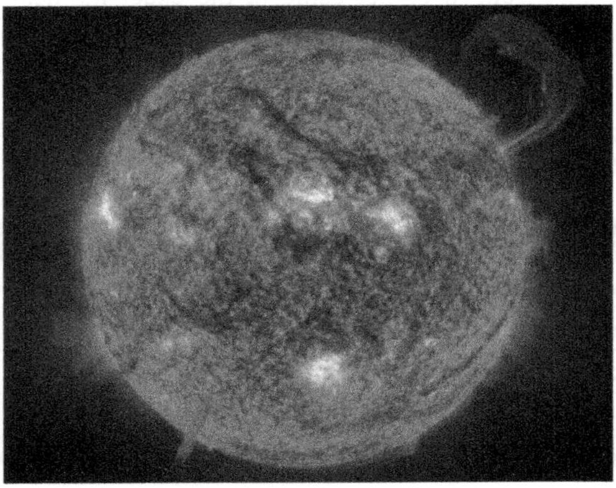

Figure 1.9. *The Sun. Source: nasa.gov.*

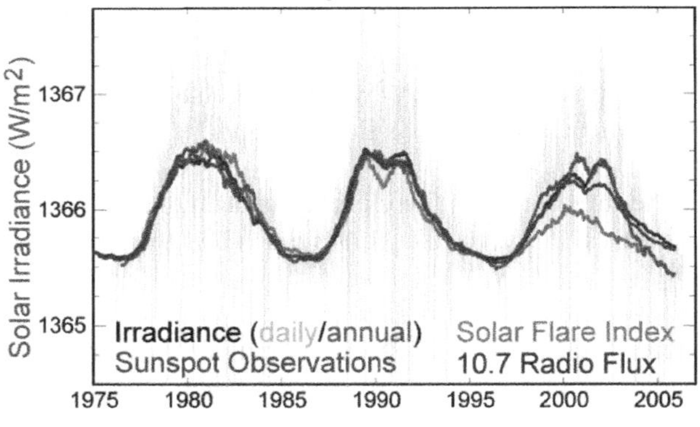

Figure 1.10. *Solar cycle variations and sunspot observation between 1975 and 2005 (Source: http://www.pmodwrc.ch/pmod.php?topic=tsi/composite/SolarConstant, International Sunspot number: http://www.ngdc.noaa.gov/stp/SOLAR/ftpSunspot number.html, Flare index: http://www.koeri.boun.edu.tr/astronomy/readme.html, 10.7 cm radio flux: http://www.drao-ofr.hia-iha.nrc-cnrc.gc.ca/icarus/www/sol_home.shtml, https://en.wikipedia.org/wiki/Solar_cycle)*

What we call the solar constant is actually not constant, and a satellite can measure the variations of solar irradiance. The internal motion in the Sun has an eleven-year cycle.

In Figure 1.10, the blue line corresponds to sunspot observations. The surface temperature of the Sun is 6,000 K. At 1,366 Wm^{-2}, we see what is captured by the Earth. The solar constant gives the amount of energy that crosses the Earth.

As for sunspot observations, there are moments during which the Sun surface is cooler and emits a bit less.

NOTE.– The Earth is visible from space because it reflects the light of the Sun.

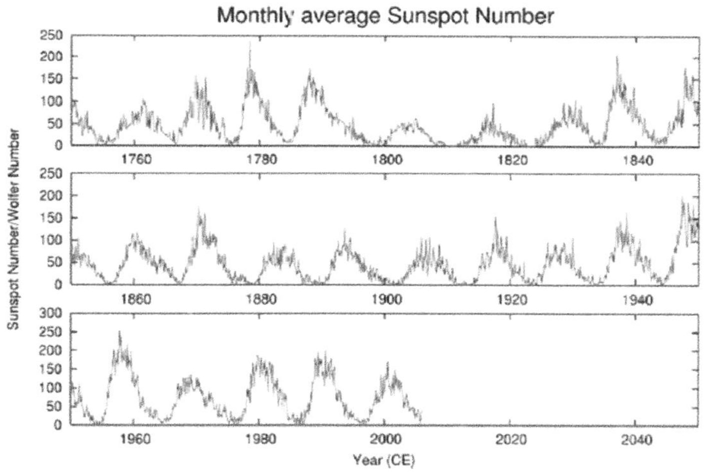

Figure 1.11. *Monthly average sunspot numbers observed since the 18th century. Source: this figure was produced by Leland McInnes using gnu plot (all data is from publicly available sources)*

Figure 1.11 shows monthly average sunspot numbers observed since the 18th century (they were measured then by looking at emission in the visible). On the upper scheme of Figure 1.11, between 1,800 and 1,820, we can see a tendency of reduction for the sunspot number: the Sun was a bit cooler due to changes in sunspots, but a volcanic eruption happening at the same time also makes things more complex to interpret.

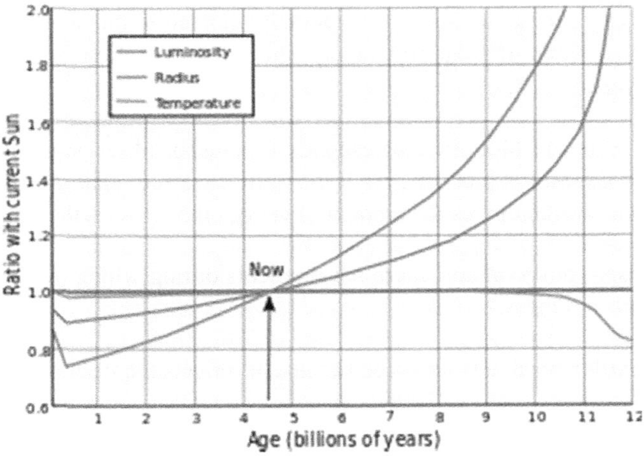

Figure 1.12. *The long term changes of the Sun "engine". Ratio with current Sun of solar luminosity, radius and temperature as a function of time (after Ribas 2010)*

Let us now consider another interesting phenomenon: 4.5 billion years ago, the Sun was younger; it had less energy. Would it have been possible to have an Earth covered with ice? We think maybe now. "The faint young Sun paradox" is the name sometimes given to the contradiction that we may find at first when considering the observations of water (in a liquid state) in the early history of our planet and the fact that the Sun's output would be 70% of the recent output at the time.

Figure 1.12 shows the luminosity, radius, and temperature depending on the age of the Sun expressed in billions of years. We can note that at around 4.5 billion years of age in the history of the Earth, the amount of radiation from the Sun increases with time.

We note a decrease of 5% in temperature from 10 to 12 billion years, while it has been stable over a period of 10 billion years.

NOTE.– A modification of solar insolation is enough to have an important impact and to disturb the system because of the form of the equilibrium. For instance, extinctions are a feature of that, a disequilibrium may raise from some cause, and if it is stable, it can have huge impacts.

Stable means, here to a first approximation, that a phenomenon lasts "long". The paradox of the young Sun asks how, if a planet was frozen, we could have gotten out of this state: one of the explanations is that, under the ice, there were submarine phenomena. We will come back to this later on.

The solar radiation is emitted in the ultraviolet (UV)/Visible/infrared (IR) domain (10–15 μ), while the terrestrial radiation is entirely in the IR domain (0–0.5 μm).

A blackbody absorbs all of the radiation it receives and re-emits it. This phenomenon can be expressed as a function of temperature:

$$E = \pi B = \sigma T^4 \qquad [1.1]$$

The term on the right side of this formalism that we call the Boltzmann equation represents the energy emitted per square meter summed up over all wavelengths; it is expressed in $W.m^{-2}$.

This phenomenon can be represented graphically as a function of wavelength in Figure 1.4, which shows the existence of two different domains of energy with the normalized blackbody curves.

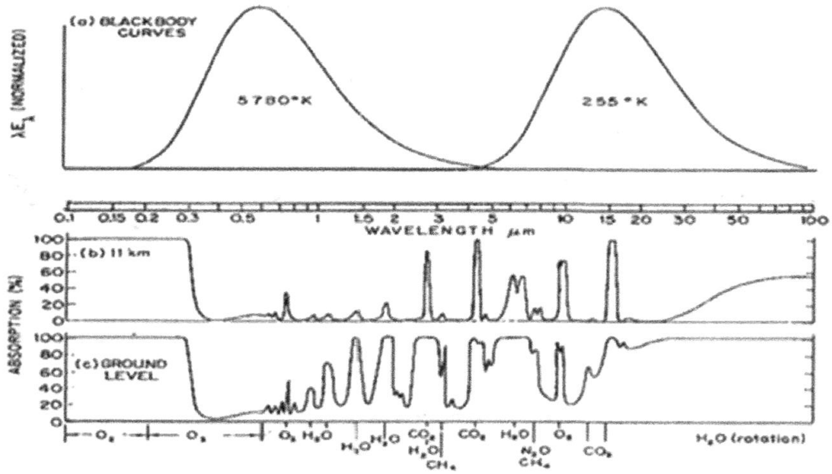

Figure 1.13. *Normalized blackbody curves for the Sun and for the Earth (Source: Brunetti and Prodi (2015)*

The Sun emits a radiation:

$$E_{Sun} = \sigma T_{Sun}^4 \qquad [1.2]$$

with T being the temperature of the Sun near the Surface, with σ = 5,67 10-8 W m^{-2} K^{-4}.

Neither the Sun nor the Earth are actually perfect black bodies, but we can do the approximation that the Earth emits a radiation that we can note:

$$E_{Earth} = \sigma T^4_{Earth} \quad [1.3]$$

On Earth, we receive only a small part of the incident solar radiation (caption of a very small part of the solar energy coming from the Sun). The method to make the measurements can be simplified as follows.

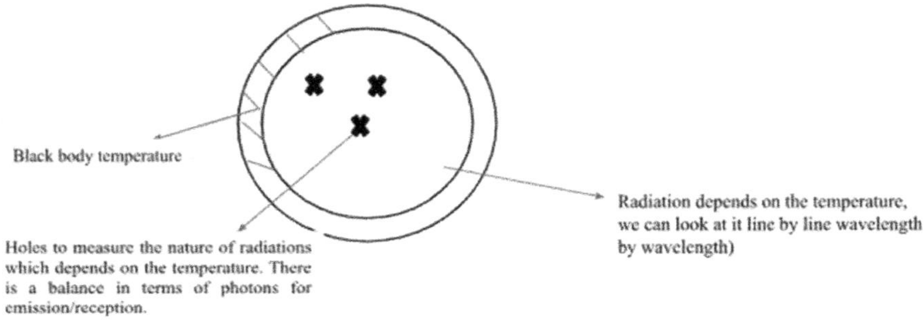

Figure 1.14. *Simplified scheme for the measurement of blackbody radiations*

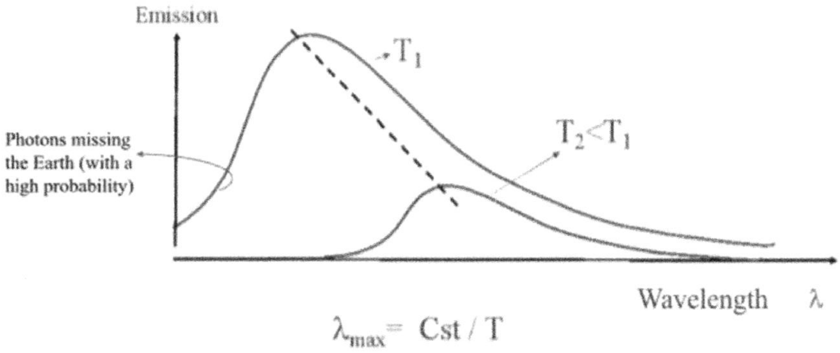

Figure 1.15. *Scheme comparing the emission of a blackbody as a function of wavelength for a given body of temperature T_1 and a cooler one with a temperature $T_2 < T_1$ (Source: adapted from Young and Freedman (2012))*

A blackbody behavior can be studied thanks to the simplified model of a black cavity with a hole and walls opaque to radiation. If the cavity is kept at a temperature fixed T, the radiation trapped inside the cavity is at thermal equilibrium with the enclosure.

Both the Earth and the Sun emit electromagnetic radiation in rather different wavelengths. Only a very small part of the solar energy is captured by the Earth, and the Earth emits energy in another domain at longer wavelengths. A radiative equilibrium is reached at different wavelengths. We can make a budget as they represent different sorts of radiations. Figure 1.15 shows the different shapes of emission curves for the energy emitted at different wavelengths. We note that for a cooler body, the maximum is a longer wavelength and the energy at a longer wavelength is less than the energy of a body that is a bit warmer. This is the principle for the Wien's displacement law.

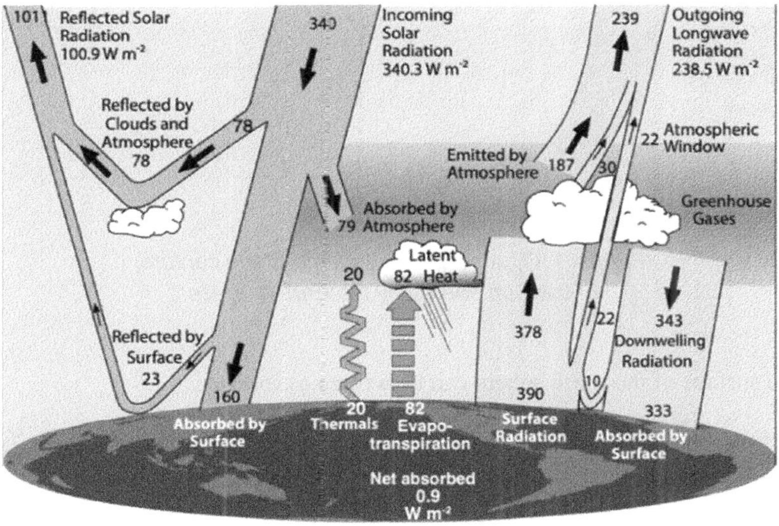

Figure 1.16. *Earth Global Radiation Budget (Trenberth and Cheng 2022)*

The Earth Global Radiation Budget depends on temperature. The unit of radiation is in W.m^{-2}. We can write the following equation:

Solar radiation incident = Reflected solar radiation + Radiation emitted in the infrared

$$340 = 101 + 239 \quad [1.4]$$

with:

Reflected solar radiation = Albedo + Outgoing reflected longwave radiation

$$107 = 77 + 30 \quad [1.5]$$

What arrives on Earth is not 1,366 $W.m^{-2}$ because the Earth is a sphere and the Earth surface that captures the solar energy is a disk; therefore, it is necessary to apply a division of this incoming energy by a factor of 4 as described in the following:

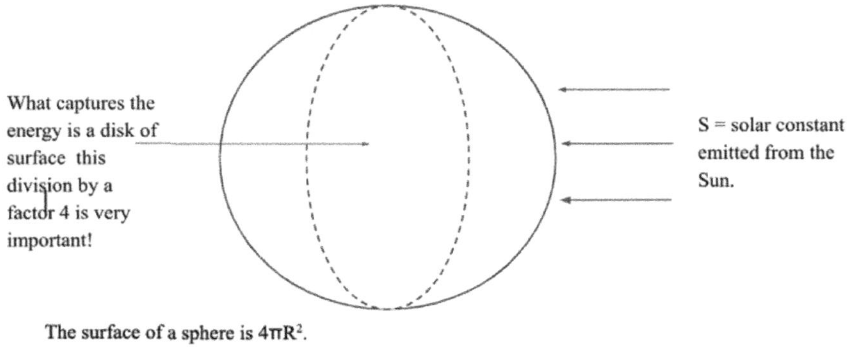

Figure 1.17. *A simplified scheme of the capture of the Sun energy by the Earth surface*

The surface of the Earth sphere can be written as follows:

$$S_{earth} = 4\pi R_{earth}^{2} \quad [1.6]$$

Therefore, if we note R_s, the solar incoming radiation captured at the Earth's surface, we have:

$$R_s = \frac{S_{earth}}{4} \quad [1.7]$$

We must be careful: the solar cycle is often misused by the climate-skeptical community members, who often forget this division by a factor of 4 parameter.

A simplified global equation can be written as follows:

$$C_t \frac{dT}{dt} = R_S (1 - \alpha) - R_t(T) \quad [1.8]$$

with T being the global earth temperature (of the largest heat storage component: the ocean), C_t being the global heat capacity, R_S being the solar incoming radiation, R_t being the outgoing terrestrial radiation, and α being the albedo.

The simplified global equation [1.7] contains much information about the complexity at stakes.

The left side of the equation represents the energy stocked in the ocean and is almost always at equilibrium. In what sense is this equilibrium stable? With an albedo of 0.3 generally, but this changed with past climates, what is absorbed is not exactly equal to what is reemitted: locally, we are not at equilibrium.

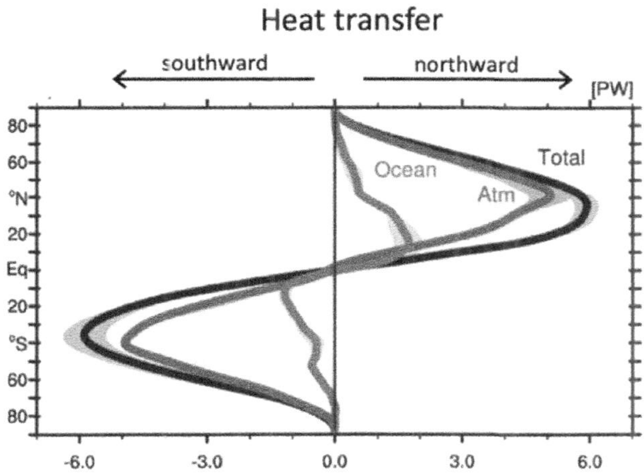

Figure 1.18. *Required heat transfer (in PW, i.e. 10^{15} W) to balance the differences of net radiation at the top of the atmosphere (black curve) and distribution of this transport among oceanic contributions (in blue) and atmospheric ones (in red). A strictly positive value on the x-axis refers to a northward transport (Source: Fasullo and Trenberth (2008))*

In Figure 1.18, we can see the dynamics of the ocean and of the atmosphere. The fact that temperatures are not warm enough in the poles is balanced because some excess heat is being brought away by the dynamics of the ocean and of the atmosphere, which take extra heat toward the poles.

1.3. Past climate history

Figure 1.19. *Picture of Louis Agassiz, first scientific proposal of past ice ages (1837), from an unknown author (public domain)*

Louis Agassiz was a paleontologist; he studied glaciers in the Alps. He found rocks named “Moraines”, organized by layers in rivers; they gave names of rivers to these past ages, and they are extremely difficult to date. Those blocks are easily recognized by geologists, as they were present very far from the glaciers. Agassiz’ first scientific proposal of past ice ages was made in 1837 at Neufchâtel. He notably stated that ancient glaciers also notably covered Europe, Asia and North America.

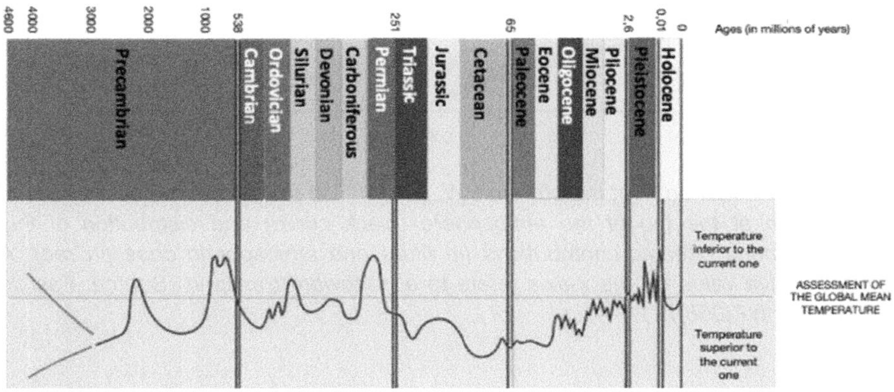

Figure 1.20. *Temperature trends over the last 4.5 billion years before present (Source: Le Treut (2013))*

Different geological indicators allow us to reproduce the global shape for the curve of temperature changes of the earth since its origins, 4.6 billion years ago. They point to a climate that is generally warmer than the one of the most recent two million years.

We are essentially interested in the Holocene represented in Figure 1.23, the period during which our human ancestors started to develop in terms of civilization.

In Figure 1.20, we can note that the evolution of the temperatures over time is not linear. The oceans were formed during the Precambrian, and the atmosphere changed its composition then. The Pleistocene is featured by cold temperatures, with oscillations.

The Cretaceous, the Jurassic and the Triassic represent a warm era compared to the Primary era.

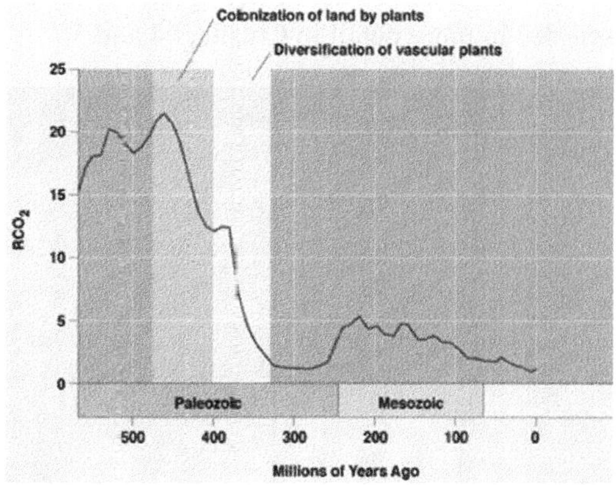

Figure 1.21. *Global temperature trends over the last 4.5 billion years before present, from an unknown author (Le Treut 2013)*

The Sun continuously brings energy to the Earth, so why is the Earth not getting indefinitely warmer and warmer? In 1824, Joseph Fourier answered with this explanation: because it emits some "Chaleur Obscure" (Invisible Energy). That is how the greenhouse effect was discovered by Fourier, who explored the heat transfers in a body. Fourier realized that for balance, the Earth has to get rid of what he called an "obscure heat", invisible to the human eyes, which must exist for equilibrium.

According to the scientist James Lovelock, a member of the British Royal Society during the 20th century, the Gaia Theory can be defined as follows: the climate of the Earth is regulated by life, which ensures an equilibrium between the increasing solar emissions, and the decreasing greenhouse effect. This theory is valid at the scale of (hundreds of) millions of years.

The RCO_2 parameter is defined as the ratio between the atmospheric CO_2 mass at a given time in the past over the current value of the atmospheric CO_2 mass. In Figure 1.21, we note that RCO_2 decreased of CO_2 since around 459 million years before present; it is very much because of the appearance of life. For Lovelock, if the Earth has been stable for such a long time compared to Mars or Venus, for instance, it is because of life maintaining life. However, it is important to note the scale of time, which is very important: it works at the scale of millions of years.

Therefore, it would not be realistic to count on that effect, as human beings would be replaced in the meantime! This time scale is so long that this phenomenon cannot be considered as a solution to avoid acting on tackling global warming, a topic that will be discussed in more detail in Chapters 4 and 5.

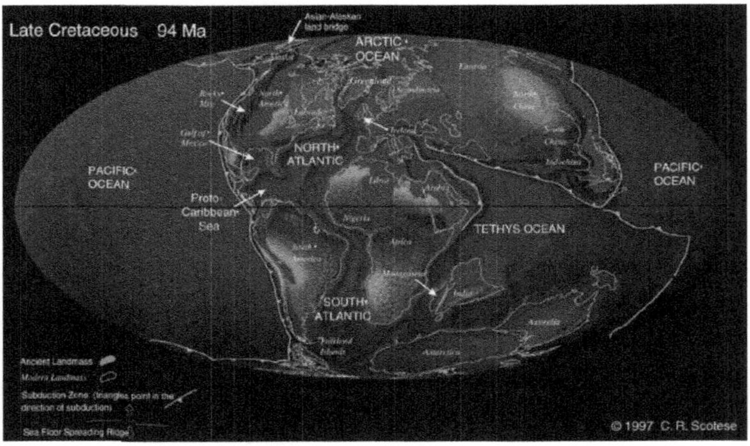

Figure 1.22. *The breaking up of the Pangea: more than 100 million years ago (Source: Scotese (1997))*

Figure 1.22 illustrates another feature: the continental drift, which produces an acceleration of the cooling. This phenomenon occurred at the same scale of time as the cooling during the Quaternary era.

The South pole was then covered with ice and the albedo participated in the cooling.

If we name α the albedo, we can ask ourselves how the climate is changed when we affect the albedo and if the effects of solar inputs are direct or indirect.

A first example of the complexity is the demonstrated dependence of the albedo α on the temperature T, which we already discussed with the following equation:

$$C_t \frac{dT}{dt} = R_S\,(1 - \alpha) - R_t(T) \qquad [1.9]$$

The cryosphere modulates α and takes three forms:

– snow cover (quick response, from days to seasons);

– sea ice (quick response, seasonal time scale);

– continental glaciers (very slow response).

When considering those cycles, there are things that cannot be explained solely by the solar input. Let us work very briefly with a few equations, even though it is not possible to solve those equations until the end.

However, those types of equations have been used to try to show why this is more complex than a simple dependence on solar input. Also, this has implications for the future climate. The equation we will consider first is equation [1.9]: if α varies in a simple way, we have three equilibrium states (see Figure 1.23):

– one is unstable;

– two are stable.

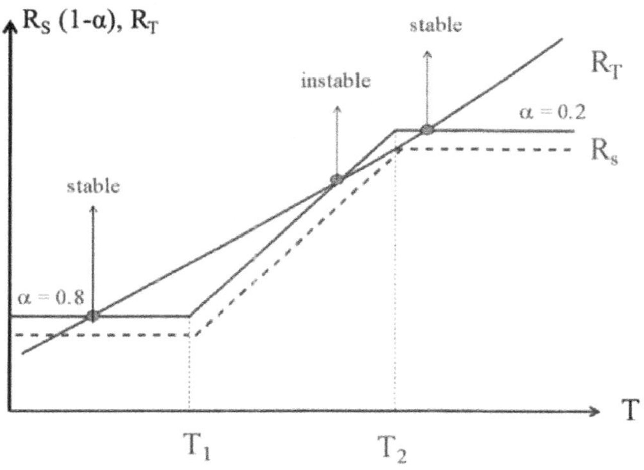

Figure 1.23. *Representation of R_S (1-α), R_T as a function of temperature*

In Figure 1.23, we can observe three points of equilibrium represented in green. Those points are either stable or unstable, found by using the equation detailed previously and by applying stability conditions (the aim here is not to deliver calculus details).

Graphically, a point of equilibrium is observed when the red curve representing the solar radiation crosses the blue curve representing the terrestrial radiation.

For the unstable point of equilibrium, it corresponds to the situation where the absorbed Sun energy brings it to a higher level: it gets out of range, and therefore, it goes out of equilibrium. This simplified model provides us with precious information about the complexity of the system.

A simplified explanation of the methods used to study past climates and oscillations between glacial and interglacial ages can be proposed as follows.

Water isotopes are essential tools to reconstruct the sequence of glacial–interglacial ages, as there is a link between the temperature for snow formation and its isotopic composition.

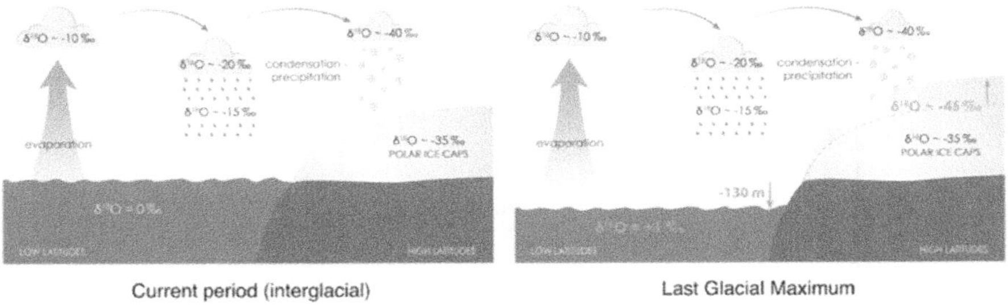

Figure 1.24. *Simplified scheme illustrating the principle at the origins of the isotopic thermometer (Source: e-CalPSu (Jouzel 2018))*

The number of isotopes indeed diminishes as the evaporating water from the poles loses temperature: we speak about an "isotopic thermometer", which is a precious tool to reconstitute climates from the past.

In the ice cap, there is a low isotopic content (low amount of O^{18}). This method was notably used by Lorius et al. (1985).

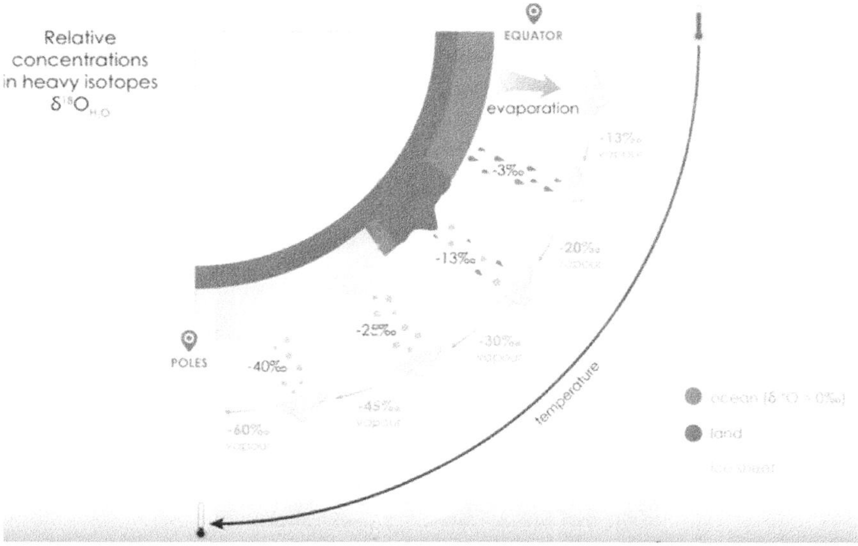

Figure 1.25. *Scheme showing the decrease of isotopes $\delta^{18}O$ as the evaporating water from the poles loses temperature (Source: e-CalPSuL, Jouzel (2018))*

More in detail, the fluctuations of the isotopic content that we can note δO^{18} at a given level *i* can be obtained thanks to the following relation:

$$\delta^{18}O = \left(\frac{\left(\frac{^{18}O}{^{16}O}\right)_{sample}}{\left(\frac{^{18}O}{^{16}O}\right)_{standard}} - 1 \right) \times 1{,}000\ ‰ \qquad [1.10]$$

with the standard $\delta^{18}O$ featured with a known isotopic composition (Vienna Standard Mean Ocean Water (VSMOW)).

Therefore, we consider the changes compared to the difference in the ocean. The method consists of assessing among O^{16}, O^{17}, O^{18}, whether the latter is more or less present. The ice caps were created by the water evaporating from the ocean (which is dense) that loses a given quantity of δO^{18}, and that is then transported in the atmosphere (less dense than the ocean) and precipitated all along its way.

By measuring the amount of O^{18} in the ocean, we can deduce the amount of ice it contains.

The level of O^{18} in the past, at the bottom of the ocean, is set at 0 when animals die: time zero is set at the time at which those animals started to sediment. The sediments correspond to past ages composed of skulls of animals containing $CaCO_3$.

The calcite shells on Figure 1.26 of the foraminifera give access to ocean water composition at the time they were living.

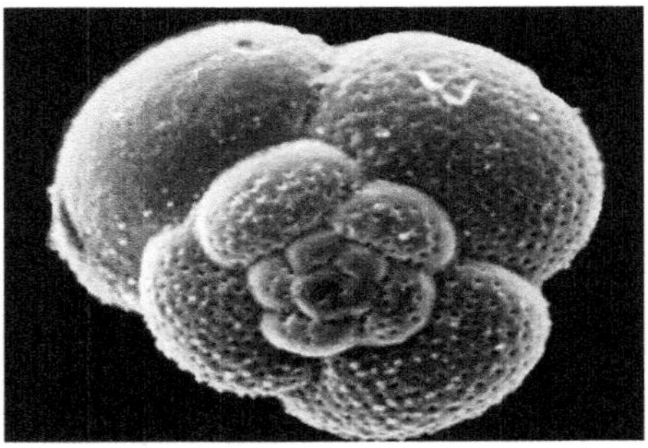

Figure 1.26. *The calcite shell of a foraminifera (Source: UCL Collection)*

It has taken decades of work in order to map out how climate has changed and compute these results in models. Different researchers made holes in the ocean floor in order to assess the variations in the volume of continental glaciers over the past 10^6 years.

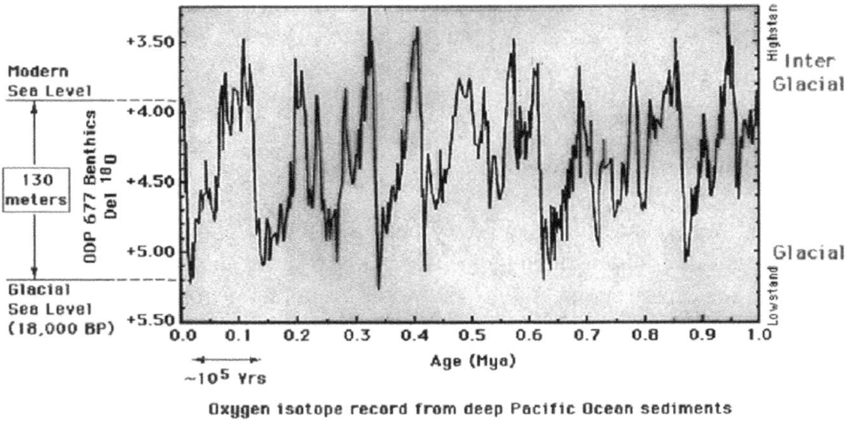

Figure 1.27. *Variation of Volume of Continental Glaciers over the past 10^6 years (Source: Denton et al. (2010))*

The accumulated ice is also able to provide climate records over almost a million years, as shown in Figure 1.27.

As mentioned previously, the isotopic thermometer is a very accurate way to measure the temperature. Glaciers are made of snow which have been circulating for several million years, and the gravity of ice layers is at play too. Jean Jouzel worked on this topic with Claude Lorius, among others, in the 1980s. In a recent book published in 2019, "Climats passés, climats futurs", Jouzel detailed the steps of this exceptional research journey.

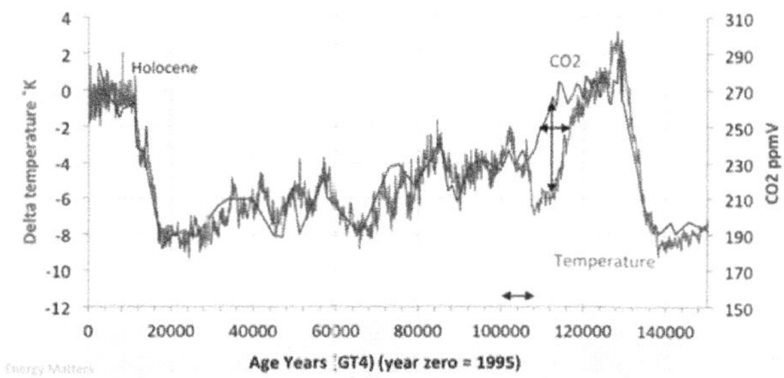

Figure 1.28. *Vostok Ice Core Temperature and CO_2 over the past 150,000 years (Source: Energy Matters, www.ncdc.gov/paleo/icecore/antartica/vostok/vostok.html)*

Figure 1.28 shows that temperatures fluctuations cannot be explained by solar fluctuations only. The most recent interglacial step was a bit warmer than now. The fluctuations that we note at around 20,000 years before present have nothing to do with the Sun. We can also note that in recent times, in less than 2,000 years, there are two temperature maxima. Without solar forcing, the system can oscillate by itself. A forced nonlinear oscillator tends to take the periodicity of the forcing. Some phenomena cannot be explained by solar effects.

Processes which can be detected from deep sea ice cores that give the isotopic content of the sea water in the past will be detailed here. We can indeed deduce the amount of ice in Antarctica, Northern Europe and North America mostly, and Greenland from deep sea cores, thanks to the isotopic contents of sea water. We can then detect carbon dioxide and methane in the bubbles of the ice.

Figure 1.29 presents carbon dioxide concentration, temperature relative to present climate expressed in thousands of years before present with time going from left to right, over the past 400,000 years. This work by Lorius et al. (1985) was pioneering for showing that carbon dioxide and methane – the two main greenhouse gases – were more or less in phase with the temperatures of Antarctica.

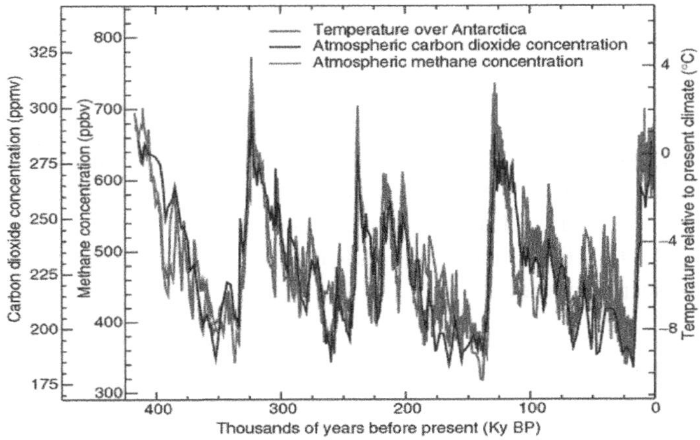

Figure 1.29. *Vostok Ice Core Temperature and CO_2 over the past 400,000 years (Source: Energy Matters; www.ncdc.gov/paleo/icecore/antartica/vostok/vostok.html)*

An interesting question is to know whether the temperature is being driven by carbon dioxide and methane or if carbon dioxide and methane are driven by temperature. This is different from the present situation: we have a climate, more or less at equilibrium, and we introduce carbon dioxide and methane in the atmosphere: therefore, we expect a direct impact of those gases over temperatures.

Instead, Figure 1.30 shows a cycle with a complex combined role of carbon dioxide and methane. Let us consider the Earth mean temperature, plus an addition of carbon dioxide and methane, like in the present situation. What existed in the past was a bit different, and we think that the solar input was probably playing a role which would be antisymmetric between the two hemispheres.

In Figure 1.30, we consider a climate more or less at equilibrium, in which two greenhouse gases, carbon dioxide and methane, are introduced: on the one hand, there is a direct impact of those two gases. On the other hand, solar input also induces temperature increase, and then CO_2 and CH_4 increase, and as a result, the temperature in the other hemisphere increases: this is probably the phenomenon occurring from one hemisphere to the other, which means that the role of CO_2 and CH_4 in the past was mainly to phase the two hemispheres, knowing that in fact the real common energy input was the change of energy from the Sun, which is itself coming from the seasonal changes.

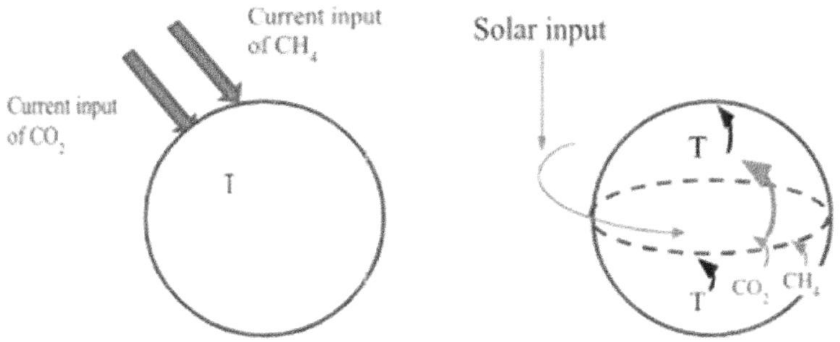

Figure 1.30. *A scheme of the combined role of carbon dioxide and methane*

There are actually three ways for modifying this energy input from the Sun at the top of the atmosphere:

1) Eccentricity, which is related to the shape of the ellipse: it changes the energy input.

2) Obliquity, which concerns the inclination of the axis of rotation of the Earth and has 40,000 years of periodicity: it does not change the total energy input.

3) Precession of equinoxes, which does not change the energy input.

This leads us to make some astronomical considerations and a short digression about the Milankovitch theory (see Figure 1.31).

Milankovitch was a Serbian climatologist who built up the theory of astronomical forcing of Quaternary era climate change in the 19th century. The discovery of Neptune by the French astronomer Le Verrier is widely regarded as a critical validation of celestial mechanics during the same century. This theory explained that 20,000 years ago, the Earth was nearer from the Sun in the summer due to the precession of solstices and equinoxes.

To come back to our considerations summarized in Figure 1.30 about the combined role of carbon dioxide and methane, it means that the way humans affect the climate through greenhouse gases in the present time and the way greenhouse gases may have played a role in the past is largely different because of the complexity of solar input going through seasonal cycles, mostly through the differences between the equator and the pole. Also, as will be detailed further in this book, the fact that CO_2 and CH_4 are well-mixed in the atmosphere creates a situation in which they play a very active role (mostly afterward).

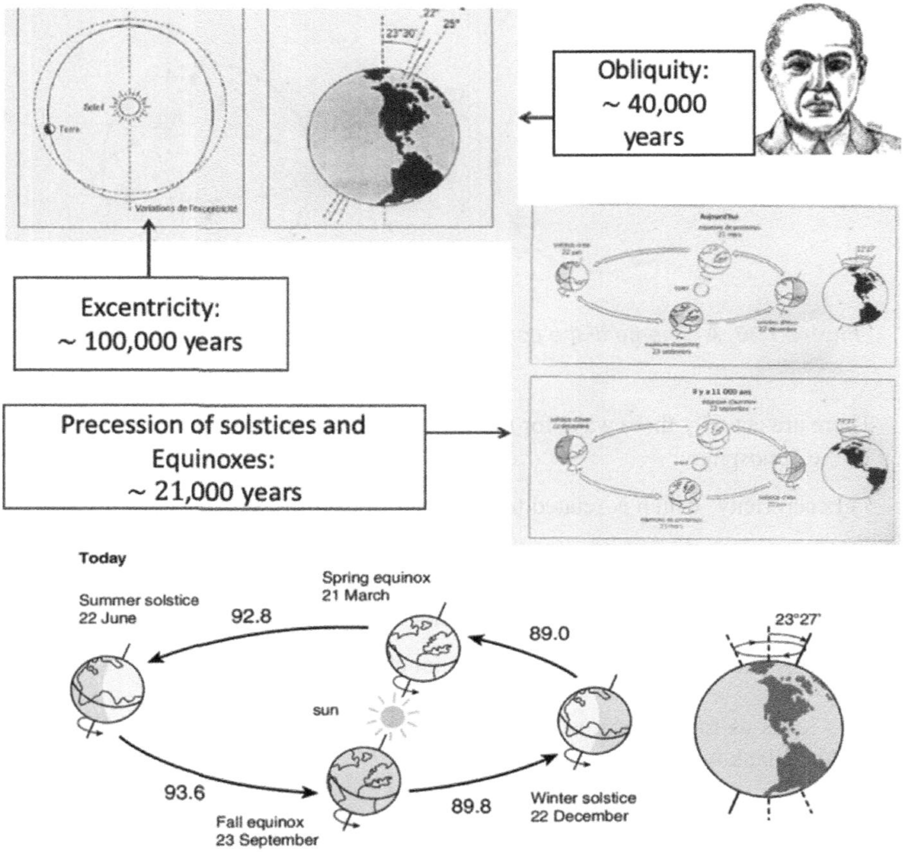

Figure 1.31. *Milankovitch theory: astronomical forcing of Quaternary era climate changes (Source: Astronomical theory and orbital forcing, Yin and Berger (2012))*

There have been a number of debates on this topic. For example, in the documentary by Al Gore "An inconvenient truth", at the time he is climbing over a ladder and shows that the greenhouse gases have been increasing, the question is raised: what will temperature do? The answer is actually not that simple: temperature will increase, but the reason why it will increase is not as obvious as it may seem. Using the past to demonstrate that temperatures will increase is not something completely correct. Beyond the simple scheme on Figure 1.30, attention must be paid.

The demonstration that CO_2 plays a role between the two hemispheres is maybe one of the most important results of Jean Jouzel.

If we go back to Figure 1.29 and observe CO_2, we may anticipate the temperature effect which we expect, and also the reverse, which was not completely expected. There were actually many debates around the following question: why does CO_2 change sometimes before, and sometimes after a temperature change? The answer is that the mechanism at play is complex. Jouzel's results show that indeed there is a link between the two and greenhouse gases are part of the problem. Temperature will increase but past reasons are not completely correct.

We already mentioned that there are two kinds of results to document past climates: the ones from the deep-sea cores, and the ones from the glaciers. We also have a lot of continental records, and the largest number by far of scientists currently working on past climate is focusing on continental records.

Palynologists (scientists studying pollen) work all around the Earth to study where different typical types of vegetation can be found, and from that, infer temperature and humidity records, as they can be linked together. The same work is also largely done for the ocean.

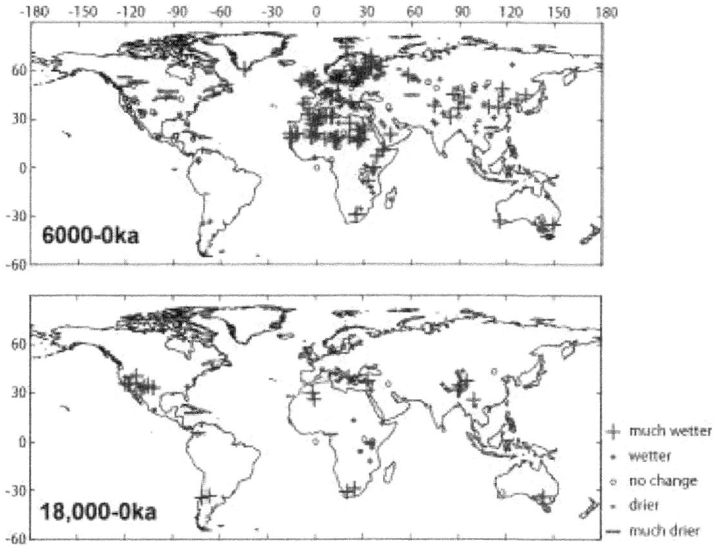

Figure 1.32. *Diagram for humidity at two different timescales based on continental records (Source: Le Treut 2018)*

Figure 1.32 shows that there was a large humidity 6,000 years ago when the North African Sahara was green. This much wetter situation is documented by a large number of records, and we know that it is due to the precession of the equinoxes.

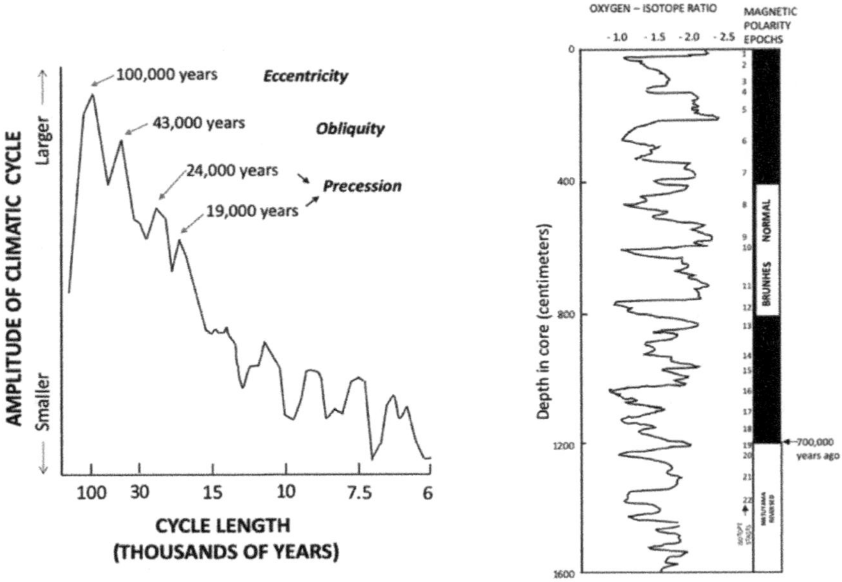

a) b)

Figure 1.33. *a) Spectrum of climatic variation of the past half-million years. This graph – showing the relative importance of different climatic cycles in the isotopic record of two Indian Ocean cores – confirmed many predictions of the Milankovitch theory; b) The "Rosetta stone" of glacial eras (Source: schemes adapted from Imbrie and Shackleton, 1974–1976)*

Figure 1.33(a) and (b) shows results by the Scottish geologist, oceanographer and military John Imbrie's teams. Imbrie (1925–2016) was very vocal and wrote a book that he entitled the "Rosetta Stone" of the glacial eras.

This so-called "Rosetta Stone" was the first core that could be dated. For this core first, the Fourier's transform calculations provided the same kind of fluctuations as the ones computed from astronomical forcing.

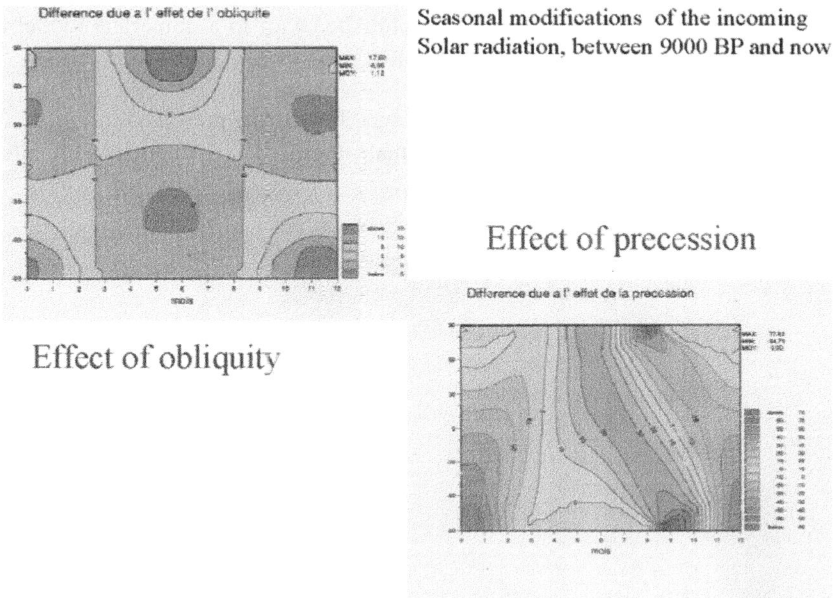

a) b)

Figure 1.34. *Seasonal modifications of the incoming solar radiation between 9000 BP and now. a) Effect of obliquity); b) Effect of precession (Source: Berger and Loutre (1991))*

The astronomical effects on the Sahara are shown in Figure 1.34, which shows the kind of seasonal modifications of the incoming solar radiation between 9000 BP and now.

The effect of obliquity (see Figure 1.34(a)) pertains to the axis of rotation of the Earth being more or less bent over the plane of the ecliptic, which is the plane of the motions of the Earth around the Sun. This obliquity is the parameter making the difference between summer and winter. When the obliquity is larger, differences between summer and winter are also larger and this configuration mostly changes the amount of energy near the poles, as shown in Figure 1.34(a).

The effect of precession (see Figure 1.34(b)) is also illustrated for one given year, and it shows the differences over this one year of rotation of the Earth around the Sun. The observation of those differences over the same period of time shows that the effects of precession induce a displacement of the warmer period during the course of the year. It is possible to impose this behavior to a numerical model and compute the climate of the past.

Such simulations were done many years ago. The results presented in Figure 1.35 show the differences between the past climate (9000 BP) and present climate in terms of summer temperatures.

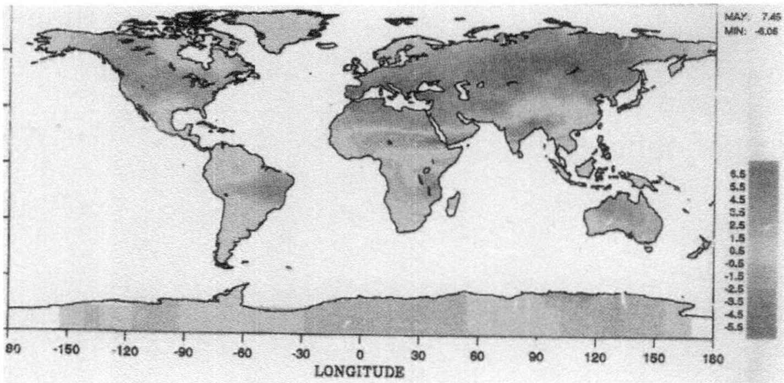

Figure 1.35. *Surface summer temperatures between 9000 BP and now (01.06–30.08). Temperature differences on the ground during June, July and August in Celsius (Source: Berger and Loutre (1991))*

In Figure 1.35, we note that the past climate was warmer during the summer (cooler places are represented in blue.). We can observe a big monsoon all over central Africa/India/Bangladesh; this was a consequence of the solar input being different. A change of precipitation in summer with a strong monsoon over central Africa and India/Bangladesh was observed 9000 BP, and it lasted until 6000 BP. It is important to understand that some changes in the past were due to solar effects; they played a role at scales of thousands of years for which they have been acting very effectively. However, this long scale of time is not comparable to the scale of time of human life, and therefore, this is not something we can expect to drive the system in the next decades.

1.4. A global evaluation of climate stability

Let us come back to equation [1.8]:

$$C_t \frac{dT}{dt} = R_S (1 - \alpha) - R_t(T) \qquad [1.8]$$

with α being the albedo, T being the local surface temperature, C_t being the local heat capacity, and R_t being the terrestrial radiation to space.

This equation has been the starting point for our discussion about past climates. We will now consider the effects of solar inputs (direct effects, and effects that are a bit less direct with a demonstrated link between the solar input and the response of the climate). This simple equation contains a lot of complex information and is still not completely solved.

1.4.1. *Let us consider the first term of equation [1.8]*

Let us consider the first term of equation [1.8]. When considering those cycles, some phenomena cannot be explained by the solar input. Let us now work very briefly with a few equations that cannot be completely solved but are nonetheless used to show a more complex behavior than a simple dependence on solar input, with implications for the future climate. In equation [1.8], if the albedo α varies in a simple way, there are three equilibrium states shown in Figure 1.36:

– one is unstable;

– two are stable.

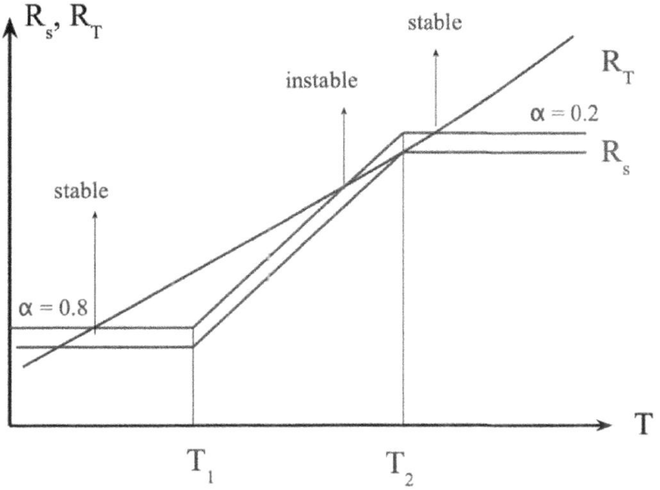

Figure 1.36. *A simplified representation of R_S (1-α) and R_T as a function of temperature*

If we note M the mass of ice, we can write another equation similar to equation [1.8], and the balance in a simplified expression would be:

$$C_t \frac{dM}{dt} = \text{Precipitation} - \text{Melting} \quad [1.11]$$

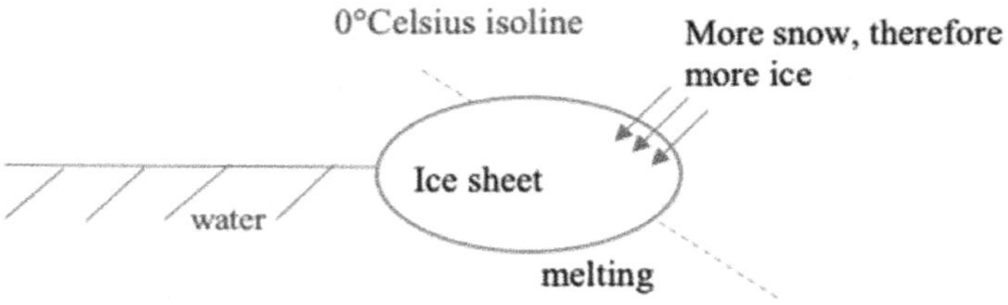

Figure 1.37. *Scheme with the water, an ice sheet and the 0°C isoline for the Arctic*

Let us consider the Arctic, for instance. We can represent it in the scheme on Figure 1.37, showing the water, the ice sheet and the 0°C isoline. Some snow and therefore more ice is added, and some of the ice is melting. There is a balance between addition of ice and melting, which is also dependent on how much the ice sheet may sink into the ground. In fact, one third of the ice at isostatic equilibrium is getting down under the current layer of the water. We also know that the weight of the ice sheet – because it is a viscous elastic body – is linked with its mass. Therefore, we can write a new equation taking that into consideration. A non-obvious phenomenon is that precipitation tends to increase with temperature, because the evaporation of the ocean increases with temperature, and the melting is also increasing with temperatures, but at a different pace. Therefore, we anticipate that when we write those more complete equations, we may have some kind of internal cycle of the climate system.

For example, if we consider the albedo parameter α, reflecting more or less M (the mass of snow/ice), we have two effects:

– If α increases, so the temperature decreases (albedo effect).

– T increases, inducing more precipitations, and therefore α decreases.

This represents what we call a nonlinear system with two opposite feedbacks, characterized mathematically by the fact it may oscillate freely. Therefore, we have the possibility to have, with the glaciers, simple oscillations of the climate system that can be demonstrated with simple equations.

This is a phenomenon that we think is happening, and there was a demonstration of that in a case which is a bit different, done with the direct role of the Sun: climate records do not only reflect the Milankovitch forcing, as illustrated by Figure 1.36. This requires some careful diagnosis.

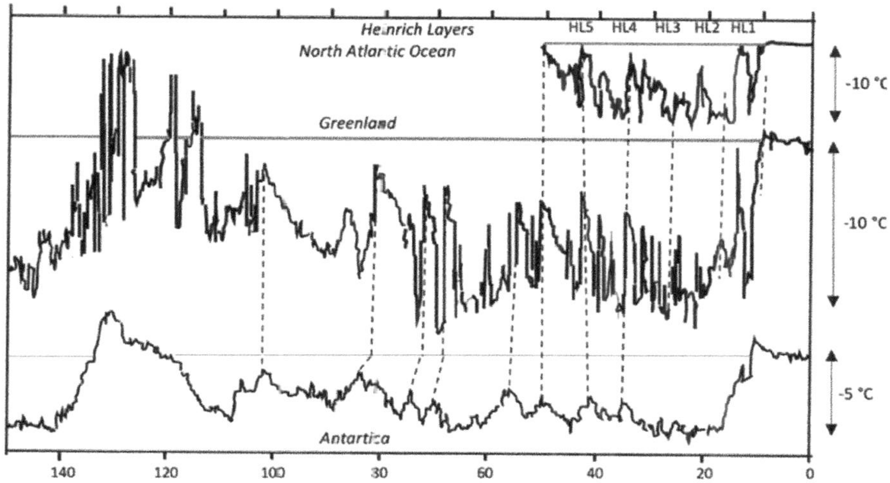

Figure 1.38. *Scheme of reconstructed climate records showing rapid changes in North Atlantic and in Greenland (ground temperatures) (Source: adaptation from Jouzel et al. (1994))*

Figure 1.38, adapted from Jouzel et al. (1994), is based on a work which radically changed the views on past climates. Reconstructed climate records showing rapid changes in North Atlantic and in Greenland are represented: the corresponding events indicated by thin dashed vertical lines are damped in the Antarctic records. Temperature (at ground) changes are estimated from the isotopic content of ice (Greenland and Antarctica) and from faunal counts (North Atlantic). "HL1" to "HL15" refer to different "Heinrich" layers. The time goes from left to right, and we see the last interglacial events: 120,000 thousand years ago, the present Holocene (warm episode).

We can also distinguish fluctuations with an indisputably astronomical origin. As for the quick deglaciations, we note that they are difficult to be linked as solar input. There are also very rapid fluctuations which are quite different and also difficult to explain directly by solar radiation. The measurements were made over

Greenland, the local temperature deduced from isotopes. In Figure 1.38, we can see a warm period, followed by a period which is cooler, various stages, and then it becomes warm again. This is a behavior similar to what we observe in the south with a larger range. This larger range is also a well-known phenomenon: the Northern Hemisphere is reacting to changes more quickly than the Southern Hemisphere. In this figure, we can also observe a large number of small events: at first sight, people thought that it was a wrong measurement. However, there were actually two kinds of measurements from an American group and a European one, working at some distance, and they got exactly the same results.

This observation is now demonstrated to be true, and subsequently, people found out that when looking at the sediments in the North Atlantic Ocean, variations at the same time scale are also found. Those very short events are still however being disputed. There are a large number of oscillators: this one is a global one, but local ones also exist with the Northern Atlantic and Greenland glaciers, which can oscillate.

Those results from Jouzel et al. (1994) show that there are one or several nonlinear free oscillations. Those oscillations generally take the size of a limit cycle with a specific range/amplitude already given by the shape of the equations. It is not similar to a pendula, when put aside; it is revolving with an amplitude that may change depending on how much impulse is given: linear oscillations have a fixed amplitude and may (or may not) have enough energy to oscillate.

Solar input serves as a clock which gives a timing of events within the system. This is fairly consistent with a lot of observed phenomena. These oscillations are simple to translate into equations (note that there are actually many other possibilities to write equations, leading to free oscillations in the system). When we talk about tipping points, the fact that the system may have properties we have yet to master is a phenomenon present in all minds.

Global models generally have no complete ice physics included in them. They involve local ice physics, but they are not able to include motions of the ice at the time scales we consider here because the physics of sea and continental ice are very complex to be described. However, intermediate models with simplified dynamics of the atmosphere, simplified dynamics of the ocean and the ice shelves exist, as well as conceptual models. The integration of ice physics in global models has started with some examples of the inclusion of feedbacks with glaciers. The slow evolution of the glaciers is one of the many reasons which may explain internal oscillations of the climate system.

1.4.2. *Let us consider the second term of equation [1.8]*

Let us now consider the second term of equation [1.8], which anticipates Chapter 2 on radiative processes.

We will describe this as an introduction which will be developed, and we will spend more time on this in Chapter 2.

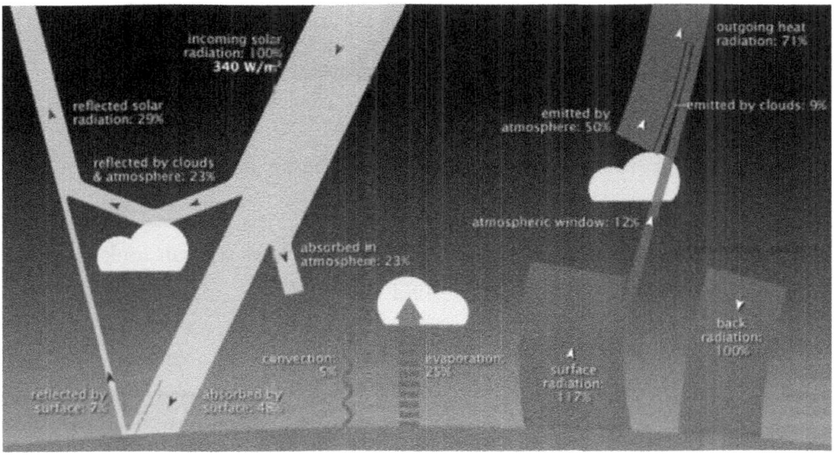

Figure 1.39. *Earth Global Radiation Budget. NASA illustration by Simmon, adapted from Trenberth et al. 2009, using CERES flux estimates provided by Norman Loeb*

Figure 1.39 shows the solar energy entering the climate system, the energy leaving it through diffusion, and what the surface emits in terms of radiation. Only a little part of this terrestrial longwave radiation goes directly to space, it being absorbed and reemitted, partly downward, partly upward. The difference between what is emitted and what actually goes to space is really what we call the greenhouse effect. It means that the surface has to really make a strong effort to be much warmer to send energy to space.

Figure 1.40 shows two sister planets: Mars and Venus. We may wonder whether the future of the Earth is more likely to be like cold, like Mars or extremely hot, like Venus. Venus is hot not only because it is close to the Sun, nor Mars cold only because it is far from the Sun; there are other explanations detailed in Table 1.2.

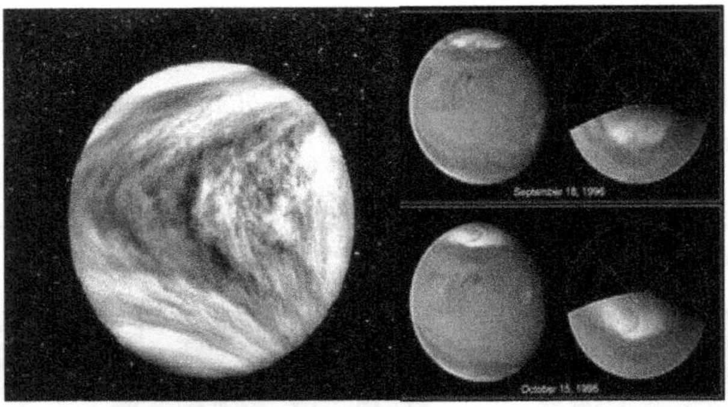

a) b)

Figure 1.40. *a) Venus. b) Mars. Source: nasa.com*

We would like to provide some indications about how we can look at the Earth when we try to compare it with neighboring planets.

Let us describe it line by line:

	Venus	Earth	Mars
Distance from the Sun (in Astronomical Unit (AU))	0.72	1	1.52
Solar Flux (in $W.m^{-2}$)	2,643	1,370	593
Albedo	0.8	0.3	0.22
Temperature without greenhouse effect (K)	220	255	212
Real temperature (K)	730	288	218
Greenhouse effect (K)	510	33	5

Table 1.2. *A comparison between three sister planets. For a color version of this table, see www.iste.co.uk/letreut/energetics.zip*

The distance between the Sun and the Earth defines a reference unit that we call one astronomical unit (AU). Table 1.2 shows that Venus (0.72 AU) is closer to the Sun than the Earth (1 AU) and that Mars is further away than the Earth (1.52 AU).

Table 1.2 also delivers the corresponding solar fluxes, and the albedo that may vary a lot with the state of the planet.

This occurs in a way that is different from the Earth for those two planets:

– Venus is very warm but has a very strong albedo (0.8) because the gases over Venus are reflecting a lot. Venus has a very deep atmosphere, with very toxic gases, and large temperatures compared to Mars and the Earth.

– The present albedo of the Earth (0.3) is smaller than that of Venus (0.8).

– Mars has changed a lot in the past; now, it is a planet with a few ice caps, which are made of CO_2. Its albedo is not very strong (0.22) because of its lack of water and because of its very thin atmosphere.

Table 1.2 also presents those three planets' temperatures without the greenhouse effect. It is interesting to note that if we just take the albedo effect and the solar input, then those three planets are mostly equivalent, with temperature differences only between 8 to 33 K.

The greenhouse effect comes from the fact that the atmosphere absorbs part of the energy emitted by the ground. This part of energy emitted by the ground may largely vary from one of the three planets considered here to another, and the energy that has to be emitted to balance out the incoming solar radiation and the albedo effect may completely change. Therefore, when we consider the respective greenhouse effects of Venus, the Earth and Mars, there are consequently huge temperature differences between those three planets.

For the case of Venus, there is a very strong greenhouse effect, leading to high temperatures (hundreds of degrees). For the Earth, we will discuss it in the course of this book. For Mars, there is a small greenhouse effect and its temperatures remain rather cool (unlike what it may have been in the past).

The figures included in Table 1.2 raise the following question: why are those three planets so different?

First, the differences come from the fact that the greenhouse effect plays a big role, and an important derivative question is: can the greenhouse effect drastically change the Earth's temperatures in the future?

When coming back to equation [1.8] and given the fact that the terrestrial radiation depends on T, we note it R_t (T). The first greenhouse gas is actually water vapor.

Therefore, we can infer that a potential feedback would reasonably be due to some kind of change due to water vapor.

The link between water vapor and temperature can be made as follows: what determines the amount of water vapor in the atmosphere is precipitation, which happens when the condensation level is reached, a level directly dependant on temperature.

The dependence of the water vapor on temperature is represented in Figure 1.41, also known as the Clausius–Clapeyron graph.

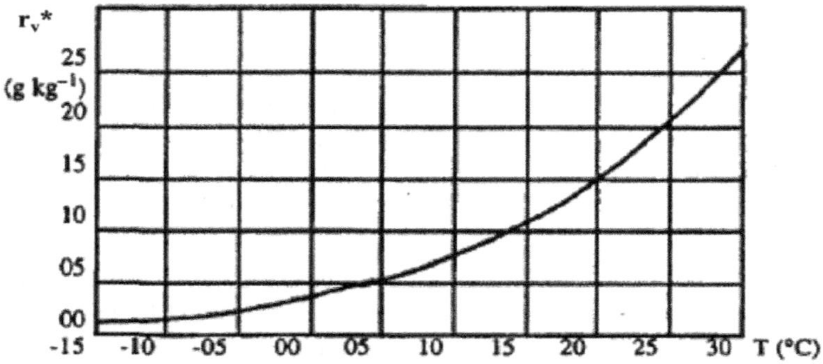

Figure 1.41. *Clausius–Clapeyron ratio of water vapor in the atmosphere at saturation. The saturation level for water vapor increases very quickly with temperature*

r_v* is the ratio of water vapor in the atmosphere at saturation. Each point of the curve refers to the value of this ratio for a given temperature where the condensation point occurs. It is expressed in grams of water vapor per kilogram of air.

We observe a very strong increase with temperature. We have negative temperatures on the far left, and we get to positive temperatures near the equator. If we place our attention on those cool temperatures, we note that r_v* is bordering 1 g.kg^{-1} at –10°C, while, for instance, at 25°C, we have: r_v* = 20 g.kg^{-1}.

This means that at a temperature of 25°C, the amount of water in the atmosphere is 20 times larger than at a temperature of –10°C. Therefore, with 35°C degrees of difference, there is a factor 20 more or less in the amount of water vapor that can stay in the atmosphere, which actually represents a very large change.

In fact, it is even more than a factor 20 because the temperatures can get very easily up to –40°C in the higher atmosphere and in tropical areas; temperatures can be occasionally higher than 35°C degrees. Therefore, there is a huge spread of water vapor content between low and high temperature values.

We now want to find a criterion for stability, and for that, we will consider a simplified version of the climate system with an ascending motion near the equator – unlike what was discussed for the albedo, what may modify the climate in terms of the greenhouse effect is occurring more strongly near the equator.

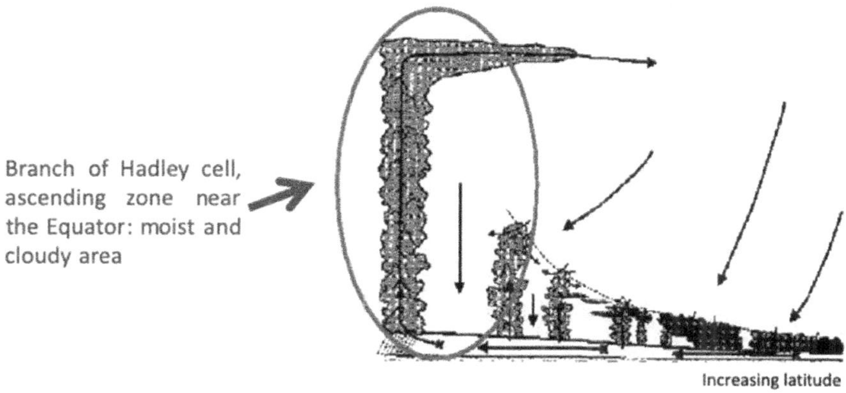

Figure 1.42. *A simplified scheme of the intertropical climate*

In Figure 1.42, an ascending region near the equator is highlighted in red, then the air is being brought to high latitudes, and afterward, it gets down. In the gray area on the left, a branch of Hadley cell is represented, and in the descending part of the Hadley cell, there can be low clouds or deserts, depending on if there is an ocean at this place.

It is possible to write a simple equation to describe this phenomenon.

1) Let us first write a simple representation of the emissions for the case of a blackbody radiation: $\boldsymbol{\sigma}\mathbf{T}^4$.

2) However, the Earth is not an actual blackbody and the terrestrial emission is less than the one that a blackbody would be, as the greenhouse effect brings back radiation to the ground: we can note this greenhouse effect *G(T)*.

3) Therefore, we can write the two components of the terrestrial emissions as the sum of the blackbody radiation minus the greenhouse effect:

$$R_t(T) = \boldsymbol{\sigma}\mathbf{T}^4 \times (1 - G(T)) \qquad [1.12]$$

4) Of course, the greenhouse effect G(T) increases with temperature, which is a source of main potential destabilization of the system.

Let us now just consider stability–instability criteria for such a system. In the case of an equilibrium, we can consider that for a fixed point that does not change, the temperature T_{eq} can be written as:

$$T = T_{eq} + \mathit{T} \quad [1.13]$$

We can write:

$$\frac{dT}{dt} = f(T) \quad [1.14]$$

Therefore, the equation becomes:

$$\frac{dT}{dt} = f(T_{eq} + \mathit{T}) \quad [1.15]$$

Also, we may develop it when we take a first development as:

$$\frac{dT}{dt} = f(T_{eq} + \mathit{T}) = f(T_{eq}) + \frac{df}{dt}\mathit{T}\ldots \quad [1.16]$$

with $f(T_{eq}) = 0$ because it is equilibrium; therefore, we can write:

$$\frac{dT}{dt} = \frac{df}{dt}T \quad [1.17]$$

To determine the stability or instability condition of a system, the following considerations can be done:

$$\frac{dT}{dt} = f(T)\frac{dT}{dt} = f(T_{eq} + \mathit{T}) = f(T_{eq}) + \frac{df}{dt}\mathit{T}\ldots$$

with $f(T_{eq}) = 0$

$$T = T_{eq} + \mathit{T}$$

with T_{eq} referring to a fixed point.

Box 1.1. *Stability condition of the climate system*

A little increase in temperature would mean that if the term:

$$\frac{df}{dt} < 0 \quad [1.18]$$

then it will bring the system back to equilibrium.

Therefore, to know whether the system is stable or instable, we should look at the sign of $\frac{df}{dt}$.

1.4.3. *A global estimate of climate stability considering another source of complexity: the greenhouse effect*

If we rewrite the energy budget equation, we have:

$$C_t \frac{dT}{dt} = R_S (1 - \alpha (T)) - \sigma T^4 (1 - G(T)) \qquad [1.19]$$

G(T) is a "greenhouse index". We then obtain, as a stability condition:

$$-R_S \frac{d\alpha}{dT} + \sigma T^4 \frac{dG}{dT} - 4\sigma T^3 (1 - G(T)) < 0 \qquad [1.20]$$

After derivation of the equation, the stability of the equilibrium according to the criterion of equation [1.18] would therefore be:

$$-\left(-\frac{DG}{DT}\right) \sigma T^4 - 4\sigma T^3 (1 - G(T)) < 0 \qquad [1.21]$$

From equation [1.21], we found that the criterion of stability that we need to have is:

$$\frac{dG}{dT} < \frac{4}{T} (1 - G(T)) \qquad [1.22]$$

NOTE.– Difficulties to get temperatures from satellite data come from the fact that for oceans, surfaces are mostly black bodies: they emit depending on their temperatures, but in cloudy areas, measurements must be done in between the cloud in clear-sky holes.

$R_t(T)$ is the radiation that gets out to space, this is the most obvious satellite data: the satellite collects all of the radiation in the infrared that gets out and $R_t(T)$ is known. This type of work is largely done by teams of Laboratory of Meteorological Dynamics (LMD) of the Institute Pierre Simon Laplace (IPSL) who have been designing such instruments.

Also, as we have the value of σT^4, $R_t(T)$, and we know how to derive G(T), we therefore can solve the equation.

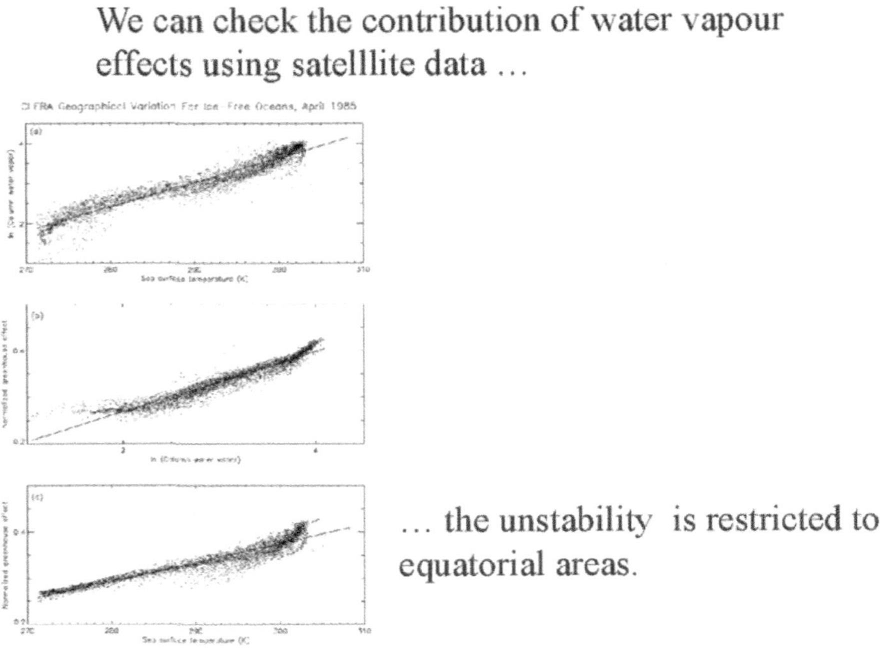

Figure 1.43. *Satellite data can be used to check the contribution of water vapor, with instability restricted to equatorial areas. CLERA project (Source: British Meteorological Office)*

Figure 1.43 shows the results from the project CLERA run by the British Meteorological Office. It shows regressions. On the top diagram, for example, a column of water vapor as a function of sea surface temperature is represented: the water vapor is increasing with sea surface temperature, with a trend comforting our previous computational results.

The middle graph of Figure 1.43 shows the greenhouse index G represented as a function of a column of water vapor. This index increases with water vapor: the greenhouse effect is largely dependent on water vapor. Mostly, this work has been done in locations allowing measurements on clear-sky conditions to get rid of clouds, observing the holes of clear-sky between clouds if necessary.

The bottom graph of Figure 1.43 emphasizes the main result by displaying the normalized greenhouse effect as a function of sea surface temperature. This is the element that has to follow some criterion for the respect of a stability condition of the equilibrium.

This criterion is quite easy to define as based on two parameters that are already known thanks to the diagram:

– the slope with a value of $\frac{4}{T}$ which depends on the temperature T;

– multiplied by (1 – G (T), and we can measure G (T) at any point on the graph.

Therefore, we have the value of $\frac{4}{T}$ (1 – G(T)), and we are able to compare this result with the value $\frac{dG}{dT}$.

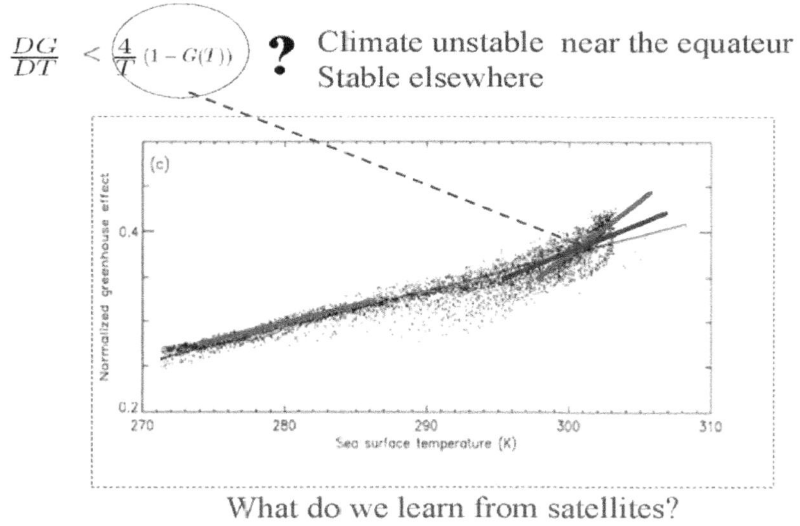

Figure 1.44. *Zoom on the regression of the normalized greenhouse effect as a function of sea surface temperature (in K) (Source: CLERA project, British Meteorological Office)*

Regarding the graph of Figure 1.44, we see temperatures going from 270 K to 310 K; the regression line in black shows a good correspondence between the normalized greenhouse gas effect and sea surface temperature.

If we imagine a zoom around 273 K of the graph in Figure 1.44 representing the normalized greenhouse gas effect as a function of sea surface temperature, we can draw the scheme of Figure 1.42 of greenhouse effect as a function of temperature G (T).

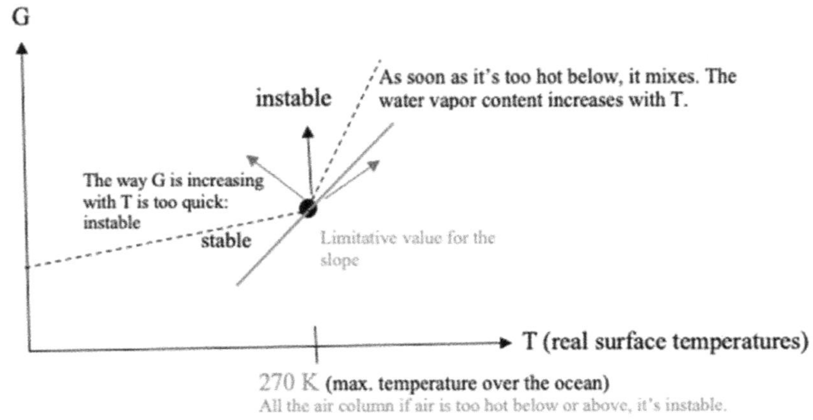

Figure 1.45. *Scheme of the regression of the normalized greenhouse effect as a function of sea surface temperature (in K) (Source: Hervé Le Treut)*

At around 270 K, the point of the curve takes on a curvy shape. When analyzing the dotes of G(T) in more detail, there is actually a limit value for the slope represented in red in Figure 1.45, meaning that where the slope is smaller than this limit value, there is a stable condition.

This shows that a simple equation can give much information, because this in fact means that 270 K represents more or less the maximum temperature over the ocean on average: between 30 N and 30 S, the surface temperature of the ocean is almost always 270 K, meaning that the way G increases with temperature at this point is too quick, and the equation is unstable. Whereas, before this point, the way G increases with temperature is smaller, so the condition is stable.

If we go back to Figure 1.42, we note that the atmosphere is ascending near the equator and descending north of the equator, with a similar situation in the other hemisphere. Where the atmosphere is descending, there is a stable climate, while where it is ascending near the equator, the climate is unstable: it means that, in fact, the climate near the equator is very warm, and when the amount of water vapor is large, the associated water vapor feedback is proportionally growing; therefore, the enlarged amount of water vapor at the equator can bring about a phenomenon named the "runaway greenhouse effect" in those regions, as it may escape out of control.

Another important feature is that near the equator, on the descending areas of the Hadley cells, on the contrary, there is a stabilizing effect: indeed, very dry air is descending in those locations, with little water vapor capacity to send back infrared radiation to space through the atmosphere because of the very little amount of greenhouse gases there. Additionally, there is a phenomenon of albedo reemitting much solar energy because of the deserts in parts of those territories.

Figure 1.46. *Are we on Mars or on Venus? Source: nasa.gov*

To the question, "are we closer to Mars or Venus' situation?", a very simple answer can be found through basic physics: we are closer to Venus near the equator, but contrary to Venus, the Earth system is not really in a situation of water vapor runaway effect because there is a stabilizing effect on each side of the equator that manages to keep the system stable.

When considering the possible versions of the Earth relative to the features of Mars and Venus, this stabilizing effect on the Earth is fundamental and has played a role for thousands of years, despite the change in the shape of the continents, as this phenomenon does not really depend on the position of the continents, but instead on the gradient of temperature between the equator and the poles – and the dynamics establishing this gradient is mostly driven by the Earth's rotation speed. Therefore, it is a rather permanent feature of the Earth system.

1.5. Conclusion

In the first part of this introductory chapter, we have seen that life is a first possible reason for a rather stable climate over years.

Another reason explaining the Earth's climate is due to the Earth's rotation upon itself and the fact that this rotation is quick enough to bring deserts very close to the equator, over an area named the desert belt. For example, over Mars, the Hadley cells go to the Poles and the situation is very different in terms of stabilizing effect.

Other factors favoring the stability of the Earth climate have been discussed among the scientific community, which we did not discuss in this chapter: one of these is the moon influence, which has kept the variations of the Earth orbit within a solar range and that may also have played a role in the climate of Earth system stability.

In Chapter 2, we will discuss radiation processes in detail.

2

The Vertical Dimension of the Energy Processes of the Climate System: Radiative Processes and Radiative–Convective Models

In this chapter, we will consider the vertical dimension of the climate system. The dynamical aspects of the system will not be much considered here. Instead, we will rather focus on the energy processes and on the way the energy transport acts in the climate system.

We will first deliver global considerations, and we will enter into more details progressively, addressing small-scale processes that are linked with radiative and radiative–convective processes.

Historically, climate models were developed after World War II with the emergence of bigger computation capacity through the development of computers. At the origins, there have been two very different designs: radiative–convective models (considering the vertical structure of the atmosphere, the types of clouds that develop, etc.), which will be discussed in this chapter; and the other type of model design, which corresponds to the first dynamical models. The mix between both models happened rather late, around the 70s/80s.

For a color version of all of the figures in this chapter, see www.iste.co.uk/letreut/energetics.zip.

2.1. Radiative processes: generalities

The diversity of the radiation budget at the top of the atmosphere highly depends on location, as shown in Figure 2.1. The differences between the poles and the equator are particularly important. For complete analyses of the problem of energy fluxes and radiations, quantum physics would be required, as well as optics and undulatory optics.

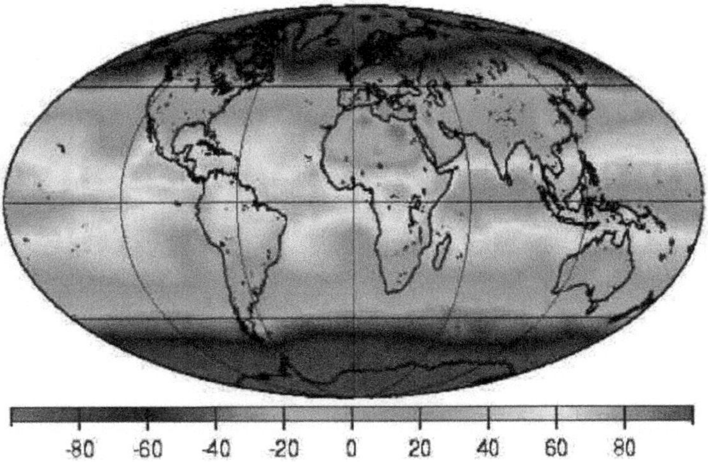

Figure 2.1. *The radiation budget at the top of the atmosphere (annual mean) in W.m^{-2} (Source: The Earth Radiation Budget Experiment (ERBE))*

In this chapter, the approach is, however, voluntarily focused on energy fluxes and thus, we will not go through all of the physics involved on a course on electromagnetic radiation. The aim here is merely to discuss how the energy fluxes connect different areas within the Earth system, using the minimal physics to describe those phenomena.

Quantum physics is necessary here as it delivers the distribution of the radiation emitted by a blackbody as a function of temperature and frequency. Full mathematical demonstrations will not be conducted here and results will merely be admitted.

The radiance of a blackbody, expressed in watts per unit surface per solid angle per wavelength [W.m^2.str^{-1}.nm^{-1}], is delivered by Planck's law:

$$B\ (\mu, T)d\mu = \frac{2h^3}{c^2(e^{\frac{h\mu}{\kappa T}} - 1)} d\mu \qquad [2.1]$$

where:

– λ is the radiation wavelength;

– c is the radiation speed, depending on if it goes through vacuum or not;

– $\mu = \frac{c}{\lambda}$ is the frequency. [2.2]

The energy of photons can be written as follows:

$$E = h\mu \qquad [2.3]$$

with h being Planck's constant $h = 6.62607015 \times 10{-}34$ J·Hz^{-1}

The radiance of a blackbody can be described and demonstrated thanks to a few equations, based on the exponent term $\frac{h\mu}{\kappa T}$ in equation [2.1].

By integrating over all solid angles and wavelengths using the Stefan–Boltzmann law, we obtain:

$$L_{total} = \sigma T^4 \qquad [2.4]$$

with $\sigma = 5.67\ 10^{-8}$ W. m^{-2}.K^{-4}

When μ is large and λ is small, photons in the system are considered very energetic. Equation [2.2] delivers information about the fact that the transferred energy strictly depends on the frequency μ/the wavelength λ.

By applying Wien's displacement law, we obtain:

$$\lambda_{max} = \frac{2900}{T} \qquad [2.5]$$

By applying equation [2.5] to the Sun and the Earth, for instance, we obtain:

$\lambda_{Sun\ max} = 0.5\ \mu m$ for $T_{Sun\ max} = 5{,}800$ K

$\lambda_{Earth\ max} = 11\ \mu m$ for $T_{Earth\ max} \approx 255$ K

Figure 2.2 shows the Plank's radiation law, using logarithmic scales for both x- and y-axes, showing the energy emitted by an element of surface in Wm^{-2} for radiation at different temperatures and as a function of wavelength. The curve at

T ≈ 6,000 K represents the solar radiation, while the one at T ≈ 300 K represents the Earth radiation. The Wien displacement law, also represented and emphasized in Figure 2.2 with dashed lines, is responsible for the fact that the summit of the curves is different depending on the temperature, as the wavelength of the emission maximum is inversely proportional to temperature.

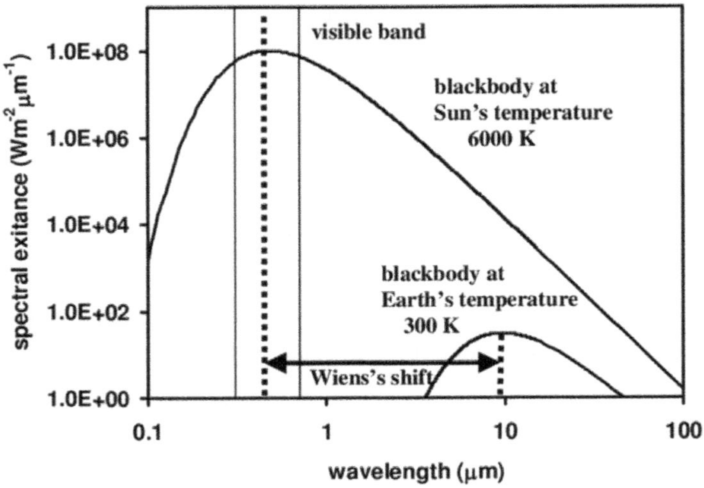

Figure 2.2. *Planck's radiation law for the sun (T = 6,000 K) and the average earth surface temperature (t = 300 K) with logarithmic scales for x- and y-axes. The wavelengths corresponding to emission maxima for these two temperatures are emphasized with dashed lines. Consistent with Wien's law, the radiation maximum is shifted towards longer wavelengths as temperature decreases (Source: Tempfli et al. (2009))*

There is no point for the purpose of this course to go into the detailed calculations. Basically, the two important computational steps are an integration over all wavelengths through Stefan–Boltzmann's law and the picking-up of the maximum wavelength using the Wien displacement law.

2.2. Absorption and greenhouse effect

Figure 2.3 shows that the atmosphere is not really a blackbody and behaves in a more complex way.

At approximately T ≈ 300 K, the curve represents the emission in the energy domain corresponding to the Earth.

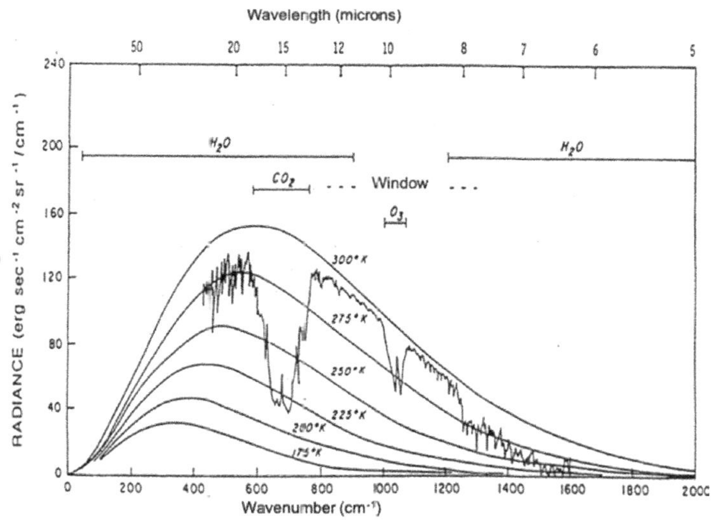

Figure 2.3. *The "real atmosphere" infrared radiation (Source: Lloyd (2005))*

The different lines show what would be the theoretical emission of a blackbody at 175 K, 200 K, 225 K, 250 K, 275 K and 300 K.

There are wavelengths where the actual Earth radiative emission is really aligning with the theoretical blackbody curve, while for others, there are important differences.

Figure 2.3 shows that when CO_2 absorbs radiation, the emission to space occurs at higher altitudes, which correspond to lower temperatures.

We can also see visually in Figure 2.3 that the water vapor absorption signal is more continuous than that of CO_2 and O_3 in terms of location of absorption. This diagram is useful to show that the infrared (IR) absorption signatures of species of the real atmospheres are measured by satellites and the relationship with temperature, illustrating that the complexity of the real Earth is more complex than a real blackbody.

Let us now write the basic energy conservation equations for the propagating energy, this time incorporating some of the physics introduced in Box 2.1.

Box 2.1 presents some important equations including the derivative of the energy in equation [2.7]: the energy I is emitted in a range depending on the frequency μ. It is emitted in a solid angle with a direction given by the vector $\vec{n}$, with s being the distance along $\vec{n}$.

From blackbodies to the radiation of real bodies

All bodies radiate and absorb electromagnetic radiation.

We describe the equilibrium conditions of this body with its environment.

Then, its emission profile **ε(μ)** is proportional to its absorption profile **κ(μ)**.

The Kirchhoff law can be written as follows:

$$\varepsilon(\mu) = \kappa(\mu)\, B(\mu, T) \qquad [2.6]$$

B(μ, T) is universal.

For a blackbody, **κ** = 1, therefore:

$$\varepsilon(\mu) = B(\mu, T) \qquad [2.7]$$

Kirchhoff law and the radiative transfer equation

In conditions of local thermodynamic equilibrium:

$$\frac{dI(\mu,\vec{n})}{ds} = -\kappa(\mu)\left[I(\mu,\vec{n}) - B(\mu,T)\right] \qquad [2.8]$$

In equation [2.8], $I(\mu,\vec{n})$ refers to disappearing energy, while B(μ, T) corresponds to reappearing energy, and **-κ(μ)** to the absorption parameter for a given layer of the atmosphere.

If we note:

$$d\tau(\mu) = \kappa(\mu)ds \text{ (optical thickness variation)} \qquad [2.9]$$

Then:

$$\frac{dI(\mu,\vec{n})}{ds} = -\left[I(\mu,\vec{n}) - B(\mu,T)\right] \qquad [2.10]$$

Box 2.1. *From blackbodies to the radiation of real bodies. Kirschoff's law and the radiative transfer equation*

The aim here is to study the energy that is propagating in a certain direction with a certain frequency: some of it is "disappearing" and some is "reappearing".

Two main processes are in effect involved:

– A process of absorption along s in the $\vec{n}$ direction, and dI is the emitted energy expressed in $W.m^{-2}.str^{-1}.nm^{-1}$.

– A process of re-emission that may take different forms: in the domain of terrestrial radiation, it corresponds to the emission from the atmosphere mostly and from the ground (if this equation is used for solar radiation, as we will do later, the result will be different because in that case, no energy is re-emitted in the same range of wavelengths).

If we consider terrestrial radiations only and we do the summing-up over terrestrial frequencies, we can write:

dI = – absorption + ***re-emission*** [2.11]

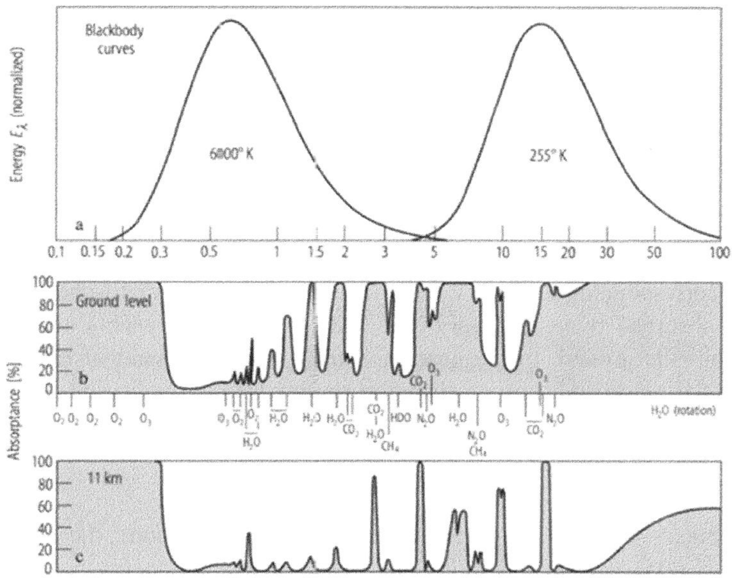

Figure 2.4. *Scheme of blackbody curves corresponding to temperature and wavelengths for downward solar radiation and for terrestrial radiation at the top of the atmosphere (a), of the absorptance of the earth radiation as seen from its ground (b) and higher up, at an altitude of 11 km (c). The x-axes stand for wavelengths in microns (Source: Ohmura and Raschke (2005), adapted from Goody and Yung (1989))*

Figure 2.4 is a scheme showing, on the left of panel (a), the blackbody emission at a temperature of about 6,000 K corresponding to solar radiation with associated wavelengths between 0.2 and 4 microns, and, on the right of panel (a), the blackbody emission at about 255 K corresponding to the terrestrial heat radiation, with a range of related wavelengths between 4 and 100 microns. Those two different energy domains are quite distinct and have almost no overlap except a very small one around 4 microns.

The emitting heat body is the Sun, the body capturing its radiative energy is the Earth. But only one-fourth of the solar irradiance is received by the Earth ($\approx$ 342 $W.m^{-2}$), corresponding to the solid angle between the Earth spherical surface and the solid angle from the Sun radiation. On a yearly scale, the amount of energy entering the terrestrial system and the amount of outgoing energy are more or less at an equilibrium. Therefore, it can be considered that the Earth is in a radiative balance with space on an annual basis.

In parts (a) and (b) of Figure 2.4, the absorptance is expressed in percentages of different species: O_2, O_3, H_2O, CO_2, CH_4 and N_2O mainly. The spectral transparency in clear-sky atmospheric conditions for the downward radiation is represented for those different molecules, respectively, as seen from the ground of the Earth in panel (a), and as seen at an altitude of 11 km in panel (b). Those two panels show well the absorption bands related to each species at various wavelengths from the shortwave ultraviolet to the thermal IR in the atmosphere.

We note that the terrestrial radiation when crossing the atmosphere is almost completely absorbed, but in the solar part, there is a complete absorption of the radiation corresponding to the UV radiation absorbed in the stratosphere. All of the energy is absorbed above, and very little energy is left. And visibly nearby, a large part reaches the ground. In the near-IR domain, the same kind of molecules act, absorbing the energy. Absorption always corresponds to the excitation of molecules.

Basically, there are two kinds of processes at play here, which can be described as follows for simplification:

– A process of diffusion (alternatively known as reflection): the basic idea of diffusion is just to consider a target and a photon with an energy hμ, μ being the frequency and h being Planck's constant. The main point of diffusion is that the path of the photon is changed, not its frequency. Therefore, its energy does not change neither along the way.

Diffusion or scattering is a process that modifies the direction of the photons but does not (directly) involve absorption (and heating). It applies mainly to solar radiation (diffusion of terrestrial radiation will be neglected) and takes three main forms in the atmosphere:

– Molecular diffusion also named the Rayleigh scattering (~ 6% of the incident solar radiation): this phenomenon explains the blue color of the sky (and the oceans or thick ice) but has no climate impact.

– Particular diffusion related to clouds and aerosols, named Mie scattering (~ 20% of incident solar radiation): this phenomenon has no colored effect but a strong climatic impact.

– Ground and ocean reflection (~ 4% of incident solar radiation): this phenomenon has a strong climatic impact.

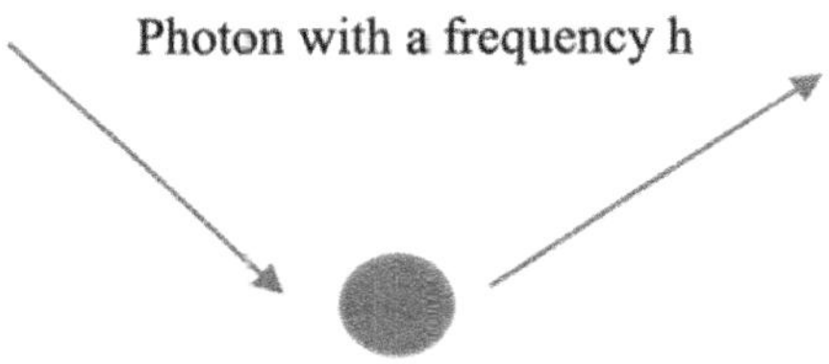

Figure 2.5. *Scheme of diffused photon, with a frequency h being unchanged along a changed path*

– Another important process is absorption. If we consider a molecule “excited”, it is actually brought to different levels of energy. The photons lose their “excitation” from one energy level to another energy level that brings the energy to another level. The distribution of these energy levels actually depends on temperature, and then this energy is lost again when the molecule de-excites. A molecule is either excited or it loses its excitation, meaning that it is respectively warming or cooling.

Therefore, the absorption of a radiation means a warming, while when there is a relaxation after the absorption, it implies a cooling. We will not enter here into the details of this physics, but experts working on radiation have to look at this effect in detail. Spectroscopy specialists especially compute in detail the absorption of each molecule in the atmosphere and how it affects each ray of energy. Establishing all of the properties of all of the molecules for the atmosphere in terms of absorption has been a huge work that was necessary for the use of satellite data among many other things.

This simple equation delivers a simple look at how complex processes are framed near local equilibrium. No more computational details are necessary at this stage for this part of the course.

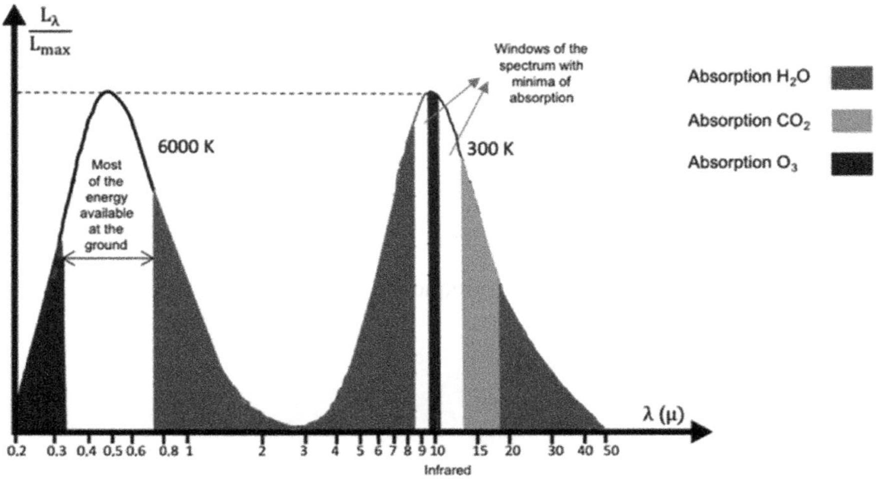

Figure 2.6. *Scheme of blackbody emissions at 6,000 K and 300 K, using a simplified view of the separation between solar and terrestrial radiations*

Figure 2.6 represents another view of the difference of absorption with some indication on the wavelength and some representation of the main atmospheric gases; a little more simplified in comparison to Figure 2.4(a).

We distinguish the terrestrial ***IR***, the absorption from the top to the bottom of the atmosphere: in some places, there is much less absorption in what are called "windows of the spectrum". This is a way we can see the bottom of the Earth from the top of the atmosphere in the ***IR*** (almost all satellite images we see on TV are captured using those channels). They allow us to see the bottom of the Earth from the top of the atmosphere in the ***IR*** domain.

We can also see the ice that roughly sees the light between 0.3 and 0.6–0.7 micrometers: it really depends on how much energy is available.

We aim to apply this to simple cases in order to derive basic physical insights into the vertical distribution of temperature.

When we talk about blackbody emission:

– the frequency of maximal emission is a function of the temperature T;

– B is proportional to T^4;

– an important result is that when we sum over all frequencies and all directions (in half a sphere), the emission per unit surface can be written as follows:

$$E = \pi B = \sigma T^4 \quad [2.12]$$

with $\sigma = 5.67 \times 10^{-8}$ W. m^{-2}. K^{-4} being the Boltzmann constant.

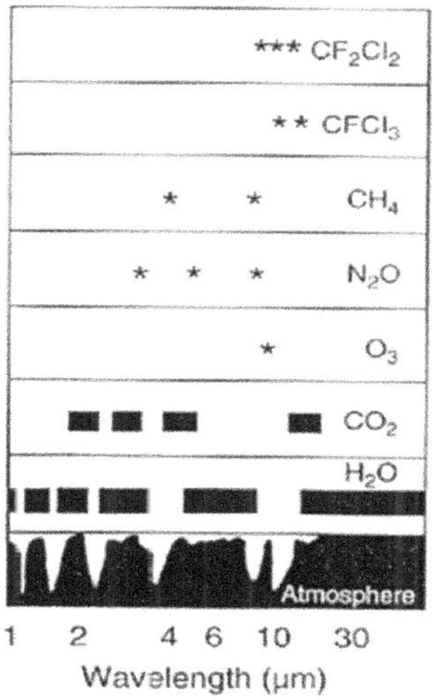

Figure 2.7. *Scheme of absorption bands for greenhouse gases*

Figure 2.7 is another view of the absorbing bands for the main greenhouse gases. Let us work on a very simple equation to deliver orders of magnitude concerning the Earth before more complex computations.

If we consider a system without the greenhouse effect, eventually no atmosphere, the equilibrium equation for the Earth climate system can be written as follows:

$$\sigma T^4_{no\ atmosphere} = R_s (1 - \alpha) \quad [2.13]$$

with $\sigma = 5.67 \times 10^{-8}$ W. m^{-2}.K^{-4} being the Boltzmann constant.

From equation [2.13], we obtain a temperature $T_{no\ atmosphere} = 254$ K, which roughly corresponds to ~ -30°C, while the value of the actual current annual mean global temperature is $+15$°C. In this virtual configuration, the Earth's mean temperature is therefore about -45°C under the real mean global temperature.

Let us now consider a simple representation of what the greenhouse effect may be, as represented in Figure 2.8.

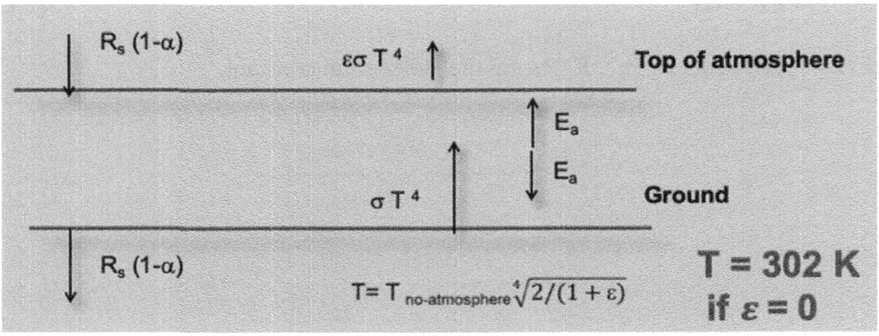

Figure 2.8. *A simple representation of the greenhouse effect*

The solar radiation is entering the system at the top of the atmosphere. We will now simplify as follows. There are two important processes: diffusion and absorption.

To perform a detailed analysis of the system's physics, let us consider each of the four boxes in Table 2.1.

	Solar energy	Terrestrial energy
Diffusion	×	
Absorption		×

Table 2.1. *Diffusion and absorption for solar and terrestrial energy*

If we want to simplify, the two very important processes are the boxes with crosses in them in Table 2.1:

– The first aspect of importance is the diffusion of solar energy. Solar energy is also absorbed in the lower atmosphere, though not significantly; alternatively, it is absorbed backward in the stratosphere. The major absorption of solar energy happens on the ground, and the absorption by the lower atmosphere is not very large: even though it is an important phenomenon, we will not look at it in much detail.

– Another important process refers to terrestrial energy, for which the reverse applies: in that case, absorption is very important (even though there is existing diffusion, that is, a smaller effect in comparison).

Therefore, we will focus on those two processes to make simple computations to explain in great lines this system mechanism.

Therefore, there will be some similarity between the two equations related to solar and terrestrial radiations.

In the solar domain, the absorption is replaced by a change in the direction: photons that are not originally directed in the $\vec{n}$ direction are afterwards directed in the $\vec{n}$ direction.

Let us now consider the context of the absorption of back terrestrial energy. Part of the incoming solar energy is reflected to space (this is the albedo), a small amount of energy travels through the atmosphere without being absorbed ($\varepsilon\sigma T^4$), and what is absorbed is reemitted to the surface.

Two equations that represent the phenomenon can be written in order to describe the energy entering the system and the energy leaving it, and there is an equilibrium between these two energy fluxes at the top of the atmosphere. There is also an equilibrium between the two at the bottom of the atmosphere.

From the numerical resolution of the equations related to Figure 2.8, we very easily obtain two equilibriums, one above and one below the atmosphere, and we find that the true temperature at the bottom is the temperature when there is no atmosphere with this value:

– If $\varepsilon = 1$, which means that there is no absorption and re-emission and that the incoming energy is getting through the atmosphere, we found that the two equilibrium values are equal.

– If $\varepsilon = 0$, which means, on the contrary, that all of the incoming energy is absorbed, and the energy is being reemitted to the ground and reemitted to space, leading to a much warmer earth temperature of 300 K (30°C), which is high.

This computation, simply based on the two equations described in Figure 2.8 and leading to a result in terms of temperature coming from the sum of both, shows that a merely radiative approach with only the greenhouse effect leads to a hot global earth temperature that is contradictory to the real one. We can also show that with a merely radiative equilibrium, the system would not work for the description of Hadley cells and the associated circulation. Therefore, an additional process is needed to realistically lower the temperature, as the values predicted by the model in Figure 2.8 are warmer than observed.

2.3. The role of convection

This additional necessary process must be accounted for in order to explain the real earth temperature is convection, emerging from the fact that when the air becomes hot enough near the ground, it goes up and mixes with higher layers, as it will be detailed later. Therefore, the equilibrium of temperatures in the atmosphere is not radiative but rather radiative–convective.

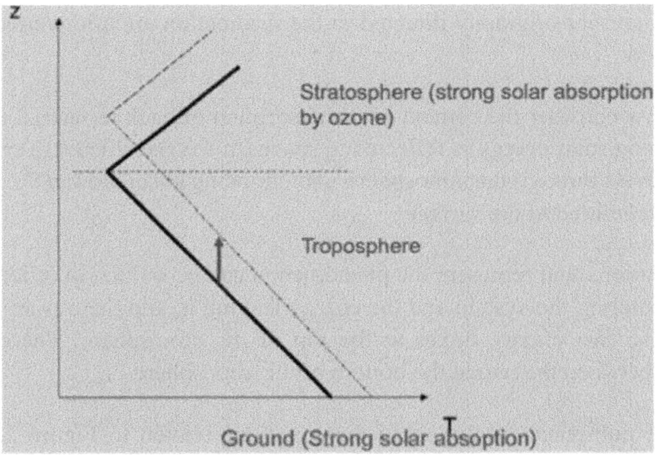

Figure 2.9. *Scheme illustrating the role of stratification in climate change*

Figure 2.9 shows the shape of the atmosphere as a function of temperature T and height z. We can see that the temperature decreases from the ground and then increases again because of the tropopause, a layer in between the troposphere and the stratosphere. The reason for this simple shape is the heating of the atmosphere by solar input, which is done mostly in two areas: near the ground and also in the stratosphere due to the ozone absorption. Therefore, if we mostly consider the source of heating due to the ground, we note that the more the height is away from the source of heating, the more the temperature tends to decrease with higher heights. The slope of this curve is very important. We see that with an increased greenhouse effect, this curve tends to be displaced upwards, and whether this curve displaced upwards manages to the same late rate, meaning the same derivative in terms of temperature to altitude and whether this derivative remains parallel with an added greenhouse effect and slightly lifts the distribution upwards, is very important. The aim is to identify what determines this decrease of temperature from the ground. For this purpose, we will examine equations of radiation to reconstruct the gradient of temperatures through this formalism. Let us first consider a simple classical equation for modeling the greenhouse effect. We make it continuous, however, two levels must be considered: the top of the atmosphere and the ground:

$$\frac{dI}{ds} = -\kappa\,(I - B) \qquad [2.14]$$

We will sum up all of the parameters to get one simple equation, and we will make a two-stream approximation, which is usual in climate models.

We can write I^+ for a radiation that goes up, and I^- for a radiation that goes down. To compute this, we integrate over a solid angle upwards and respectively over a solid angle downwards, over frequencies, in order to obtain the result of this summing up.

We define the parameter τ, the optical thickness:

$$\mathrm{d}\tau = \mathrm{K}\,\mathrm{ds} \qquad [2.15]$$

When we rewrite the equation as such, without worrying about the direction and by integrating the term of absorption, in τ which is dimensionless, we obtain:

$$\frac{dI}{d\tau} = -\kappa\,(I - B) \qquad [2.16]$$

If we forget the re-emission, we can write the following simplified equation:

$$\frac{dI}{d\tau} = -I \qquad [2.17]$$

In this case, the solution would be:

$$I = I_0 e^{-\tau} \qquad [2.18]$$

This gives us some kind of vertical dimension, which is in fact dimensionless and is a metric indicating how the radiation is absorbed and a vertical description of the system.

If we sum up the equations, we must consider the direction in which we go:

– upward equation:

$$\frac{dI^{+}}{d\tau} = -I + B \quad [2.19]$$

We have to integrate that equation over all of the system. This integration must be done over solid angle and over frequencies (we saw that absorption strongly depends on frequency), using the following simplified equation:

$$\frac{dF^{+}}{d\tau^{+}} = -F^{+} + \sigma T^{4} \quad [2.20]$$

We then obtain the flux of energy, which is summed up over frequencies and over the complete solid angle upward F^{+}.

$d\tau^{+}$ is the convention used here to signify that the study is delivered in one direction, which is an upward motion.

A similar equation can be written for the radiation going down:

– upward equation:

$$\frac{dF^{-}}{d\tau^{-}} = -F^{-} + \sigma T^{4} \quad [2.21]$$

$d\tau^{-}$ is the convention used here to signify that the study is delivered in one direction, which is a downward motion. F^{-} is the flux of energy, which is summed up over frequencies and over the complete solid angle downward.

We need to account for all of the upward and downward fluxes.

2.3.1. *Where does the - $F^{\pm}$ comes from?*

When we write the equation in the direction of propagation, τ gives the direction of propagation, F^{-} propagates downward and is absorbed downward. Of course, the direction of propagation in the two equations [2.20] and [2.21] above are not the same. Therefore, if we want to have the same τ in the two equations, knowing that the equations correspond to propagation in the same media, we need to write:

$$d\tau^{+} = -\, d\tau^{-} \quad [2.22]$$

Therefore, we need to consider the fact that we must study how the flux changes along the direction of propagation. This leads to the following simple system of equations:

$$\begin{cases} \frac{dF^+}{d\tau} = - F^+ + \sigma T^4 & [2.23] \\ \frac{dF^-}{d\tau} = + F^- - \sigma T^4 & [2.24] \end{cases}$$

This system is a basic tool to learn a little about cloud radiation in the system. The principle for solving those equations is simple and typical with this kind of symmetry: we consider the sum divided by 2 of both fluxes (upward and downward), as well as their difference:

$$\Delta = \frac{F^+ + F^-}{2} \qquad [2.25]$$

$$A = \frac{F^+ - F^-}{2} \qquad [2.26]$$

In terms of physics, if we do not assume solar absorption, it then leads to a sort of conservation law.

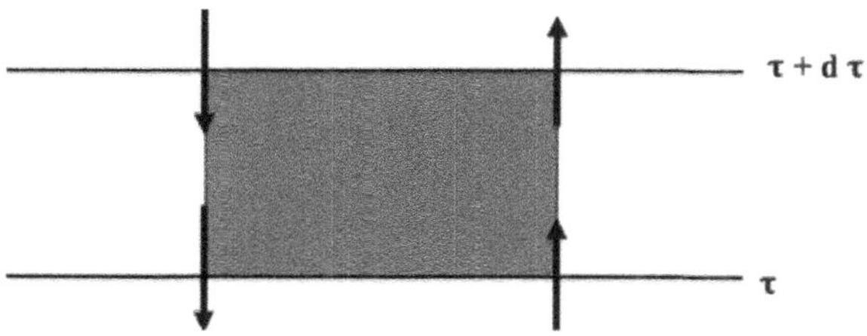

Figure 2.10. *Scheme illustrating upward and downward fluxes for two different levels corresponding to τ and $\tau + d\tau$*

Let us consider the scheme of Figure 2.10 with levels τ and $\tau + d\tau$.

From equation [2.25], if we look at $\frac{d\Delta}{d\tau}$, what enters the "box" between the levels τ and $\tau + d\tau$ has to leave the box, except if there is solar absorption. In order to have an equilibrium without solar absorption, there must be an equilibrium between those areas, and we thus obtain:

$$\frac{dA}{d\tau} = -\Delta \qquad [2.27]$$

$$\frac{d\Delta}{d\tau} = -A + \sigma T^4 \quad [2.28]$$

In addition, if there is no solar absorption, $\frac{d\Delta}{d\tau} = 0$. This important, as the absence of solar absorption actually creates the balance between what goes up and what goes down.

Therefore:

$$A = \sigma T^4 = -\Delta\tau + \text{constant} \quad [2.29]$$

This allows us to develop these equations and to have a profile of temperature as a function of τ.

Other important elements to be considered are boundary conditions: in fact, we must determine what escapes at the top of the atmosphere, yielding a value known as the "temperature of equivalent emissions", which represents the energy leaving the system and must, ultimately, balance the incoming solar radiation.

There are also conditions near the surface to be accounted for: the surface temperature provides the upward energy in the system. This happens at the bottom, near the ground.

After computing, we will then discuss the limits, why it is not completely correct, and especially why we need to have a convective redistribution of temperatures.

In the troposphere, the absorption by the sun is limited. We study the lapse rate along the vertical to see if something else is needed to describe the system. The "missing ingredient" is convection.

The equivalent temperature can be written as follows:

$$R_t(T) = \sigma T_e^4 \quad [2.30]$$

with $R_t(T)$ corresponding to the global solar absorption that goes to the Earth and is reemitted by the Earth.

The equivalent temperature can be defined as follows: it refers to the temperature of a body that would emit to space the same temperature as the one on Earth. A term releasing some external condition from the sun is also needed.

For the boundary conditions, with τ_0 being the optical thickness at the top of the atmosphere, and τ_0 - 1 the optical thickness at the ground, we can write the following.

For $\tau = \tau_0$, at the top of the atmosphere:

$$F^+ = \sigma T_e^4 \text{ (definition of } T_e) \quad [2.31]$$

$$F^- = 0 \quad [2.32]$$

For $\tau = 0$ at the ground:

$$F^+ = \sigma T_s^4 \text{ (emission)} \quad [2.33]$$

Hence, for $\tau = \tau_0$ at the top of the atmosphere:

$$A = \Delta = \frac{1}{2}\sigma T_e^4 \quad [2.34]$$

This represents what gets out.

For $\tau = 0$, with the term in italics representing the condition near the surface (what is absorbed):

$$A = F^+ - \Delta = \sigma T_s^4 - \frac{1}{2}\sigma T_e^4 \quad [2.35]$$

For $\tau = \tau_0$ at the top of the atmosphere:

$$\frac{F^+}{2} = A = \Delta = \frac{1}{2}\sigma {T_e}^4 \quad [2.36]$$

For $\tau = 0$, some of the descending radiation goes back to the ground $F^- \neq 0$, and therefore:

$$A = F^+ - \Delta = \sigma T_s^4 - \frac{1}{2}\sigma T_e^4 \quad [2.37]$$

For τ at any height:

$$T^4 = \frac{1}{2} T_e^4 \,(1 + \tau_0 - \tau) \quad [2.38]$$

It is possible to measure temperatures at all levels because the height is measured by τ_0.

The temperature reached at T_s is a temperature that serves to emit. Temperature decreases with τ, meaning that temperature decreases with height.

Two conclusions can therefore be taken from those results: this gives a simple continuous model of greenhouse gases, and there is a large rate of decrease of temperature with height.

If the temperature at the ground is too large, there is convection, because the vertical lapse rate of temperature is too large to prevent from convection. There is a very large decrease in temperature with the source of heat, a nonpermanent state that is modified by convection.

The equivalent emission level is the level where the temperature is T_e, and we need to have:

$$\tau = \tau_0 - 1 <=> T^4 = T_e^{\ 4} \quad [2.39]$$

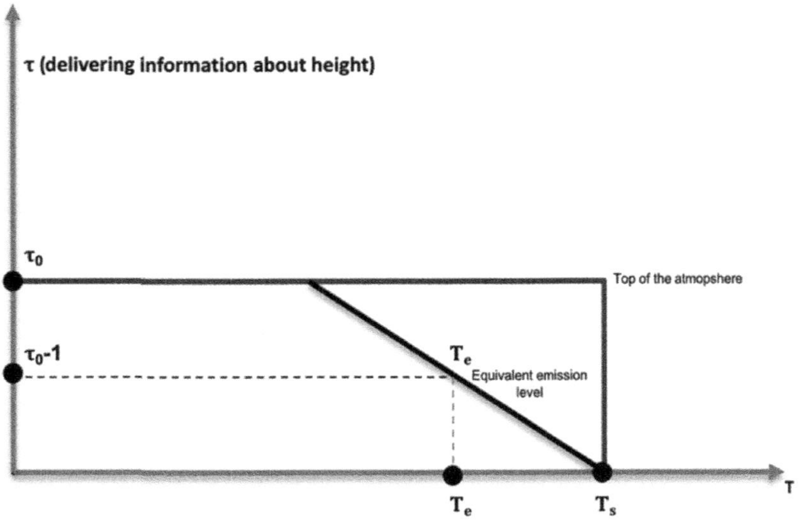

Figure 2.11. *A simple scheme illustrating temperature as a function of height in the atmosphere and the notion of "equivalent emission level"*

The level at which the atmosphere is emitting to space is well defined with more CO_2 (i.e. with a more opaque atmosphere: the level where the atmosphere is absorbed is then higher with a stronger greenhouse gas effect). The higher the CO_2 concentration in the atmosphere, the more energy coming from the ground is absorbed by this CO_2. All of the absorbed energy is then reemitted at different heights and temperatures. Convection is a main related issue for the locations affected by this phenomenon.

If we want to discuss a simple approach to stratification of the atmosphere, we may ask ourselves what will happen with a strong vertical gradient due to the absorption of solar radiation. For that, some thermodynamics tools are needed:

– Using the first principle of thermodynamics, let us look at the atmosphere with a few parameters to provide a simplified view of the phenomenon. When considering a gas such as the atmosphere (bi-variant), we can describe it by two parameters: the pressure P representing the vertical height and the entropy S. There is conservation of energy if a motion is not quick enough. Therefore, we can write the density of the atmosphere ρ as a function of two independent thermodynamic parameters, the entropy per mass unit S, sensitive to warming, and the pressure P linked to height with height through the hydrostatic equilibrium:

$$\rho = \rho\,(P,S) \qquad [2.40]$$

The pressure equilibrates very quickly close to the hydrostatic equilibrium:

$$\frac{\partial P}{\partial z} = -\rho g \qquad [2.41]$$

This equation is a representation of the vertical scale. It gives a good approximation of the height of the particles and equilibrates very quickly.

The entropy S is a representation of the energy exchanges through the well-known equation:

$$TdS = \dot{Q} \qquad [2.42]$$

with dS being the variation of the entropy S and $\dot{Q}$ the heating rate. For a short-term motion, S is conserved, meaning that the exchange does not have enough time to develop largely (as an example, in winter, the air coming from the Arctic remains cold).

We can also define the potential temperature in the atmosphere as follows:

$$\theta = T(\frac{P}{P_n}) \qquad [2.43]$$

The temperature of air mass may be modified by adiabatic dilatation (without heat exchanges).

Indeed, when we consider an air mass, between the top of the atmosphere and the ground, there are about 40 degrees of difference. The pressure of this air mass traveling from bottom to top varies, as does the temperature because:

$$P = \rho RT \quad [2.44]$$

Actually, pressure varies very quickly with height: we take away the part of the modification due to adiabatic exchanges, and we consider only the part due to heat exchanges.

Entropy is, in fact, a quantity that will not change when the air mass comes from Siberia to France, the potential temperature of an air mass does not change because of conservation of entropy and we can write the following equation:

$$\theta = T\left(\frac{P}{P_{Surface}}\right)^{R/C_P} \quad [2.45]$$

with R being the perfect gas constant and C_P being the calorific constant, which measures how much heat the Earth is able to take.

For stable conditions, if we raise the parcel up by δ, then it will go down:

$$\rho\,(P + \delta P, S) > \rho\,(P + \delta P, S + \delta S) \quad [2.46]$$

What makes the difference in this term is the entropy S. Knowing that we have been looking at an upward motion, we have:

$$\delta S > 0 \quad [2.47]$$

It depends on the sign of $\frac{\partial P}{\partial S}$: if $\frac{\partial P}{\partial S} < 0$, then ∂S will decrease.

Therefore,

$$\delta S > 0 \quad \Longrightarrow \quad \frac{ds}{dz} > 0 \quad [2.48]$$

because if the conditions are stable, the parcel of air has to be denser than ambient air, so it goes back to its original state (stability condition):

$$S = C_P \ln \ln (\theta) + \text{Constant} \quad [2.49]$$

Equation [2.48] shows that the potential temperature has to increase with height.

Let us now complete equations for a simple approach to stratification of the atmosphere:

$$\rho \cong \rho\,(P_0, S_0) + (s)(P_0, S_0) \times\ (P -\ P_0) + (P)(P_0, S_0) \times\ (S -\ S_0) \qquad [2.50]$$

when S increases, θ increases.

$$\delta P = \frac{\partial P}{\partial z}\delta z \qquad [2.51]$$

$$\delta S = \frac{\partial S}{\partial z}\delta z \qquad [2.52]$$

$$\rho_{\text{ambient air (environment)}} \cong \rho\,(P_0, S_0) + (s)(P_0, S_0) \times (\delta P) + (P)(P_0, S_0 \times (\delta S) \qquad [2.53]$$

$$\rho_{\text{particle}} \cong\ \rho\,(P_0, S_0) + (s)\,\rho\,(P_0, S_0) \times (\delta P) \qquad [2.54]$$

The stability condition for equations [2.53] and [2.54] is:

$$\frac{\partial \theta}{\partial z} > 0 \qquad [2.55]$$

and with the hydrostatic approximation, we have:

$$\frac{\partial (C_P T + gz)}{\partial z} > 0 \qquad [2.56]$$

$$\theta = T\left(\frac{P}{P_{\text{Surface}}}\right)^{-R/C_P} \qquad [2.57]$$

For *hydrostatic equilibrium*, we have:

$$\frac{1}{\theta}\frac{\partial \theta}{\theta z} = \frac{1}{T}\frac{\partial T}{\partial z} - \frac{R}{C_P}\frac{1}{P}\frac{\partial \mathbf{T}}{\partial \mathbf{z}} \qquad [2.58]$$

$$\Longrightarrow$$

$$\frac{1}{\theta}\frac{\partial \theta}{\theta z} = \frac{1}{T}\frac{\partial T}{\partial z} - \frac{R}{C_P}\frac{1}{P}(-\rho g) \text{ with } \rho g = \frac{P}{RT} \qquad [2.59]$$

$$\Longrightarrow$$

$$\frac{1}{\theta}\frac{\partial \theta}{\theta z} = \frac{1}{T}\frac{\partial T}{\partial z} + \frac{1}{T}\frac{1}{C_P}\,g > 0 \text{ (this term has to be positive)} \qquad [2.60]$$

and:

$$\frac{\partial T}{\partial z} > \frac{-g}{C_P} \quad [2.61]$$

With $C_P \approx$ 1,000 S.I. (1,000 joules per unit of mass), this equation gives the convective lapse rate, i.e. the rate at which temperature decreases with height:

$$\frac{g}{C_P} \approx -10°\text{C. km}^{-1} \quad [2.62]$$

This computation may not sound intuitive if hot air goes up. What makes the balance is the pressure between the bottom and the top of the atmosphere: hot air is defined by the potential temperature, with which we have taken away/corrected the changes of pressure by adiabatic dilation. Therefore, there is a competition in the atmosphere between changes of temperature in height due to convection. This is the reason why low temperatures are reached.

Radiative equilibrium does not really exist because the vertical gradient decreases with height, and therefore, the temperature of the surface goes much quicker than the one allowed from the laws of convection. The example of the Hadley cells, which take the energy from downward to upward, illustrates that the general mechanics does not work without this stability criterion.

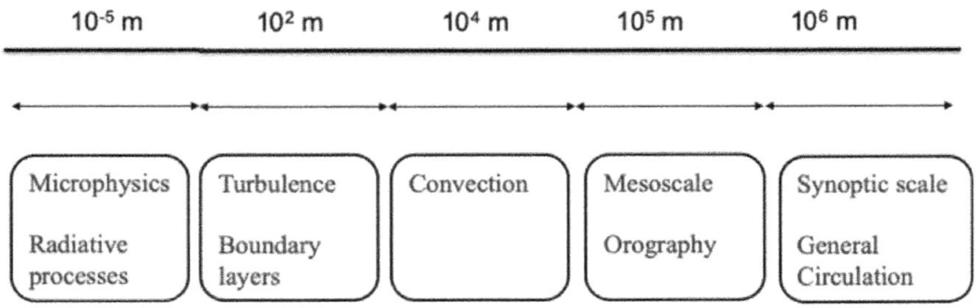

Figure 2.12. *The atmospheric motions: a large range of dynamical scales*

Figure 2.12 shows the different orders of magnitude of the atmospheric motions, which affect and change the radiative processes. The synoptic scale is associated with general circulation. The mesoscale is related to winds such as Mistral, associated with orography. Turbulence and convective phenomena are associated with hot air near the surface/in the boundary layers. As for the smaller scale of microphysics, a large part of the greenhouse gas effect is due to clouds.

We can write simplified equations for virtual temperatures by using parameters (moist air thermodynamics) by setting P_a, the pressure of dry air and P_v, the pressure of water vapor, both expressed per mass:

$$P = P_a + P_v = \rho_a R_a T + \rho_v R_v T \quad [2.63]$$

The virtual temperature is the temperature of the gas to which we add components that give a perfect gas.

Let us introduce a new stability condition, with S_a as the dry air entropy (mostly referring to nitrogen and oxygen), S_v as the entropy per mass unit of water vapor and r_v as the density of the total mass of water vapor:

$$\frac{\partial}{\partial z}(S_a + r_v S_v + r_l S_l) > 0 \quad [2.64]$$

which means that an increase in entropy with height is necessary for stability.

$$r_v = \frac{\rho_v}{\rho_a + \rho_v + \rho_l} \quad [2.65]$$

We make the following simplifications: ρ_v is small, so equation [2.65] is approximately equal to:

$$r_v \cong \frac{\rho_v}{\rho_a} \quad [2.66]$$

If we write r_l, the density of total mass of liquid water that stays in the cloud and is associated with the air, we can write the following sum to obtain the entropy S:

$$S = S_a + r_v S_v + r_l S_l \quad [2.67]$$

One way to describe the increase in height is to make a small modification in equation [2.67], with the difference S_v - S_l reflecting the change of water phase from vapor to liquid. At saturation, this ratio can be written as $\frac{L_v}{T}$, and we can therefore write the following equation for the latent heat:

$$\frac{L_v}{T} = S_v - S_l \quad [2.68]$$

This quantity is conserved, meaning that the total mass of water is not changing for non-precipitating clouds (it only changes with precipitating clouds).

A description of how it will increase with height is proposed below, with r_v^* as the saturation level of water:

$$\frac{\partial(S_a+\frac{r_v^* L_v)}{T}}{\partial z} > 0 \qquad [2.69]$$

The hydrostatic approximation gives:

$$\frac{\partial(C_{pa}+gz+r_v^* L_v)}{\partial z} > 0 \qquad [2.70]$$

$$\frac{\partial T}{\partial z} = \frac{\partial(C_P T+gz)}{\partial z} > 0 \qquad [2.71]$$

with $\frac{\partial(C_P T+gz)}{\partial z}$ representing the energy in large part transported by the atmosphere.

We suppose that L_v is constant and that the temperature does not change much. For clouds, we have:

$$\frac{\partial(C_{pa}T+gz+r_v^* L_v)}{\partial z} > 0 \qquad [2.72]$$

where C_{pa} is a term decreasing with height, gz increases with height, and $r_v^* L_v$ decreases with height. We therefore have both terms decreasing and increasing. This situation is a source of complexity.

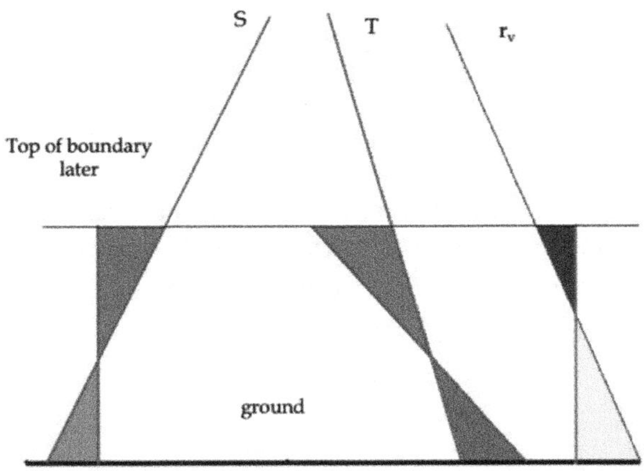

Figure 2.13. *Scheme of low cloud formation, a mechanism reinforced by the interaction with radiation*

Figure 2.13 shows a scheme of low cloud formation, representing the vertical gradient of the lower part of the atmosphere:

$$\frac{\partial(C_{pa}T+gz+r_v^* L_v)}{\partial z} > 0 \qquad [2.73]$$

with $r_v^* L_v$ decreasing with height. L_v is almost constant and disappears when we derive it, but r_v^* strongly decreases with height.

The total mass is not precipitated; it is conserved. The liquid entropy does not change much either.

r_v^* only depends on the temperature level of water vapor at saturation level r_v^* when the air climbs, it loses little by little its water vapor until it becomes very dry at the top of the atmosphere. Therefore, we consider extremely dry clouds with $r_v^* \sim 10^{-2}$ near the tropopause and $r_v^* \sim 10^{-6}$ very high in the troposphere.

This situation makes the criteria delivered by equation [2.73] harder to meet. Therefore, it is harder to stay stable.

For moist instability:

$$\frac{\partial T}{\partial z} > -6.5\ °C.km^{-1} \qquad [2.74]$$

There will therefore be a difference between the gradient of temperatures with or without clouds.

Let us ask ourselves how we can understand the main cloud regimes.

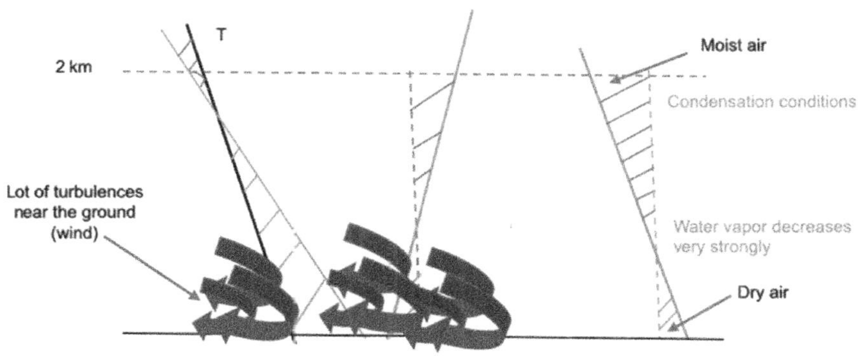

Figure 2.14. *Scheme of low clouds, in the lowest part of the atmosphere (below 2 km)*

What is the effect of turbulence? Turbulence is a phenomenon that actually mixes quantities that are conserved in their air motion. The potential temperature θ is conserved (it was designed for this purpose and exemplified with the case of Polar air). However, in the case of the presence of turbulence, the temperature will not be conserved. But the actual total amount of water will be conserved. We know that for turbulence represented in the blue-colored areas on Figure 2.13, the potential temperature θ can be written as follows:

$$\theta = T\left(\frac{P}{P_0}\right)^{\frac{R}{C_P}} \qquad [2.75]$$

θ and T change their sign at the same height, meaning that at a given height, the sign of change of temperature T will be the sign of change of potential temperature θ.

For water vapor, there is also a mixing. With a view of the situation from above, the potential temperature and temperatures are cooler: turbulence brings more humidity close to the boundary layer and tends to spread.

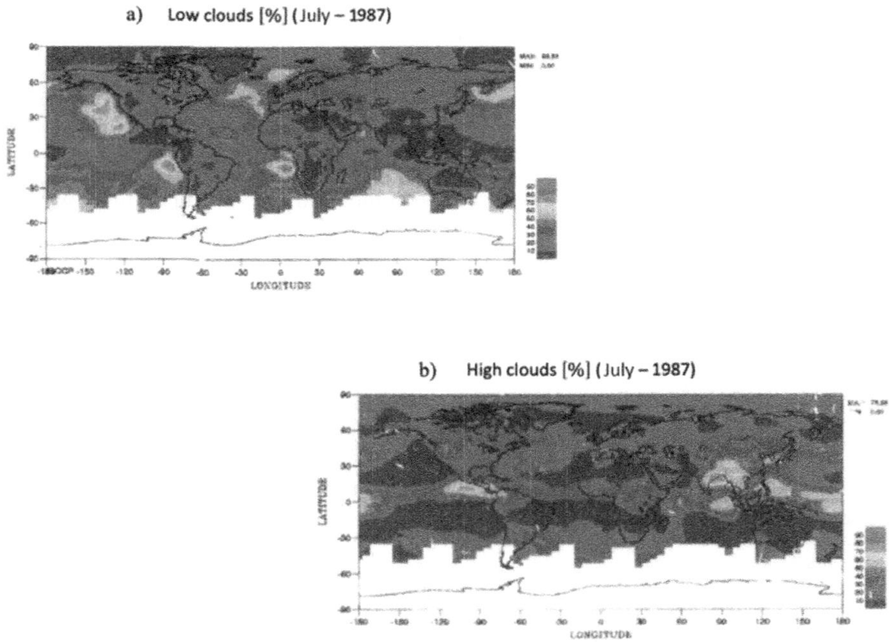

Figure 2.15. *a) Satellite views of low clouds; b) satellite view of high clouds*

Figure 2.15(a) shows low clouds viewed from space, thanks to satellites for a full monthly climatology. They cover large parts of the ocean. Figure 2.15(b) shows high clouds, corresponding to unstable atmospheric moves. For more details, we must look at the horizontal scale.

In order to deliver a grasp of how we understand the main cloud regimes, let us first consider what happens when we adopt the view of the situation from above.

At the top of the boundary layer, the potential temperature and the actual temperature are cooler and the water vapor ratio is larger: there is more humidity, so the turbulence brings it just near the top of the boundary layer with more humidity and cooler air. Therefore, there are condensation conditions there. Low clouds associated with this boundary layer form from the top.

The principle of convective clouds is the following: the air ascends when it becomes too hot, making condensation, which causes the air to ascend a bit quicker. The air that moves down is accelerated by the evaporation in the boundary layer: this is a different process than clouds, which are formed from the bottom. The difference is that we considered stable conditions most of the time for low clouds, which develop from the top of the boundary layer, and stable conditions to have clouds ascending very high in the atmosphere up to 16 km. Those clouds also tend to spread and form cirrus clouds, which extend very far away from the convective cell.

In Figure 2.15, the different satellite images show a full monthly climatology of the clouds, separating between high and low clouds.

Low clouds are what we can see directly from space and are near the ground. They cover a large part of the oceans, with a part of the ocean where conditions are basically stable.

Reversely, high clouds are the ones that we can see well from space, corresponding to unstable conditions. In order to go further into detail, we have to look at the horizontal dynamics based on local thermodynamics to consider the different types of air motions.

Clouds are modifying the radiative budget of the Earth for two reasons: first, they represent greenhouse objects since clouds are mostly the condensed part of the water. Second, they also diffuse the solar radiation (albedo effect), with basic processes linked to the absorption of terrestrial IR radiation from the atmosphere.

Let us now consider the solar energy that comes from space (in the end, we will mix the two approaches).

2.4. Diffusion of solar radiation

As detailed before, the radiative transfer equation for the sun is different to the one for the Earth. Indeed, the terrestrial ***IR*** radiation is not very diffused but is strongly absorbed. Reversely for the solar radiation, half of it reaches the ground. The largest part is actually diffused to space. A large part of the climate change problem is about whether this solar radiation diffusion will increase or not.

Climate models start with basic equations to solve. Let us first consider a very simplified framework of what was done for Earth radiation:

$$\frac{dI}{d\tau} = -I + X \qquad [2.76]$$

where the term $\frac{dI}{d\tau} = -I$ referring to diffusion with little absorption, while X refers to terms redirected in the same I direction.

In fact, some complexity is at stake for describing convection from moist to dry air. A universal element for describing dry air patterns is the necessity that the entropy increases with height, in order to have stable conditions. An interesting point is to know where the conditions become unstable because this really changes the pattern of the occurring phenomena. The equation for the stability condition will therefore involve the entropy of dry air only, which we can note as S_a, for mostly nitrogen and oxygen, which are by far the largest part in the atmosphere, as well as S_v as the entropy for water vapor and with entropy per mass unit water vapor and r_v as the mass ratio referring to the density of the total mass of water vapor.

As a reminder, diffusion is the redirection of photons coming from the sun. When considering the solar energy, we roughly have the following orders of magnitude:

– Diffusion: about 30% (responsible for large fluctuations).

– Absorption by the atmosphere: about 20% (mostly by the stratosphere).

– Absorption by the ground: about 50%.

As already discussed, there are different types of diffusion, including:

– Molecular scattering (radial scattering), which explains why the sky is blue: molecules oscillate, and there is a re-emission of these oscillating molecules. The diffusion will depend on the size of the molecule.

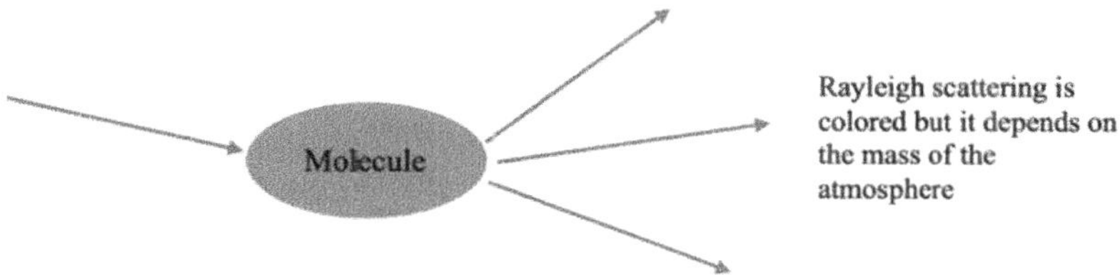

Figure 2.16. *Simplified scheme for Rayleigh scattering*

There is reflection when the molecule vibrates and transmits energy at a forced power of vibration. If an object vibrates very quickly, it transmits more energy. The molecule becomes like an antenna since it vibrates/oscillates and emits at the same wavelength as it receives energy. Therefore, photons of the different colors are not reemitted at the same frequency.

Rayleigh scattering is responsible for the strong effect of color. Scattering with a blue color is a spectacular process, important for many purposes, which reflects the mass of the atmosphere, but this phenomenon is not that important for climatologists.

Aerosols and clouds are much larger targets. They represent about 20% of the incident solar radiation. There is a big disproportion between their mass and their impact: this is one of the reasons why we do not have a great understanding of the climate functioning.

Ground and ocean reflection represent about 50% of solar radiation entering the surface, and 4% varies a lot. Local changes are very strong from the ocean depending on whether the water is in liquid form or is colored white when in ice form, etc.

A very simple experiment consists of orienting a light towards a hole and looking at the light on a large screen.

As diffusion depends on wavelengths, it brings colors. The sizes of the particles make a difference. Clouds are white without remedy because of this factor 2 symbolized in Figure 2.17.

Figure 2.17. *Scheme of light size increase on a screen after passing through a hole*

Let us consider a droplet particle with a radius that we can note as a and a light with a wavelength λ.

We can use the following simplified mathematical description for absorption and diffusion: the interaction between the radiation and the particle is described by the angular distribution function S(θ,ϕ) of the complex amplitude radiated by this particle. A key parameter is the effective cross-section for extinction, defined by the following equation:

$$\delta I = -\,\sigma_{ext}\, I_0 \qquad [2.77]$$

When scaled by the droplet geometrical cross-section, it becomes the efficiency factor for extinction:

$$Q_{extinction} = \frac{\sigma_{extinction}}{\pi a^2} \qquad [2.78]$$

with a being the radius of the particle.

We can have efficiency factors both for absorption and diffusion, respectively:

$$(\delta I)_{absorption} = -\,\sigma_{absorption} I_0, \text{ and } Q_{absorption} = \frac{\sigma_{absorption}}{\pi a^2} \qquad [2.79]$$

and

$$(\delta I)_{diffusion} = -\,\sigma_{diffusion}\, I_0 \text{ and } Q_{diffusion} = \frac{\sigma_{diffusion}}{\pi a^2} \qquad [2.80]$$

Figure 2.18 shows a graph of the efficiency factor Q_{ext} as a function of $x = \frac{2\pi a}{\lambda}$.

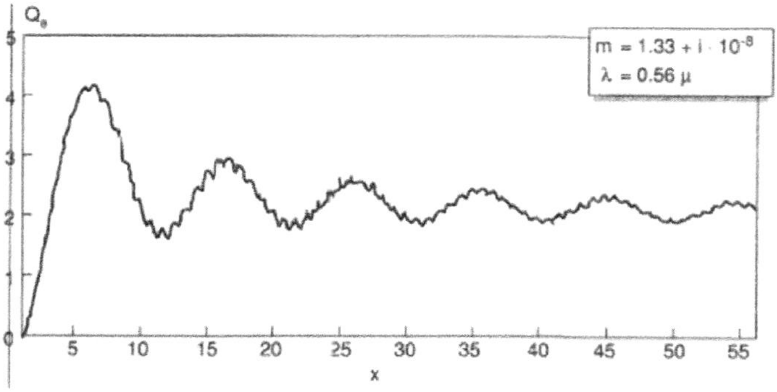

Figure 2.18. *Scheme of the efficiency factor Q_{ext} as a function of* $x = \frac{2\pi a}{\lambda}$

For the Mie scattering, diffusion depends on the wavelength.

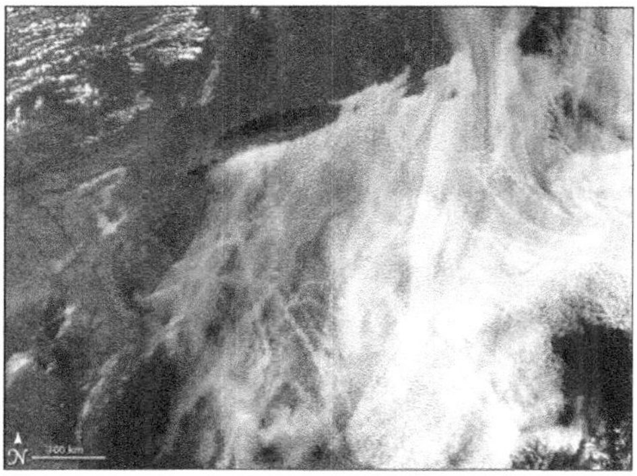

Figure 2.19. *View of low clouds and ship tracks (Source: Moderate Resolution Imaging Spectroradiometer (MODIS) on NASA's Terra satellite, on May 11, 2005, NASA image courtesy Liam Gumley, Space Science and Engineering Center, University of Wisconsin–Madison)*

Figure 2.19 shows a description of homogenous clouds with ships under the low clouds and provides a view of their path from above. In order to understand how fluxes work, we need to know how clouds are formed.

We can write:

$$Q_{efficiency} \cong 2\pi a^2 \quad [2.81]$$

This formalism can be applied to absorption, and there is a symmetry between upward and downward diffusion. Condensed water makes a huge difference for this reason. Indeed, a sky with a large number of cloud droplets is very different to a sky with a small one. Therefore, the cloud droplet distribution matters hugely.

For a large number of cloud droplets, it is possible to analyze the mass of water that does not change and is condensing: for a number N of droplets, we need to know the amount of surface covered by the clouds. Let us consider that the N droplets are from the same radius a. The evolution of the surface covered by the clouds is a function of N. If we write M the mass of water and ρ the density of the water, we have:

$$M = N\rho\frac{4}{3}\pi a^3 \qquad [2.82]$$

If we note Σ the surface covered by clouds, we have:

$$Q_{extinction} = 2\Sigma = 2N\pi a^2 \qquad [2.83]$$

When writing equation [2.83] in another form, we replace a^2 and eliminate N to obtain the mass. The clouds with small values of a have the largest impact. A larger number of droplets results in greater impact. When the droplet number increases, $Q_{extinction}$ increases at a fixed radius.

Why do we see the boat traces on Figure 2.19? These boat traces are actually a reflection of the clouds above the boat: the reflection depends on the number of droplets.

If there is an increase in the number of droplets, low clouds increase and reflect more largely into space.

The effect is reversely proportional to the size of the droplets. There is a strong effect of pollution/aerosols on radiation.

With a certain mass of water, we translate it to a $Q_{extinction}$: there is a larger diffusion with a larger number of droplets.

The way droplets interfere with solar radiation is critical.

Let us consider a "slab" of cloud, with a surface of 1 m^2, with an associated optical thickness τ.

For this slab, we can write:

$$\frac{dF^-}{d\tau} = -F^- \qquad [2.84]$$

A diminution of the radiation is occurring here.

We consider that the droplets have a radius a, and N is the number of droplets per m^2. The surface covered S is therefore:

$$S = N\pi a^2 \qquad [2.85]$$

The diminution of the radiation $\frac{dF^-}{d\tau}$ will be associated with the surface covered. The target is in fact two times the surface. What dampens the radius is the factor 2, also named efficiency factor, which helps to have simple computations:

$$d\tau = Ndz \times 2\pi a^2 \qquad [2.86]$$

Let us write dM as the mass of water. We can write the following equation combining the mass of the water and the size of droplets:

$$dM = Ndz\rho \frac{4}{3}3\ \pi a^3 \qquad [2.87]$$

Therefore:

$$d\tau = \frac{dM}{\rho\frac{4}{3}\pi a^3} \times 2\pi a^2 \qquad [2.88]$$

with Q = 2 being the efficiency factor.

$$d\tau = \frac{dM}{\rho}\frac{3}{2}\frac{1}{a} \qquad [2.89]$$

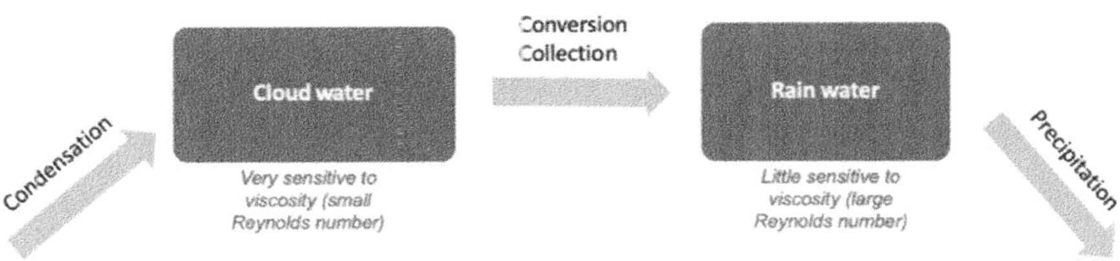

Figure 2.20. *Scheme of the cycle of atmospheric condensed water*

Figure 2.20 shows the cycle of atmospheric condensed water. The condensation phenomenon does not operate spontaneously: nuclei are necessary for the process to happen. For the anecdote, in his appendix of the *Discourse on the Method*, René Descartes added a treaty about rainbows. He was indeed puzzled by the fact that some droplets remain in the sky, while others do not. Descartes observed that the droplets were actually bubbles of varying density and that a weight function was not relevant to the phenomenon. Indeed, a key property of viscosity is the capacity of different layers of flow to be in contact and to entrain each other.

It is interesting to introduce here considerations about the Reynolds number, i.e. a non-dimensional representation of viscosity. The Reynolds number is actually one of the most important nondimensional parameter in fluid mechanics. We can write it as follows:

$$R_e = \frac{L_V}{\nu} V \qquad [2.90]$$

with L representing the characteristic length of the considered fluid expressed in meters related to the horizontal radius of the droplets in this case, V the fluid flow speed expressed in $m.s^{-1}$ and ν its kinematic viscosity, expressed in $m^2.s^{-1}$.

The Reynolds number has to be small under the conditions of a viscous flow.

We also introduce here the notion of terminal velocity, defined as the maximum speed that can be reached by a given object moving through a given fluid. Here, we consider the case of water droplets through the atmospheric fluid. If we consider falling droplets at equilibrium, the terminal velocity is constant (there is no acceleration), and we can write:

$$P = \rho_{water} \frac{4}{3} \pi r a^3 g \qquad [2.91]$$

with P being the weight of the water droplet.

In order to maintain the balance, we can also write the following equation, translating the way viscosity operates on the surface as well:

$$P = 6\pi \rho_{air} \nu V R_e \qquad [2.92]$$

Therefore:

$$6\pi \rho_{air} \nu V R = \rho_{water} \frac{4}{3} \pi r a^3 g \qquad [2.93]$$

The Reynolds number has to be small under the conditions of a viscous flow, and a small Reynolds number basically means that droplets are attached. On the contrary, a big Reynolds number means that the fluid is no longer viscous. In this case, the droplets will therefore fall down because of gravity.

We can also write the following equation for the water falling velocity:

$$V_{fall} = \frac{9}{2} \frac{\rho_{water}}{\rho_{air}} R_e^2 \, \nu^{-1} \qquad [2.94]$$

Equation [2.92] therefore makes it possible to define the fall velocity as proportional to R_e^2, defining different domains for droplet behavior. Using V_{fall} and equation [2.92], we conclude that:

$$R_e \Leftrightarrow a < 5 \ \mu m \qquad [2.95]$$

Whatever the type of cloud, larger droplets fall quicker.

When $a > 5$ μm, the viscous capacity to hold droplets in the sky disappears. This has many consequences: the condensation phase builds some droplets (through very complex mechanisms). Droplets grow by collection: larger droplets "collect" the small. At the beginning of cloud life, droplets increase by falling. They grow until they are no longer viscous, and then they fall down.

Atmospheric cloud droplet formation

Thermodynamics explain how initial droplets form: to form a droplet, the droplet state changes from water vapor to liquid water. Atmospheric cloud droplet formation is not spontaneous because of capillarity.

The computation of Gibbs energy variation when N water vapor molecules condense out a total of N to form a cloud droplet and g_{vapor}, the Gibbs energy of a molecule of water vapor corresponding to its initial state, can be written as follows:

$$G_{initial} = N g_{vapor} \qquad [2.96]$$

The surface pressure holds the droplets together around the droplet, and it is named superficial pressure.

If we write g_{liquid}, the Gibbs energy of a molecule of liquid water, A, the cloud droplet surface, g_{vapor}, the Gibbs energy of a molecule of liquid water and $\boldsymbol{\sigma}$ the capillary tension, we can write:

$$G_{final} = (N\text{-}n)\, g_{vapor} + n\, g_{liquid} - A\sigma \qquad [2.97]$$

Actually:

$$\Delta G = d\frac{4}{3}4\ \pi a^3(g_{liquid}-g_{vapor}) + 4\sigma\ \pi a^2 \quad [2.98]$$

with d being the density of droplets corresponding to a mass of water. For the droplet radius a being small, the surface effect dominates the change of phase from liquid to vapor (capillarity effect).

Therefore:

$$\Delta G = n\ (g_{liquid}-g_{vapor}) + A\sigma = d\frac{4}{3}4\ \pi a^3(g_{liquid}-g_{vapor}) + 4\sigma\ \pi a^2 \quad [2.99]$$

Saturation conditions

In saturation conditions, $g_{liquid} = g_{vapor.}$

And more generally:

$$g_{vapor} = g_{vo}(T)\ + kTLn\ (P_{vapor}) \quad [2.100]$$

Hence:

$$\Delta G = -\ d\frac{4}{3}4\ \pi a^3 kT\ Ln(S_{a)}) + 4\sigma\ \pi a^2 \quad [2.101]$$

with $S = \frac{p_v}{p_{v*}}$

Box 2.2. *Atmospheric cloud droplet formation and saturation conditions*

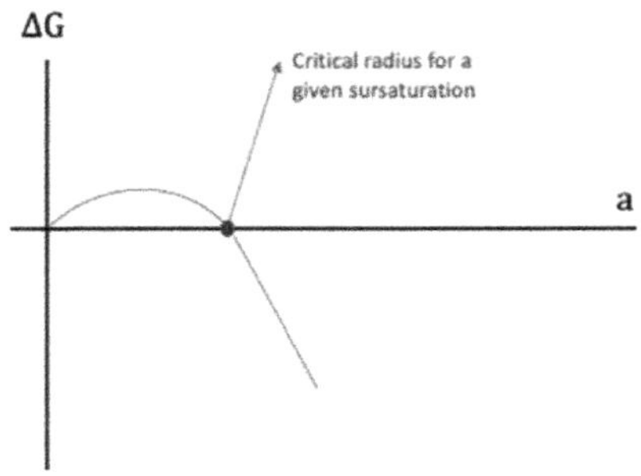

Figure 2.21. *Scheme of a graphical representation of ΔG as a function of the droplet radius a and of the critical radius value for a given sursaturation*

Figure 2.21 shows that there is a critical radius for a given sursaturation: for each target radius, there is a threshold to be reached for a saturation phenomenon, leading the droplet to fall. The result is that normal (homogeneous) nucleation does not happen.

Normal nucleation is very difficult: there is a need for a sursaturation of about 2: it is twice the amount of water vapor without capillarity. If water is condensed earlier, it makes this process inefficient. "Cloud nuclei" can help the process: real sursaturation has to be more than 1, and more precisely about 1.05. This is why we can build droplets much more easily with nuclei such as sulfates/gases that convert, i.e. with hydrophilic droplets that attract water molecules. In the atmosphere, the formation of droplets never occurs spontaneously. We need some particles to form clouds and the number of droplets depends on the way they are formed, and therefore on the cloud nuclei. All of this will initiate a chain of processes that play a crucial role in how solar radiation is utilized at the Earth's surface.

Let us now consider a few examples.

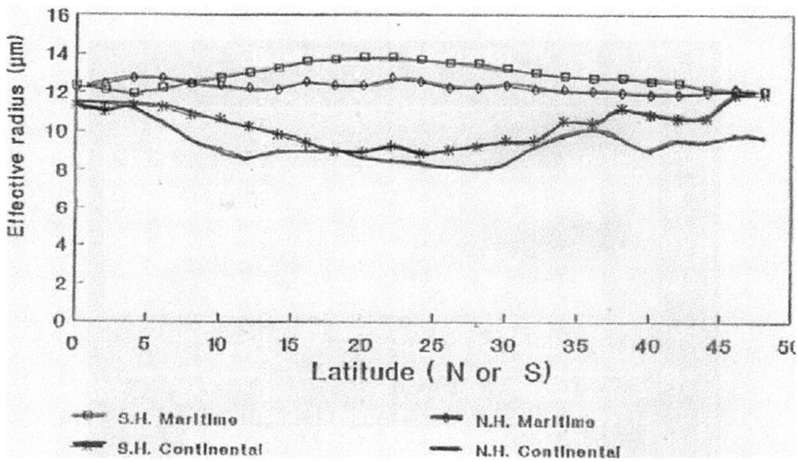

Figure 2.22. *Scheme of cloud droplet size with interhemisphere comparison (Source: Olivier Boucher 1995)*

Figure 2.22 shows an old scheme delivering an interhemisphere comparison of the effective radius in μm as a function of latitude for cloud droplet size.

The figure is an old diagram taken from Olivier Boucher's PhD dissertation, who is now the VP and director of the modeling group of IPSL and who has been the leader of the Climate and Chemistry group at the Hadley Center in Great Britain.

To assess an effective radius, we do not directly measure the cloud particles' size, but there is a distribution of several sizes in a cloud (there is not one standard size). The effective radius delivers a sort of mean of those particles. It is a bit larger than 5 μm due to this effective radius. The dependency is studied over two parameters: latitude (we note that the effective size does not depend very much on latitude), and there is a separation between the two hemispheres and between the maritime part and the continental part.

In a continental situation and in the northern hemisphere, the droplets are wider and smaller, because that represents a configuration where the quantities of cloud nuclei are the largest. In the Northern Hemisphere, we are indeed in a continent where there is industrial and transportation activity with much release of aerosols. Therefore, there are a lot of nuclei, and hence many droplets, which means a smaller radius of the droplets.

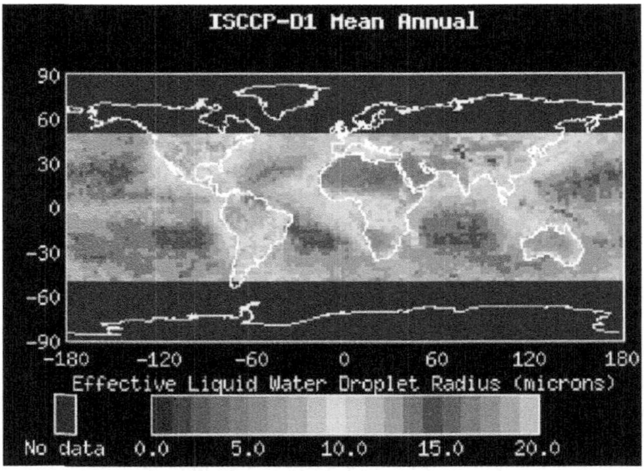

Figure 2.23. *Satellite view of liquid water droplet with associated radius in microns (Source: International Satellite Cloud Climatology Project)*

Figure 2.22 also shows the reverse condition: the Southern Hemisphere is a maritime continent, with very little human activity, which means that there are less nuclei and therefore larger cloud droplets. This marks an important difference in terms of how the radiation is affected between the two hemispheres. However, in the following chapters, we will only address the question of how human activities may affect/trigger changes of those differences for the climate in the future.

Figure 2.23 shows a satellite view with the location of liquid water droplets of different sizes. We can also see places with no clouds (0 μm water droplets can indeed be interpreted that way). The dominant color above continents is orange referring to small droplets, and at tropical latitudes, red is the dominant color, due to even smaller droplets than the one of the Northern continents. Over central parts of the oceans, there are bigger droplets because there are less aerosols and less nuclei.

The red spots in the ocean are due to dust coming from the Sahara. The amount of water that can be condensed is regulated by the atmospheric circulation. There is a convergence of water vapor in some regions, which is dictated by the motions of the atmosphere.

When considering only one vertical dimension, the availability of water is very largely due to the convergence of water vapor through the large-scale atmospheric dynamics: what is received as water has been brought by this large-scale dynamic, and this very naturally tends to end-up as a condensation or a precipitation because convergence occurs at small altitudes, then it has to go somewhere: there is an ascending motion. Ascending motions mean that levels are reached in the atmosphere which have a low capacity to hold water: 10,000 times smaller than near the ground. This is the reason why all of the water falls down. But the water will fall down very differently in terms of the number of cloud droplets if there are enough cloud nuclei. In case there are more nuclei, this condensation concerns the same amount of water vapor but distributed over a larger number of droplets. This is due to a general command of how much water will condense, first of all dictated by large-scale atmospheric motions. This is the reason why, for instance, artificial rain can be made for a few days, but it is not possible to continue this process at a large scale to maintain water. A rare place where this process is successful is the Andes in Chile, for example, where water is gathered through "cloud nets" processes by gathering the small droplets in a net that can be managed because of the size of mountains, which go from the sea up to 4–7 km of height.

Figure 2.24 shows the satellite view with the location of ice crystal particles from different sizes. In high latitudes, when temperatures are low enough, there is almost always an ice phase.

Ice crystals are very different to water droplets. They have various and complex shapes, and generally, they begin to fall down immediately. They do not have the same kind of viscous phase as water droplets, and they fall down slower due to their shape. Therefore, the analyses for water droplets do not strictly apply for crystals.

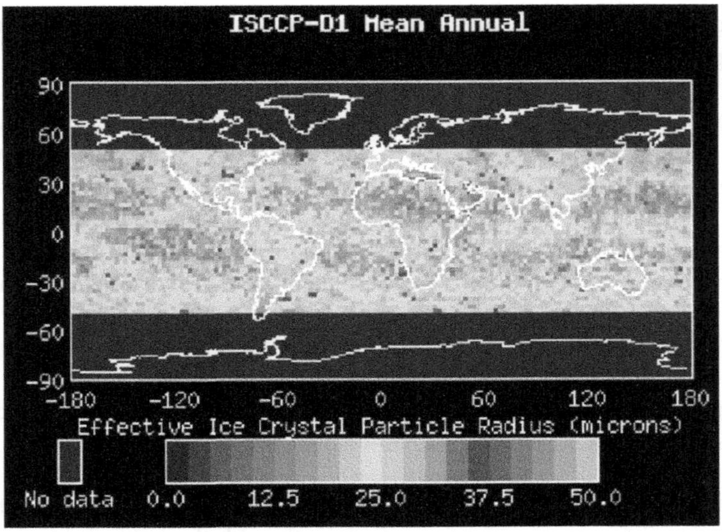

Figure 2.24. *Satellite view of Ice Crystal Particle with associated radius in microns (Source: International Satellite Cloud Climatology Project)*

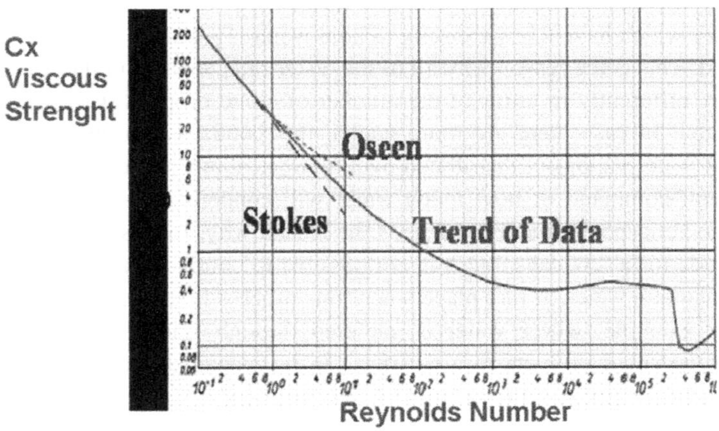

Figure 2.25. *Diagram showing the viscous strength Cx as a function of Reynolds number using a logarithmic scale*

Figure 2.25 shows a diagram discussing the Reynolds number and the viscous trend using logarithmic values. The way viscosity increases for small Reynolds numbers is extremely quick. Viscosity is exponentially increasing or decreasing depending on the Reynolds number. This metric helps discussing forces that act on particles in a fluid with a given Reynolds number, and there is a very strong-scale effect depending on this number.

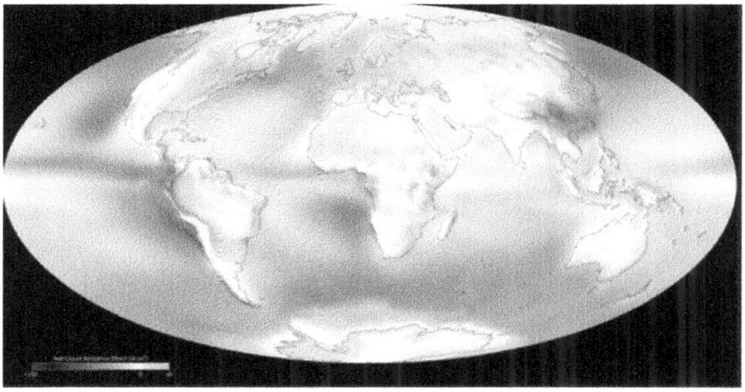

Figure 2.26. *Net Cloud Radiative Effect average 2000–2015 (Source: Clouds and the Earth's Radiant Energy System (CERES); nasa.gov)*

The understanding of liquid and ice droplet behaviors is narrowly linked to the comprehension of cloud radiative impact.

Clouds have an important effect on the Earth climate for their double effect of increasing the albedo, which cools the system, while they can also warm the Earth's climate system by decreasing the IR radiation leaving to space.

The resulting net effect of clouds is shown in Figure 2.26 and obtained by computing the sum of the two opposite clouds' effects. We can see that observations coming from satellites show that the net effect tends to be a cooling effect.

However, the future effect of clouds on the Earth climate through global warming remains one of the main sources of uncertainties in climate models.

3

The Horizontal Motion of Atmospheric and Oceanic Transport

In this chapter, we study the horizontal motion of the atmosphere and the ocean as a key phenomenon to balance the very differentiated distribution of solar energy flux between low latitudes, where it enters the system, and high latitudes, where it leaves.

So far, we have been looking at vertical energy transfer. In this chapter, we will focus on atmospheric and oceanic transport. First, we recall a few generalities.

3.1. Energy budget at the top of the atmosphere

The Earth's radiative budget, already largely discussed, is depicted in Figure 3.1.

The diagram in Figure 3.2(a) shows the measurement of radiative budget at the top of the atmosphere (TOA), as seen by a satellite. The yellow/orange/red and green areas around the equatorial belt distinguish the region where there is more solar radiation penetrating the atmosphere than outgoing terrestrial radiation leaving the atmosphere, i.e. with a gain of heat. Figure 2.1 delivers an average value over one year, with an amplitude delivered in W m^{-2}. There is a net 160 W m^{-2} amplitude of radiative forcing between regions losing or receiving heat energy, which is huge.

For a color version of all of the figures in this chapter, see www.iste.co.uk/letreut/energetics.zip.

Energetics of the Climate System,
by Hervé LE TREUT and Mounia MOSTEFAOUI. © ISTE Ltd 2025.

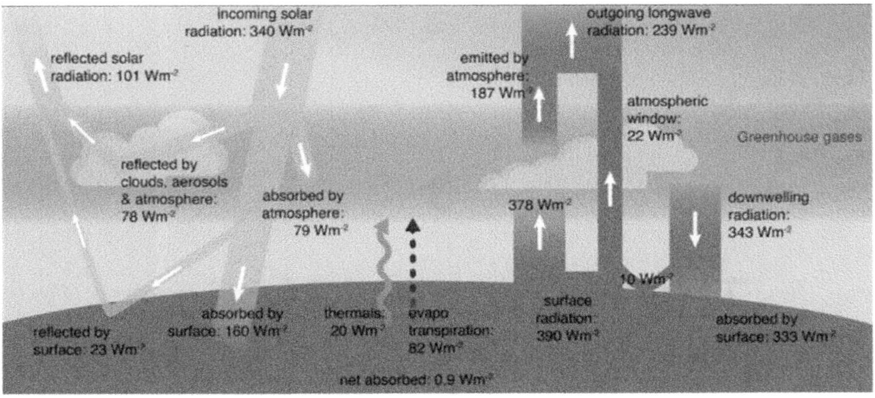

Figure 3.1. *Diagram of the Earth's radiative budget (Source: Institut Pierre Simon Laplace – Climate Graduate School/e-courses for masters 2024)*

In Figure 3.2, we can also see the reverse situation, with regions losing heat (represented by the pink to blue gradient). There is some complexity behind this diagram that cannot limit itself to this big difference between the Poles and the equator: for instance, in North Africa and in the Middle Eastern peninsula, over South Australia, South Eastern China, the Western coast of Chile and Peru. We distinguish whole regions losing heat in the Sahara and other deserts. Those regions are losing heat in equilibrium with the transport of heat by the atmosphere. The system at the scale of a few years is more or less at equilibrium.

If we can see some local disequilibrium, this is because there are sinks of heat that are the result of the motion of the heat over the terrestrial sphere, and this motion is due to the atmosphere and the ocean. The analysis of this motion of the atmosphere and the ocean is the heart of this chapter.

Figure 3.2(a)–(c) shows the contrast between respective annual mean, winter (December/January/February) and summer (June/July/August) conditions, in W m^{-2}.

If we look at an average over several seasons (see Figure 3.2(b) and (c)), the pictures are very different than the mean annual conditions of Figure 3.2(a). We clearly distinguish a reverse image between the summer and winter cases. There are also associated motions that are very different from the studies until now: monsoons. In each situation, there are transfers of energy between the hemisphere with more solar energy income to the hemisphere with less incoming energy, and not only from the equator to the Poles.

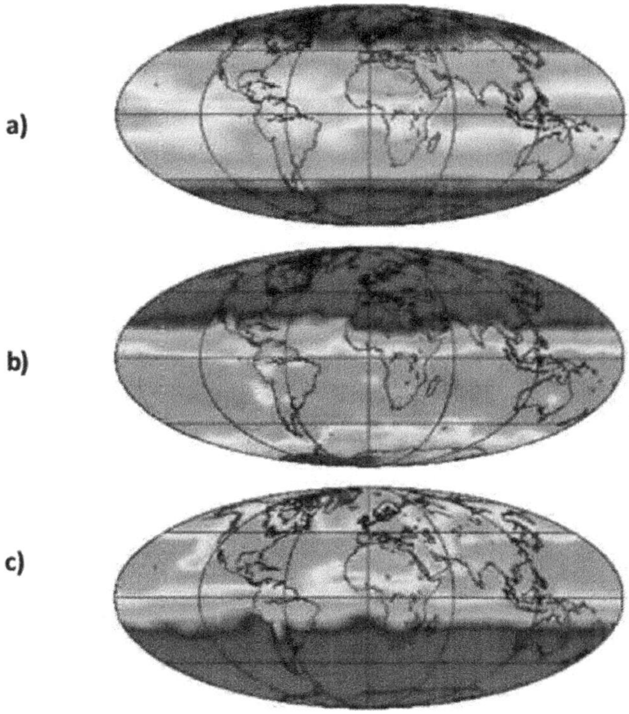

Figure 3.2. *Radiative budget at the top of the atmosphere (TOA) in W m^{-2}, a) annual mean, b) under winter conditions (December, January, February) and c) under summer conditions (June, July, August) (Source: The Earth Radiation Budget Experiment (ERBE))*

The complexity comes from the fact there are two fluids at play: the atmosphere and the ocean. When discussing transport, motion cannot be forgotten.

Figure 3.3 shows the energy fluxes from the atmosphere and ocean. The equator is emphasized with a red point, and the South Pole and the North Pole are easily localizable thanks to latitude on the x-axis. The energy is transported from the equator to the Poles. There are two layers: the atmosphere and the ocean, and TA and TOC respectively stand for transport of energy across a latitudinal wall in the atmosphere (see Figure 3.3(b): middle panel) and ocean (see Figure 3.2(c): bottom panel). We may or may not expect that the two will participate in the top panel (see Figure 3.2(a), representing TA + TOC). In order to separate those two effects, distinct budgets are established: the heat budget at the TOA, the surface heat budget, and the bottom of the ocean heat budget (with no real exchange with just a little geothermal energy).

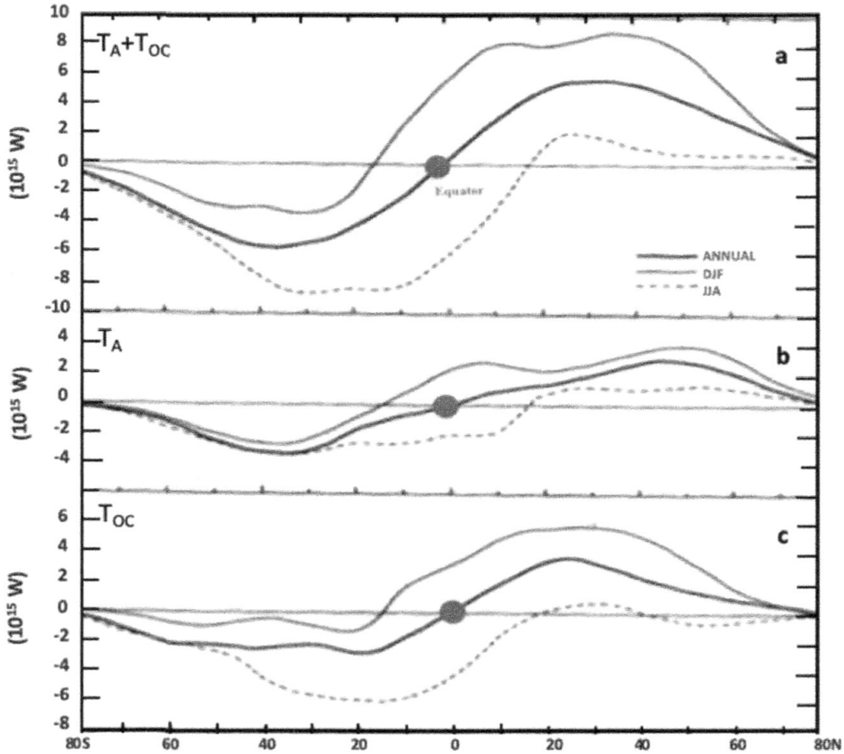

Figure 3.3. *Energy fluxes from the atmosphere and ocean are circulations necessary to maintain a global equilibrium. The red point represents the equator (Source: Carissimo et al. (1985))*

The overall principle is as follows: from the TOA heat budget, we know more or less how the energy should circulate; from the TOA heat budget, it is possible to mathematically derive the kind of transport of energy below that will balance it, with boundary conditions, from regions with an excess of heat to regions lacking heat. The energy flux due to the atmosphere and the energy flux due to the ocean must also be respectively identified and computed. The easiest way for that to happen is to know the surface heat budget; that is the reason why we will focus on energy exchanges with the surface in the following.

It is extremely difficult to measure the surface heat budget. What we call heat budget corresponds to fluxes that enter or leave the system. There are also fluxes at the interface between the atmosphere and the ocean.

Heat transfer happens between space and the atmosphere, and there is also heat transfer between the atmosphere and the ocean. From the study of its functioning, we may deduce the necessary transfer of heat from the equator to the Poles first, and then with other dimensions on the Earth's surface.

In the diagram in Figure 3.3, the main message is that from the energy budget at TOA, we can derive first in annual mean, and in DJF (December/January/February), or in JJA (June/July/August), the kind of heat transport from the South to the North Pole. In the annual mean (black bold line), the point in red Figure 3.3 is very near to the equator and has actual measurements: the transport of heat is in the direction of the North Pole located in the right part of the diagram (80N), and in the direction of the South Pole (80S) corresponding to the directions seen in the previous maps. Therefore, consistently, the total amount of heat (DJF full line and JJA dashed line) has to be consistent and re-equilibrate the lack of equilibrium of the two other radiative diagnostics. In other words, transport has to correct all the disequilibrium of the radiative process.

Also, during different seasons, this transport line is shifted a bit more southward, consistently with the changes seen in the maps between summer and winter. The negative curves (right part of the diagram relative to the equator) indicate that the energy goes from the equator to the South Pole, and after the equator point, the values become positive because the energy goes from the equator, where it is maximum, to the North Pole, with a weight function at play. In both cases, even though they look antisymmetric due to the positive/negative convention, they are in fact symmetric (through Hadley cells and processes that will be discussed in Chapter 4 of this book), but the direction of the winds is to the North from the one side and to the South from the other side. This functioning allows the equator to cool off by transferring energy to the Poles.

A striking fact (even though there are some distortions) is that the amplitudes of the curves of the upper diagram (see Figure 3.3(a)) are reduced by half in the middle diagram, representing the transport of the atmosphere (see Figure 3.3(b)), and in bottom diagram, representing the transport of the ocean (see Figure 3.3(c)). Hence, the system will be balanced more or less in equal parts by the transport of heat due to the atmosphere and by the transport of heat due to the ocean. Those diagrams are from relatively old studies, and nowadays, the transport of heat due to the ocean has been shrinking slightly, but this pioneering work stays important in terms of understanding trends.

The question now is how much energy is being transferred from the atmosphere to the ocean.

The complex Earth climate system actually has different compartments linking together the atmosphere and the ocean.

3.2. Energy exchanges with the surface

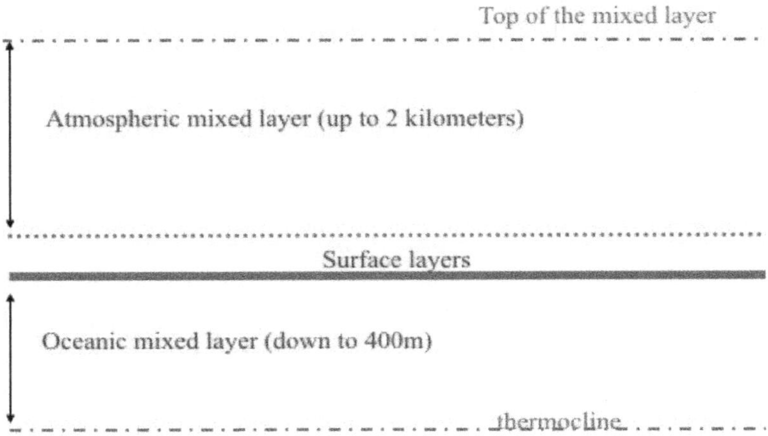

Figure 3.4. *Interface layers between the atmosphere and the ocean*

Figure 3.4 shows the atmospheric mixed layer, which stores many elements and has a critical interface role for the exchange with the ocean. There is also an oceanic mixed layer at the bottom and surface layers between the two. The exchanges of energy between the free atmosphere and the ocean go through all of the layers.

We will also discuss the possibility or impossibility of computations in those conditions.

We saw in Chapter 2 that temperature has a specific way of growing in layers where there is turbulence.

We also showed that the potential temperature has a specific shape that we can represent as in Figure 3.5.

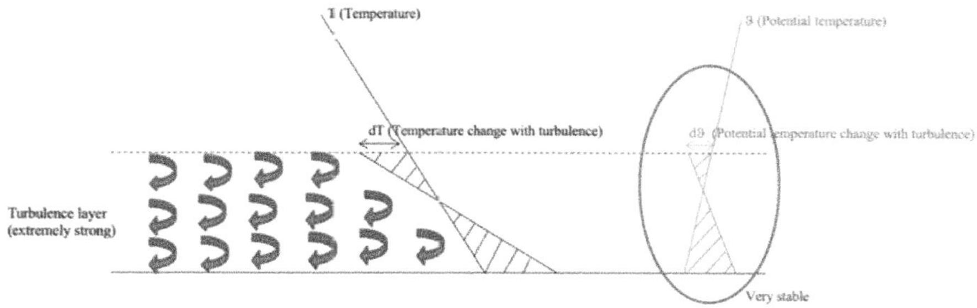

Figure 3.5. *Scheme of temperature and potential temperature changes for the atmospheric turbulent layer*

Figure 3.5 shows that the real temperature decreases with height and is not conserved in the circulation, while the potential temperature θ increases with height, and this latter quantity is mixed because it is a conserved motion. Therefore, there is some kind of load for potential temperature.

What we see in Figure 3.5 is a very large dθ, which refers to a very stable situation, as seen in Chapter 2. Another point is that, in fact, turbulence, because it produces this strong mark, puts a lead on the boundary layer. Everything happens as if there was a very clear and strong limit above the turbulent layer represented on the scheme, with turbulence not being able to go beyond it. Similarly, we also have a turbulent boundary layer in the ocean, which is also featured with a mixing in Figure 3.5.

The ocean temperature decreases from the surface in the ocean mixing turbulent layer as shown in a simplified way in Figure 3.6. Temperature decreases very quickly in locations with cold water and less easily in locations with warm water that cannot get down very easily due to their lighter mass.

In summary, we study a system that is actually featured with three levels of exchanges, all dependent upon the others:

– the exchanges between the atmospheric mixed layer and the TOA;

– the exchanges near the surface between the atmosphere and the ocean;

– the exchanges between the ocean mixed layer and the deeper ocean.

The phenomenon of energy exchange between the ocean and the atmosphere is complex because of the three kinds of energetic exchanges involved in the system.

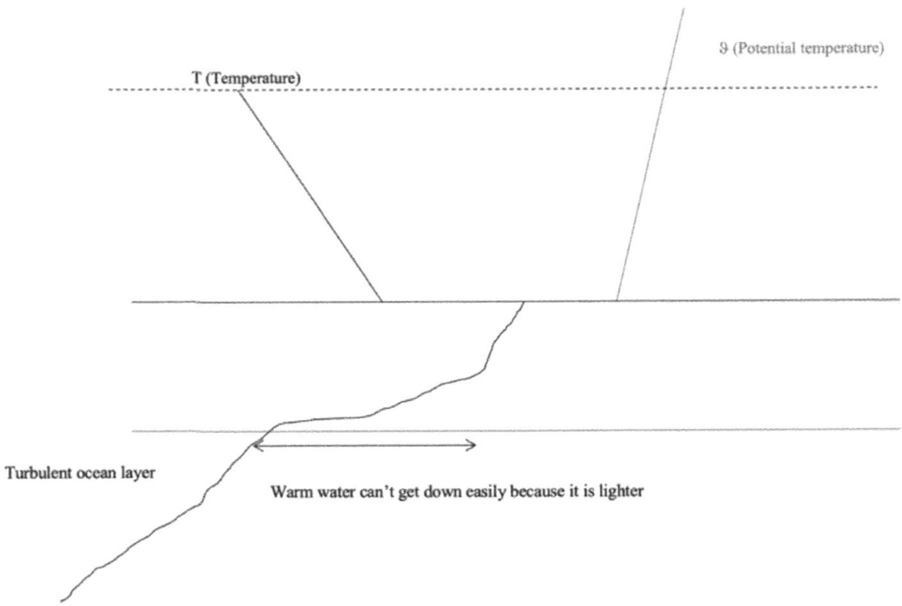

Figure 3.6. *Scheme of temperature and potential temperature changes for the oceanic turbulent layer*

Furthermore, the mixed layer amplitude is highly dependent on seasonality: the atmospheric mixed layer can extend up to 2 kilometers in summer, and to only a few hundred meters in winter. The mixed layer of the ocean near the equator represents about a few tens of meters, while in high latitudes, it can go down to more than 400 m, as shown in Figure 3.4.

The in-between layer connecting the atmospheric mixed layer to the oceanic mixed layer is represented in Figure 3.7.

The mixing delivers the value of potential temperature evolving, as represented in blue Figure 3.6 in terms of trend. If we simplify the ocean, we also obtain the curve represented in red on Figure 3.8, as the ocean is similarly mixed. The represented mixing on the bottom delivers a surface temperature for the ocean, the mixing on the top delivers a surface temperature for the atmosphere, and the energy has to be transferred with continuity. The surface layer is a third important connecting layer between the ocean and the atmosphere: in this surface layer represented in Figure 3.7, the atmosphere, which is the system that is quicker to react, will mix, as represented in Figures 3.7 and 3.4.

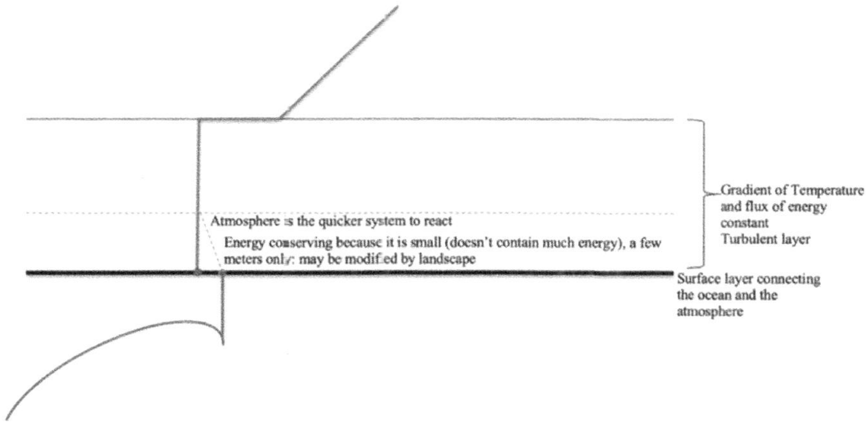

Figure 3.7. *Scheme of energy transfer configuration around turbulent boundary layer*

In the surface layer, there is an associated profile of the atmospheric temperature that matches up with the profile of the ocean temperature. Therefore, this surface layer is energy conserving because it is very small, with a size of a few meters only, and therefore, it does not contain much energy. It is modified by the buildings, the landscape and trees, for example. We know that, in fact, in this layer, there is a gradient of temperature, but a flux of energy which is not modified (constant). This gradient of temperature will be associated with a flux of energy going from the ocean to the atmosphere or from the atmosphere to the ocean that will not change very much over this surface layer because there is no possibility to conserve the energy in the surface layer, and basically everything that goes in has to go out. Hence, it represents a perfect place to try to measure this, as the flux does not change very much with height. When it comes to the ocean, those measurements must be done from boats.

Hence, measures of fluxes at the interface between the atmosphere and the ocean have been done on boats for many years, using the masts of the boat to take measurements near the water and parameters related to clouds, but nowadays buoys can be used.

An important point is that it is extremely difficult to make boat measurements, and they have actually been made for years by using empirical dimensional analysis. It is indeed possible to use simple formula that relate the difference in temperature between the ground and the temperature at the z level in the surface layer, the wind at the same level and some parameters. With those elements, we can, however, only approximately compute the heat exchange between the ocean and the atmosphere.

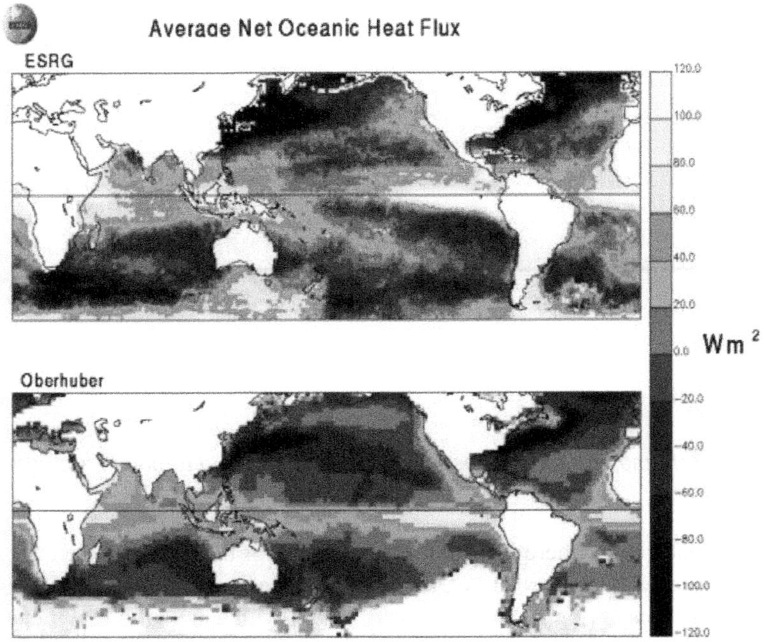

Figure 3.8. *Average net oceanic heat flux (Source: top map – satellite data from European Space Research Group (ESRG); bottom map – boat measurements from Oberhuber)*

Figure 3.8 shows a few climatologies that are now a bit old, but have been among the most referenced, and are still interesting to comment on, delivering the average net oceanic heat flux: the energy entering the ocean is positive for light orange to yellow colors, while blue/dark colors stand for regions sending energy back to the atmosphere.

The map in Figure 3.6 is from the European Space Research (ESRG), and is one of the earliest attempts to do this using satellite data. The bottom map, named after the climatologist Oberhuber, results from several years of boat measurements.

Those two maps are not completely different: for both we can see energy entering near the equator, energy leaving the system away from the equator and a global consistency.

However, when it comes to having a numerical representation, we obtain less consistency with measurements, and when we want to derive currents more or less at equilibrium, we also obtain less consistency in the results.

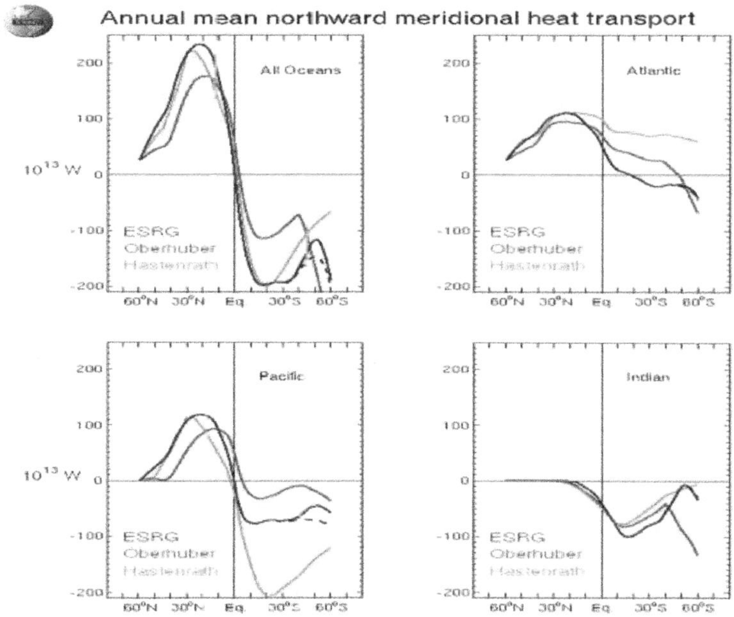

Figure 3.9. *Annual mean northward meridional heat transport in all oceans, in the Atlantic, in the Pacific and in the Indian Ocean (Source: satellite data (ESRG and boat measurements (Oberhuber and Hastenrath)))*

Figure 3.9 shows a panel of four diagrams showing annual mean northward meridional heat transport in all oceans, in the Atlantic, in the Pacific and in the Indian Ocean. In each diagram, three climatologies are represented: satellite data from ESRG (blue curve) and two curves showing measurements done by boats (Oberhuber in orange and Hastenrath in green). For all oceans, we have the reconstruction of the energy flux by the ocean that could be at equilibrium with the reconstructed heat flux from boat measurements. In other words, from the reconstructed heat flux, we derive what the ocean “should” transport to balance this reconstructed heat flux at the ocean surface.

For all oceans, the results appear correct. For the North, we can see on the left upper panel of Figure 3.9, that the transport goes to the South Pole, with consistency. But when looking ocean by ocean, it becomes, however, more difficult.

For the Indian Ocean, there is consistency because all models agree to estimate that there is not much transport of heat by the ocean when it goes to the continent. However, for the transport, for example, in the Southern Hemisphere, in the Pacific, there are differences among models in terms of estimating how much heat energy goes to the Pole, depending on the different climatologies. Even in the Atlantic, the sign of heat transport in the Southern Hemisphere is unknown: this may contradict the trends observed in other oceans with transport from the South to the North; it has been a source of debate for many years and remains very difficult to assess.

The way forward has been to rely a bit less on the ocean, even though measurements in the oceans are constantly improving, and proceed differently: we cannot reliably measure what is in the ocean; however, we can determine what is inside the atmosphere. We can therefore obtain a diagnostic of the differences seen above from the atmospheric models that serve meteorological purposes, from pure measurements of the atmosphere. Then, we can have a satellite signal that comes to the TOA, the meteorological model that gives the difference discussed above, and obtain a different version of what goes into the ocean, that is not based on oceanographic data, but that is instead based on atmospheric and spatial data.

We will now ask ourselves what kind of data we can rely on and how much we can trust the share delivered between the atmospheric transport and the oceanic transport.

Atmospheric transport and oceanic transport go in the same direction; they help each other and both have similar importance/size: in that sense, it facilitates their study. In effect, analyses of the atmosphere's functioning show that this system brings energy to the Poles. Similarly, studies of the ocean functioning from this perspective also show that this system tends to transport energy to the Poles, with a few exceptions, including in the South Atlantic.

In the following, we will discuss the dynamics of the atmosphere and the ocean, not heavily and completely, but just for the purpose of analyzing the transfer of energy from one region to the other.

As a reminder, until now, the distribution of solar radiation has been dependent on different parameters: it is very differentiated between low and high latitudes, with energy entering the system near the equator and leaving the system at higher latitudes, and this has to be balanced by the horizontal motion of the atmosphere and the ocean.

The budget can be assessed by analyzing the TOA energy fluxes. The computations are difficult because of the constraints at the North and South Poles, but computations can be done in order to obtain a flux, which is actually an implied flux with an implied current or implied atmospheric winds, by the differentiation between regions where energy is received, and regions where energy is lost.

Figure 3.9 shows the result of the total amount of energy; for balance, the flux of energy is directed from the equator to the Poles, where there is a lack of received energy. The flux of energy on Figure 3.3 (going 80S–80N with fluxes expressed 10^{13} W) is directed to the North when it is positive and to the South when it is negative.

Those global diagrams, when analyzed seasonally, are featured with some shift of the curves, from the South to the North, from winter, when there are larger fluxes of transfer, to summer characterized with less energy flux transfer in the corresponding hemisphere. The flux in the summer JJA is the summer for the Northern Hemisphere, and in fact, there is a very large negative flux that goes from the summer to the winter (big loss of energy in the winter in terms of energy received by the Earth, and this is compensated by the difference).

Let us now have a closer look at the differentiation between the energy from the atmosphere and the energy from the ocean.

The following step is to pay attention to the two intermediate layers of the atmosphere and the ocean respectively, as represented in the graphics of Figures 3.4–3.6. There is one reservoir of heat/water vapor in the atmosphere, as well as intermediate layers: one is the layer connecting the surface atmosphere with the ocean. The surface layer is more on the atmospheric side because the atmosphere goes quicker than the ocean, and the top of the mixed layer acts as a lead at the bottom of the ocean mixed layer called the thermocline. These are important processes to analyze in order to understand how heat is transferred from the ocean to the atmosphere.

Figure 3.4 shows that above the ocean, the surface layer connects an area with the top of the mixed layer and is featured by a well-mixed potential temperature θ, down to the thermocline. As there has to be continuity in terms of temperature, there has to be a connection (of about a few tens of meters). This is a part of the air column that is very specific, as it makes a link between the atmosphere and the ocean, and it is apart from a constant flux because there is no accumulation of heat within (it is too small to accumulate anything). Another way to explain it is to simply state that whatever energy enters must leave upward or downward because it cannot be stored. Therefore, the flux is constant because of the impossibility of energy storage in this small surface layer. This constant flux in the surface layer links the atmosphere and the ocean. A lot of work has been done in order to characterize the flux in this constant layer, mainly through two methods:

– Actual measurements need to precisely describe the small motions of turbulence occurring at a small scale. On land, towers are used, while in the ocean, it is a bit more complicated because using boats implies adding correcting terms.

– Another way to do it is to use some simplified laws resulting from dimensional analysis. This is a very common practice in fluid mechanics when we do not know the laws describing a phenomenon, but we know the main parameters that have to remain more or less constant.

Summary of the energy fluxes at the ocean–atmosphere interface

– Radiative contributions (solar and terrestrial).

– Turbulent contributions (sensible and latent heat).

The turbulent fluxes are computed empirically using dimensional analysis

$$\tau_x = -\boldsymbol{\rho_a \overline{w'u'}} = \boldsymbol{\rho_a u_*^2} \quad [3.1]$$

Equation [3.1] defines the turbulent velocity scale $\mathbf{u_*}$, which represents a simplified description with only u direction.

The following general law can also be written:

$$\frac{\partial \overline{X}}{\partial z} = \frac{F_X}{u_* k z} \quad [3.2]$$

with the Karman constant k = 0.4

First example: X = u

From equation [3.1], we have:

$$\frac{\tau_x}{\rho_a} = u_*^2 \quad [3.3]$$

Equation [3.2] can be integrated as follows:

$$\overline{u}(z) = \frac{u_*}{k} \mathrm{Ln}\left(\frac{z}{z_0}\right) \quad [3.4]$$

$$\tau_x = \boldsymbol{\rho_a} u_*^2 = \boldsymbol{\rho_a} C_D \overline{u}^2 \quad [3.5]$$

with $CD = \left(\frac{k}{Ln(\frac{z}{z_0})}\right)^2$ [3.6]

A limit condition is prescribed at height z0 (with rugosity parameter CD).

Second example: X = CPT

We can integrate the equation [3.2], considering FE:

$$\frac{\partial \overline{T}}{\partial z} = \frac{F_E}{C_P u_* kz} \quad [3.7]$$

with CP being the universal constant.

$$\overline{T}(z) = T_{sol} + \frac{F_E}{C_p u_* kz} \text{ Ln}\left(\frac{z}{z_C}\right) \quad [3.8]$$

We obtain:

$$FE = CPCD\overline{u}(z)\,(\overline{T}(z) - T_{sol}) \quad [3.9]$$

Box 3.1. *Equations related to the turbulent fluxes empirically computed using dimensional analysis*

In Box 3.1, τ_x is a flux of momentum, which is the result of a squared velocity multiplied by density. The study of the small fluctuation of this flux, multiplied by the fluctuation of the upward wind, delivers information about the quantity going through the surface layer. Therefore, instead of making the detailed measurements that would be required, scientists study whether there could be some general law that would help to make the description of the phenomenon.

General laws are based on parameters that can be described and deliver non-dimensional numbers. As described in Box 3.1, let us consider the flux of any quantity x, that we want to know.

What we can measure and determine is the gradient of temperature, much more easily measurable than the solar flux. We can also choose some altitude z in the surface layer and have some idea of the equivalent velocity of the small motions in the surface layer, with u_* being the characterization of this turbulent velocity associated with turbulence, and we can relate it to this parameter. Therefore, there are three identified parameters corresponding to the physics of the problem, with k being named the Karman constant (approximately constant). Researchers rely on that to make measurements in the oceans, for example, u_* can be estimated approximately as the average velocity of the wind. z is the altitude at which the measurement is made, generally at either 2 meters, 5 meters or 10 meters, depending on the boats' masts, and from that it is possible to compute 2 or 3 values of the temperatures, or other parameters like wind when going upward in the surface layer. This method has been used for years, and many theories have been developed using such dimensional analyses of the turbulence within the surface layer, which have led to the establishment of most of the flux maps.

When we look at the accuracy of those measurements by dimensional analysis, it reaches about 10^2 $W.m^{-2}$, which is very rough. When we sum it up over large areas, we may think that the accuracy will improve, but it is actually still rough, and this method has been the only estimate for years to assess flux budget at the interface between the atmosphere and the ocean.

There are map fluxes for which it is possible to integrate those equations by taking X of any quantity. It can first be the velocity, using equation [3.1]. Then, we can use it in the equation for the flux of energy at the interface. Therefore, when we solve those equations (which we will not be doing here), the logarithmic solution form corresponds to the shape of the wind near the ground. But we also need a limit condition prescribed a bit higher than the ground, at z_0, which is named the rugosity parameter.

From this equation, we get a value of u^* that can be reinjected in equation [3.7], which delivers the actual flux of energy F_E, as detailed in Box 3.1. The gradient is limited, as it is integrated over the ground up to a level z. Furthermore, in equation [3.7], $\bar{u}(z)$ is the wind at the z level that can be measured, we know the universal constant C_P and we can deduce the parameter C_D that we can compute from z_0. The representation of this equation is correct as long as the gradients of potential temperatures are not too high. It is simple and intuitive: the flux depends on the gradient of temperature; it depends on the wind (in other words, if the wind is blowing strongly over a surface, there will be more evaporation, which is one way to have heat transfer; there is also direct heat transfer). However, if this representation is correct, it is not accurate. This method has been used to deliver maps showing distributions of net oceanic heat flux made several years ago when it was the one possible way to estimate heat fluxes, as shown in Figure 3.8.

Therefore, net oceanic heat fluxes have to be taken with care, as roughly only some areas are a good fit with the data purely taken from boats, as shown in Figure 3.8. The upper diagram of Figure 3.8 indeed represents a mix with some satellite data: it delivers an average over the whole year, with various notable discrepancies with data observations from boats.

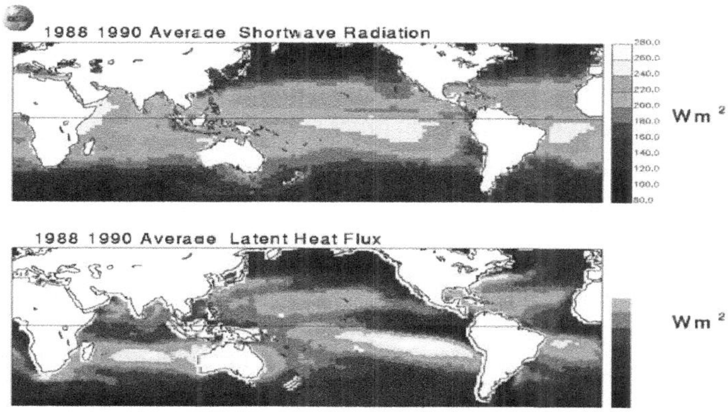

Figure 3.10. *Maps of average 1988–1990 shortwave radiation (top) and 1988–1990 latent heat flux (bottom) in W.m^{-2} (Source: satellite data (ESRG))*

The top panel in Figure 3.10 shows the average shortwave radiation, which is difficult to obtain because it is located at the bottom of the atmosphere. The shortwave radiation estimate is therefore subject to approximation. The estimate of solar flux, moving down from the atmosphere, may be corrected by its value at the bottom of the atmosphere at the interface with the ocean, by considering the clouds reflecting part of the solar radiation.

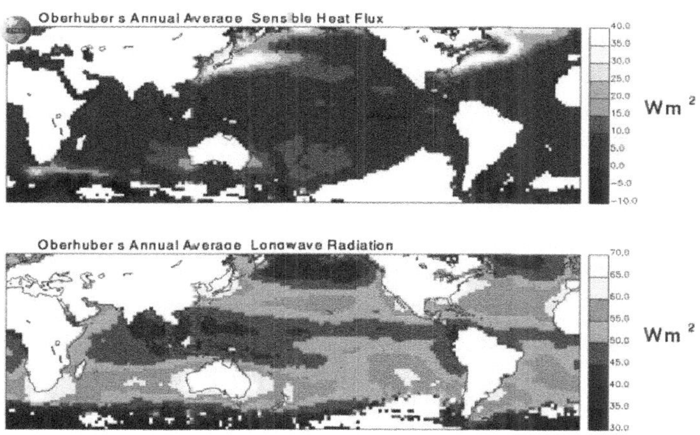

Figure 3.11. *Maps of average 1988–1990 sensible heat flux (top) and 1988–1990 longwave radiation (bottom) in W.m^{-2} (Source: Oberhuber)*

The same exercise can be done for terrestrial longwave radiation, as shown in the bottom panel in Figure 3.11: from the ocean temperature, we know the emitted quantity at the surface of the ocean, and it is possible to correct that with water vapor at the surface, and with clouds, in order to obtain an estimate of this value.

Two turbulent fluxes are also represented in Figures 3.10 and 3.11: one is direct evaporation and condensation (latent heat flux) in the bottom panel of Figure 3.11, which is the largest one, and the other is direct temperature (sensible heat flux) in the upper panel of Figure 3.11.

The flux of energy at the interface between the atmosphere and the ocean is the sum of the four contributions shown in Figures 3.10 and 3.11, two corresponding to radiative transfer (atmospheric or terrestrial), and two corresponding to turbulent transfer of heat directly (sensible heat flux) or transfer of water vapor being evaporated over the ocean (water may condensate again, delivering its heat when it condensates), named latent heat flux.

More recently, the ability to make a diagnosis has improved thanks to a constantly refining meteorological forecast. Nowadays, within the atmosphere, we can access measures of different parameters such as ratio of water vapor, temperature and winds. Those measurements are done through satellites, and there is, in the atmosphere, the capacity to check consistency through a process called data assimilation.

When we do a forecast, the data assimilation principle is used, which means that we go from -1 day to 0 day and so on. During this whole period of time, assimilation functions as follows: we let the model start at -1 day, for example, and as it starts, it takes into account all the available possible information. For one complete day, various measurements from satellites, surfaces, etc., are taken throughout an important network of data providers. The mathematics allows us to combine all that is simple in terms of general ideas, but complex in the set-up. The general idea is to minimize the distance to observations. Of course, the model is not perfect and neither is the data. We have estimations of the potential errors on data, and we try to minimize the distance between both through what is called a cost function: with this constraint, we know more or less the kind of potential errors of the measures. Basically, it is like minimizing data regression data with square distance approximation, but it is much more complex in that case as modelers use a huge matrix. This has been done at the European center since the 1980s by climatologists such as Olivier Talagrand and Philippe Courtier, who did a huge amount of work that has been integrated into models with satellite data. The result is that we know many things from the atmosphere indirectly derived from the data and from the model physics, and we can get to fluxes derived from analysis. This analysis is not perfect because it is generally now done from one day to the other, and we do not have the benefits of a long memory (for water resources in the soils, etc.), even though it has been improved over time.

Research groups tend to rely more on estimates of fluxes between the atmosphere and the ocean resulting from such a process, rather than on direct measurements in the ocean, but the two approaches exist for the budget: we know what is entering above as in fact what we measure well is the energy entering in the atmosphere, and so we can derive what quantity of energy has to leave the system at the bottom.

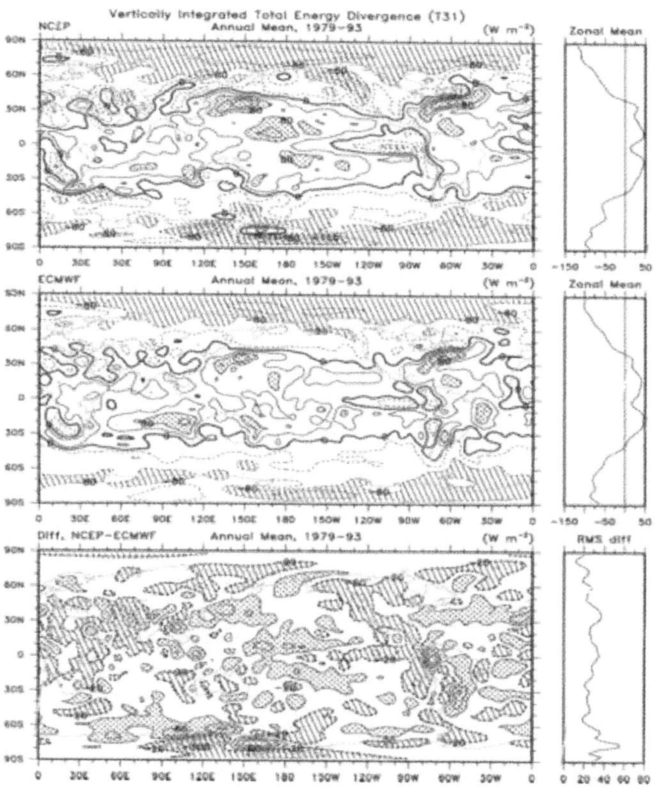

Figure 3.12. *Annual mean vertically integrated total energy divergence for 1979–1993 from NCEP (top) and ECMWF (center), and their difference (bottom) in W.m^{-2} (Source: Trenberth et al. (2001))*

The results of such a procedure are widely used, and Figure 3.12 shows other results following this procedure, with the first results of such models in the 1990s. Figure 3.12 is taken from a paper by Trenberth et al. (2001) about atmospheric energy budgets and their implications for surface fluxes and ocean heat transport. Recent IPCC reports still display figures using these methodologies.

Figure 3.12 shows the annual mean vertically integrated total energy divergence for 1979–1993 from different models: National Centers for Environmental Prediction (NCEP; top) and European Centre for Medium-range Weather Forecasts (ECMWF; center), and their difference (bottom), expressed in W.m^{-2}. Figure 3.12 shows that there are discrepancies in terms of results, but implied fluxes from the equator to higher latitudes, as expected theoretically.

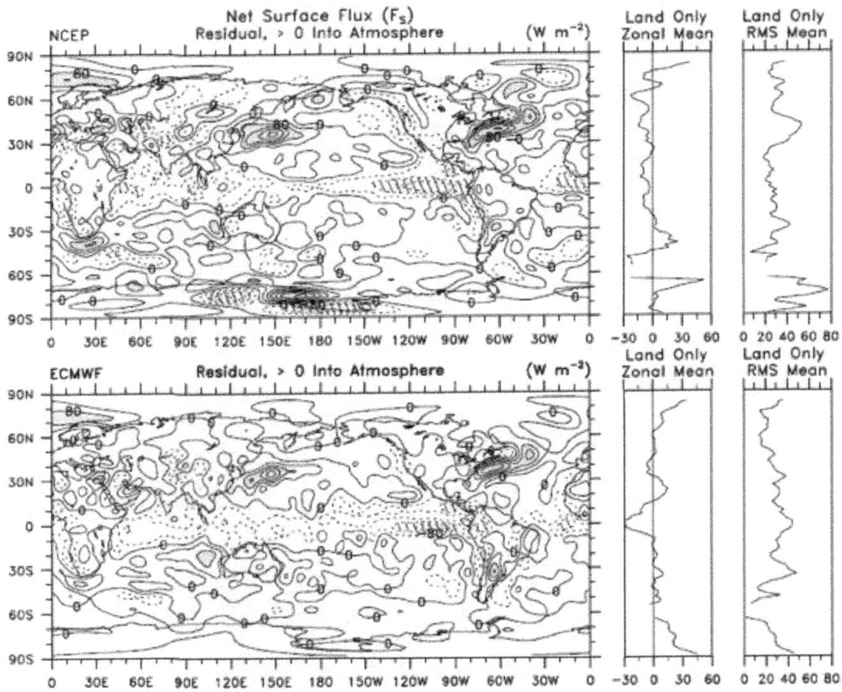

Figure 3.13. *Net surface fluxes derived from the total atmospheric energy divergence and TOA radiation for February 1985 to April 1989 expressed as the annual mean in Wm^{-2} (Source: Trenberth et al. (2001))*

Figure 3.13 shows an estimate of the net surface fluxes derived from the total atmospheric energy divergence and TOA radiation for February 1985 to April 1989, expressed as the annual mean in Wm^{-2}.

In the next section, we will more specifically focus on the transport by the atmosphere and the ocean.

3.3. Transport by the atmosphere and the ocean

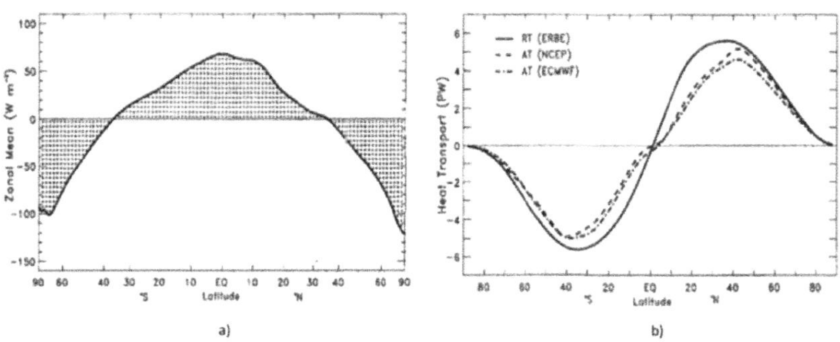

Figure 3.14. *a) Top of the atmosphere (TOA) annualized ERBE zonal mean net radiation (Wm^{-2}) for February 1985–April 1989. b) The required total heat transport from the TOA radiation RT is given, along with the estimates of the total atmospheric transport AT from NCEP and ECMWF reanalyses (PW) (Source: Trenberth et al. (2001))*

Figure 3.14 shows the results from the satellite ERBE (Earth Energy Budget Experiment), which has been in orbit for many years, but it has failed to transmit data to the ground. Figure 3.14(a) shows that there is an excess of energy within the intertropical area and a lack of energy at high latitudes.

In Figure 3.14(b), the estimates of atmospheric transport are presented, showing the differences between satellite data from ERBE, and the two sources of estimates, from different meteorological systems: one is the ECMWF and the other is the NCEP, which is part of the American Meteorological System. Figure 3.14(b) shows some consistency between ECMWF and NCEP, which estimates of the energy budget of the atmosphere and the required heat transport to balance it.

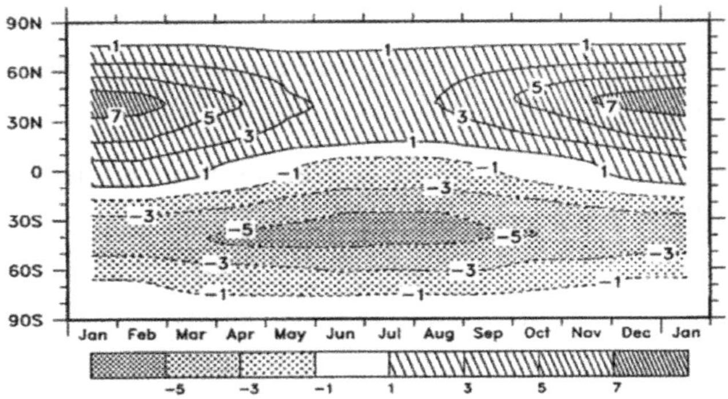

Figure 3.15. *The average zonal mean annual cycle of northward atmospheric energy transports for 1979–1998 (PW) (Source: Trenberth et al. (2001))*

Figure 3.15 shows the average zonal mean annual cycle of northward atmospheric energy transports for 1979–1998 and shows the changes with the seasonal cycles.

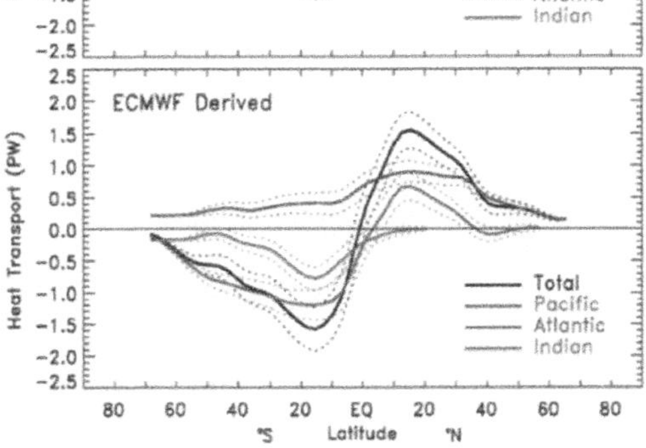

Figure 3.16. *Implied zonal annual mean ocean heat transports, based upon the surface fluxes for February 1985–April 1989 for the total, Atlantic, Indian and Pacific basins for NCEP and ECMWF atmospheric fields (PW). The 1 standard error bars are indicated by the dashed curves (Source: Trenberth et al. (2001))*

Figure 3.16 shows the kind of budget that we can get: not only in complete zonal average over the longitude, but also for different ocean basins. The top panel shows the transport of energy from the American model NCEP, and the bottom panel shows the European center model. We can see several elements consistent with the theory: an ocean transport in the Atlantic (purple line) that is always northward, i.e. with a transport of heat that goes from the Southern Hemisphere to the North. For the Pacific, on the contrary, there is symmetry between the two hemispheres, with energy being transferred from the equator to the Poles (green line).

Of course, for the Indian Ocean, energy does not really go towards the North; there is necessary transport southward. This stops at 60S where there is the Circle Polar current, and it also stops at 60N, as after this there is the Arctic Ocean. The difficulties of those implied currents are always the boundary conditions when considering the heat with its exchange vertically over the Earth and transport: the boundary values of the heat transport have to be known, and estimates must be done separately to get the sort of results presented in Figure 3.16.

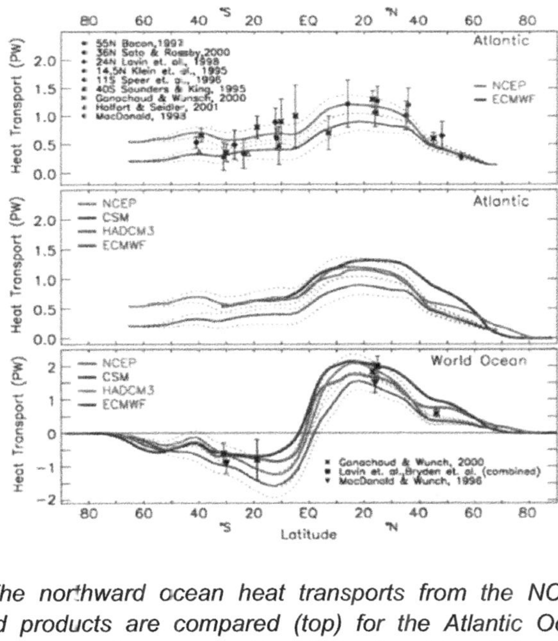

Figure 3.17. *The northward ocean heat transports from the NCEP-derived and ECMWF-derived products are compared (top) for the Atlantic Ocean with direct ocean estimates from sections, as identified in the key. The dashed curves show the 61 std err for the derived transports. Where given in the original source, error bars are also plotted and the symbol is solid (Source: Trenberth et al. (2001))*

If we compare these results to others like in Figure 3.15, for instance, it is roughly consistent: all the oceans show a transport that is towards the Poles globally. The Atlantic is different; it depends a bit on the climatologies. The Indian Ocean is to the South and the Pacific is symmetric. The agreement between the ECMWF and NCEP meteorological is better than the agreement of the values derived directly from the ocean (see Figure 3.17).

Figure 3.17 represents the heat transport for the Atlantic. A main question that it raises is why the results for the Atlantic Ocean are so different from other oceans.

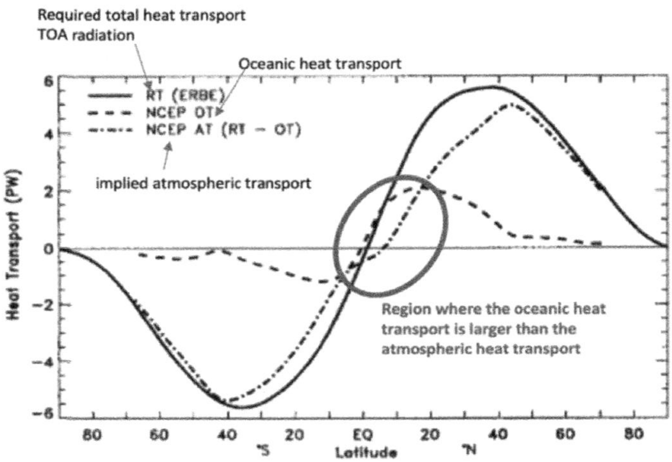

Figure 3.18. *The required total heat transport from the TOA radiation RT is given, along with the estimates of the total atmospheric transport AT from NCEP and ECMWF reanalyses (PW) (Source: Trenberth et al. (2001))*

Figure 3.18 shows another estimate of the role of the transport of the ocean compared to the atmosphere. The ocean heat transfer is represented with the dashed line "NCEP OT", the total transport, which balances the energy from the satellite, is represented with the black line "RT (ERBE)", the atmospheric heat transfer is represented with the dashed lines "NCEP AT" and the ocean is much less, except in the tropical area.

Therefore, this study shows a role of the ocean transport that is a bit limited to the tropical area and much less at higher latitudes. This is different from what was expected from older boat measurements, where the oceanic heat transfer looked equal to and sometimes even larger than the atmospheric heat transfer.

We will see later in the chapter that the ocean cannot transport the energy horizontally as much as we first thought, but the energy of climate change is stored in the ocean, which represents a very important component locally.

In the next section, we will discuss the links between atmospheric circulation and energy transport.

3.4. Vertical and horizontal transports: the links

Let us begin with the atmosphere and add a few words on the ocean later on. This requires us to consider the different energy forms to have the full landscape in mind.

Note that the following paragraph requires a bit of background on physics, but the general conclusions we will deliver at the end of this section are broadly accessible.

First, let us clarify again what is the energy in the system. There are different types of energy in the atmosphere, which will be more or less the same in the ocean, except the ocean is a bit simpler as its density does not change.

1) We first have the *internal energy* that may take two forms. The usual one is C_vT (with C_v expressed per mass unit).

2) *Potential energy*, usually represented as Φ, is another form of energy. It has two components: gravity and the work of centrifugal forces, which are linked to the rotation of the Earth.

Both terms depend only on the position of a given air mass on the sphere; therefore, they depend on the potential, and they determine the vertical dimension of the atmosphere, orthogonal to the isogravity lines, which are also lines of centrifugal forces.

But this is very small, as the Earth is a slightly distorted sphere:

$$\Phi = \text{gravity} + \text{centrifugal forces} = gz \quad [3.10]$$

with z being accounted from the surface or from the center of the Earth to the surface.

3) *Kinetic energy* is the energy of the motion when the air moves, which has kinetic energy.

4) *Latent heat* is the most important source of energy captured by the atmosphere, but we do not mention it much in this chapter.

When water vapor evaporates over the ocean, the ocean cools down and energy is earned in the atmosphere in the form of water vapor. This energy can be transferred back as direct heat (it is converted into internal energy in that case), when the water vapor condenses again.

Water vapor serves as a source of energy that can be released anywhere when the air's water vapor is condensing.

The term "latent" more broadly refers to something that may arrive, and similarly for energy, it refers to an energy that is not directly available, but may become available anytime when there is condensation of water vapor occurring. The phenomenon occurs when the wind is upwards, and it always occurs because of the smaller limit value for the water vapor content.

Therefore, to look at the transport, we should write an equation for each of those energies, as shown in Box 3.2.

Conservation of the different energy forms

Internal energy:

$$\frac{\mathbf{D c_v T}}{\mathbf{Dt}} = -\frac{\mathbf{P}}{\rho}\boldsymbol{\nabla}.\mathbf{U} + \frac{\mathbf{1}}{\rho}\,\mathrm{F_e} \qquad [3.11]$$

Geopotential energy:

$$\frac{\mathrm{D\Phi}}{\mathrm{Dt}} = \mathbf{U}.(\boldsymbol{\nabla}.\Phi) \qquad [3.12]$$

Kinetic energy:

$$\frac{\mathrm{DK}}{\mathrm{Dt}} = -\,\mathbf{U}.(\boldsymbol{\nabla}.\Phi) - \frac{1}{\rho}\mathrm{U}.\nabla\mathrm{P} + \mathrm{U.F} \qquad [3.13]$$

Summing up of the different energies

$$\rho\frac{\mathbf{DE}}{\mathbf{Dt}} = -\,\mathrm{p}\boldsymbol{\nabla}.\mathbf{U} - \mathrm{U}.\nabla\mathrm{P} + \rho\ \mathrm{F_e} + \rho\,\mathrm{U.F} \qquad [3.14]$$

$$\frac{\partial(\rho\mathrm{E})}{\partial\mathrm{t}} = -\,\boldsymbol{\nabla}.(\rho\mathbf{U}\,(\mathrm{E} + \frac{\mathrm{P}}{\rho})\,\rho\ \mathrm{F_e} + \rho\,\mathrm{U.F} \qquad [3.15]$$

Box 3.2. *Summary of the conservation equations of the different energy forms*

1) For *internal energy conservation*, equation [3.11] in Box 3.2 displays the use of Lagrangian derivatives, meaning that we are looking at the terms, which are directly the sources of heat, and the two terms can be detailed as follows:

$-\frac{P}{\rho}$ represents the work of pressure forces per unit mass, and $\nabla.U$ is the integrated divergence of the wind.

$\frac{1}{\rho}F_e$ is the heat coming from the sun or from the terrestrial radiation heating the system.

As a reminder, the difference between Lagrangian and Eulerian approaches is summarized in Figure 3.19.

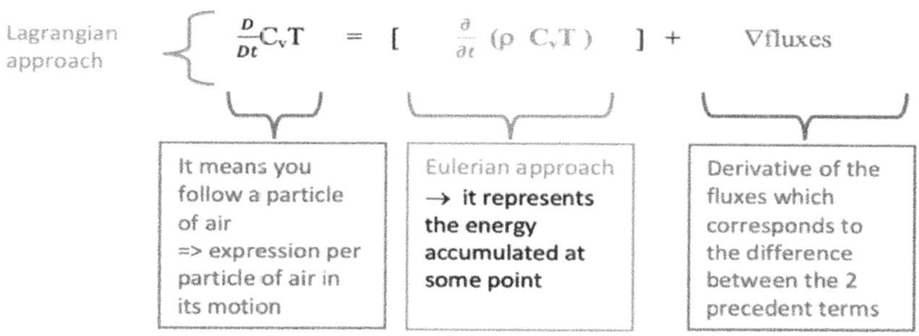

Figure 3.19. *Differences between Lagrangian and Eulerian approaches*

In Figure 3.19, the Lagrangian approach is first summarized: we consider a particle of air, and we want to know what has been accumulated at a certain point, and what is transferred.

We have to be especially cautious as we are in the context where the atmosphere is compressible, i.e. where ρ is changing.

Furthermore, **∇** fluxes stand for the motion of a particle, so we have an additional term because the particles are moving.

2) For *potential energy*, a given location does not induce any change. Therefore, the mathematical expression contains only the term $\frac{D\Phi}{Dt}$ and the flux term $.\Phi$.

3) *Kinetic energy* is obtained by multiplying the basic equation of the flow by U.

After summing up the conservation equations for these three energies, we obtain the equation [3.14], summarized in Box 3.2.

In it, we have the Lagrangian term, with the source of heat. In equation [3.14], there is one term associated with the pressure forces that create motion in the atmosphere: $-\mathrm{U}.\nabla\mathrm{P}$. Another phenomenon creating motion is viscosity, which is usually small.

Then, we obtain the equation simplified in Figure 3.20, which can be transformed using small mechanics terms put into brackets and varying locally in a flux:

$$\underbrace{\rho \frac{DE}{Dt}}_{\text{Lagrangian}} = \underbrace{\frac{\partial}{\partial t}(\rho\,\mathrm{E})}_{\text{Local energy change}} + \underbrace{\nabla(\underline{\mathrm{U}}\,\rho\,\mathrm{E})}_{\text{Term of flux}} + \textbf{sources}$$

Figure 3.20. *A simplified expression of the addition of the equations of energy conservation*

Figure 3.20 shows that the integration of ρ in the term of local energy change, where there is a derivation, which is a bit complex, must be done with care; it is based on a mass conservation that must be introduced. In the end, we obtain the equation represented in Figure 3.20 for the whole energy, distinguishing local energy change, as represented in the equation of Figure 3.20, and a term of flux plus sources. This is the method to compute the energy stored somewhere and the energy that is transferred at a given location.

After including all terms, including pressure terms, the transported heat energy per unit mass by the atmosphere can be written as follows:

$$\textbf{Heat energy transported by the flow} = \mathrm{C_vT} + \Phi + \underbrace{\frac{1}{2}\mathrm{U}^2}_{\text{Kinetic energy}} + \frac{P}{\rho} + \underbrace{\mathrm{LH}}_{\text{Latent heat}}$$

Figure 3.21. *A simplified expression of heat energy transported by the flow*

Let us focus on the terms of the equation translating the heat energy transported by the flow. Note that there are different kinds of latent heat for ice, clouds and water clouds.

Because of the perfect gas law, $P = \rho RT$, and when we sum up terms, we obtain:

$$C_v T + \frac{P}{\rho} = C_v T + RT + \Phi + \frac{1}{2} U^2 \quad [3.16]$$

with $C_P = C_v T + RT$, with C_v as the thermodynamic constant at constant volume and C_p as the thermodynamic constant at constant pressure. [3.17]

We neglect latent heat and add kinetic energy as follows:

$$\mathbf{E_t = C_pT + \Phi + \frac{1}{2} U^2} \quad [3.18]$$

Very small

If U =10 m.s-2

At 10 km height,
we have 10^4 × 10 J

Then as:

$$\Phi = 10^5 >> \frac{1}{2} U^2 \times 10^2 \quad [3.19]$$

and with $C_p \sim 1000$ and $T \sim 300$ K; therefore, it comes that:

$$C_pT >> \frac{1}{2} U^2 \quad [3.20]$$

and,

$$\Phi >> \frac{1}{2} U^2 \quad [3.21]$$

Therefore, the atmosphere is a good engine to transfer energy quickly from one side of the Earth to the other. *The winds are quick but transfer little mass, and in fact, little energy.* Therefore, the atmosphere does not store kinetic energy, but it does transport kinetic energy. Hence, we can write the following simplified equation for the energy transported by the atmosphere E_t:

$$\boxed{\mathbf{E_t = C_pT + gz}} \quad [3.22]$$

2 main terms

This is a very precise analysis, which is computed with deep care in climate models.

We have seen that the energy in the atmosphere is transported, and the next question to address is: how is this transport functioning?

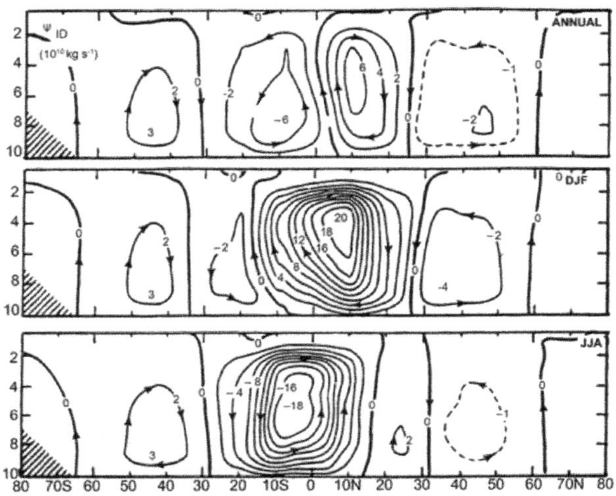

Figure 3.22. *What is the expected energy transport by the meridional cells? (Source: Peixoto and Oort (1992))*

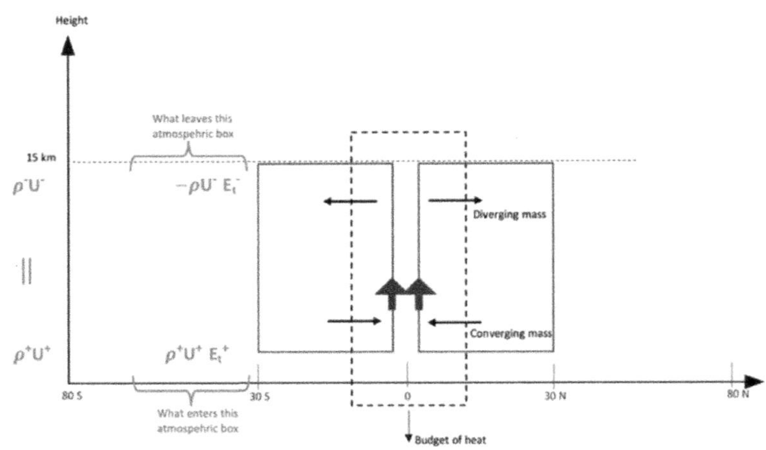

Figure 3.23. *Scheme of heat budget within Hadley cells for an axisymmetric system at rest*

Figure 3.22 shows the motion by meridional cells, and the question is: how does energy transport by those cells create motion to the Poles? Several issues are at play in this matter, and one of them is the link with the vertical gradient of the atmosphere. Figure 3.23 shows an illustrative scheme of Hadley cells 80S–80N.

Figure 3.23 shows the heat budget functioning within Hadley cells. A negative level refers to energy leaving the atmospheric box, while a positive level refers to energy entering, with a view of a system that is axisymmetric and at rest, meaning that the mass is conserved within the flow.

This roughly means that the energy that enters at the bottom of the atmosphere and the energy that leaves at the TOA are equal. Whether a system is gaining or losing energy basically depends on the comparison between the following two terms, and the system will lose energy if $\underline{E_t}^- > \underline{E_t}^+$, which also means, in other words:

$$\frac{\partial E_t}{\partial z} > 0 \qquad [3.23]$$

Equation [3.23] also recalls another important phenomenon: the stratification of the air. The related important feature is that because the air is stratified, the atmosphere is characterized by a radiative-convective equilibrium instead of a simple radiative equilibrium.

We can write the following two equations marking the stability of this radiative-convective equilibrium:

Regarding the transport to the Poles, the stability condition for stratification is met if:

$$\frac{\partial(C_p T + gz)}{\partial z} > 0 \qquad [3.24]$$

with

$$gz = \Phi \qquad [3.25]$$

We may also add here latent, which leads to the expression:

$$\frac{\partial(C_p T + gz)}{\partial z} + LH > 0 \qquad [3.26]$$

There is almost an identity between those two criteria for equilibrium because the water vapor becomes more complex due to the Hadley cells' ability to transport energy to the Pole, and in fact, there is a stable stratification.

3.4.1. *What is stable stratification?*

Stable stratification is not obvious and is a phenomenon obtained by *permanent convection* occurring, which is also necessary for the Hadley cells to transfer energy to the Poles.

Without those permanent *convective clouds* near the equator, the Hadley cells system would not function and could not transfer energy to the Poles. The first global climate system models that were set a few decades ago used to function without the addition of convection and did not work, as they were set up in the wrong dimension.

The scientific community then had to deal with the idea of integrating small-scale processes; it was first done by ensuring that the atmosphere was stable.

Therefore, algorithms were progressively added in the climate models *in order to stabilize the atmosphere*: the first climate models just used that as a representation of sub-grid scale processes, which are quite complex.

The consequences of the conservation of energy forms are as follows:

– If the stratification is stable, it means that $\frac{\partial E_t}{\partial z} > 0$.

– The direction in which the energy is transported is directed by the motions of the wind in high latitudes.

Let us consider the following cell.

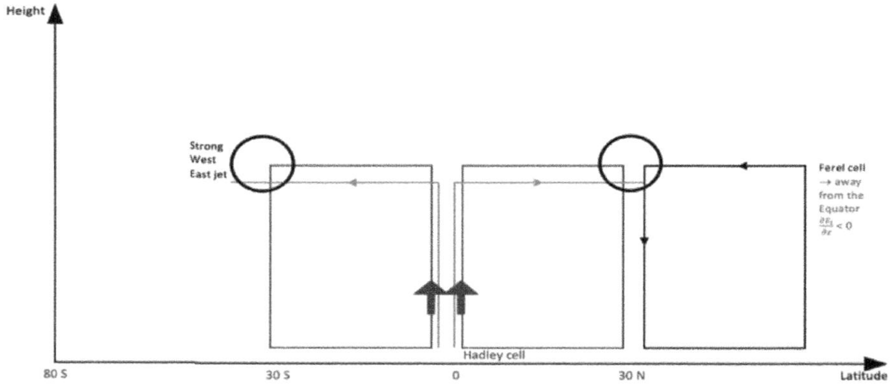

Figure 3.24. *Scheme of Hadley and Ferrel cells*

In Figure 3.24, there is no equilibrium when the flow goes from the Poles to the equator for Ferrel cell. On the contrary, there is equilibrium in the configuration shown in Figure 3.24 when the Hadley cell transports energy from the equator to the Poles.

The diagram represents the motion of the atmosphere averaged over time and over longitude for quantities that do exist for Hadley cells. There is really air descending near the equator, moving away from the equator into the higher layers of the atmosphere: this is not a mathematical trick. It differs from the situation at higher latitudes, as on the contrary, the Ferrel cells are instead a "mathematical trick" and only mathematical results give some continuity to the flow. But when observing the actual circulation, we do not see "Ferrel cells", but only instead horizontal motions going in one direction.

Figure 3.24 shows a quite simple functioning: stationary in time and abscissa metric in longitude. This is more or less true for the Hadley cells, but not at all for the mid-latitude area of the system. Therefore, the phenomena occurring from the equator must be considered a bit differently.

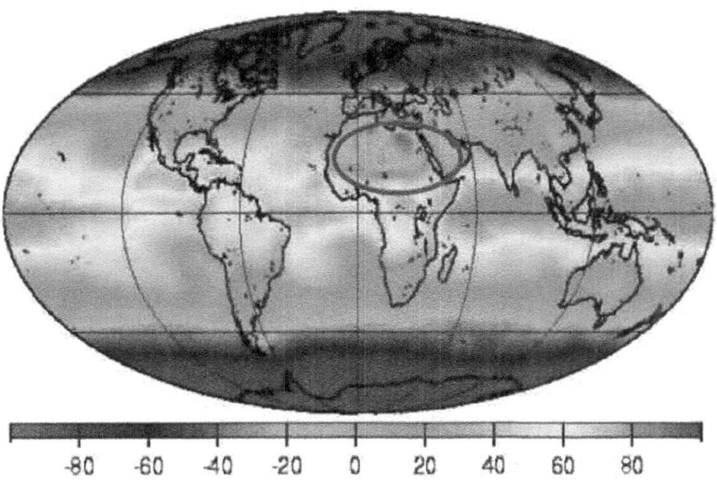

Figure 3.25. *Radiative budget at the top of the atmosphere (TOA), annual mean in W m^{-2}. The desert area with a negative radiative budget is circled in red (Source: Earth Radiation Budget Experiment (ERBE))*

Figure 3.25 shows the radiative budget at the TOA from a very high-resolution model. Pictures are extremely difficult to separate between satellites and models.

We see something occurring near the equator, and the small motions in the Indo-Pacific region are the ones that allow the energy to be transferred to the Poles.

If we focus our attention on the region highlighted with a red circle in Figure 3.25, we do not see very much when the air is dry, as there is no cloud to visualize it. The warm air moves up from the equatorial region, then it moves out from this region, and then moves down and marks the desert location. The heat is transferred as follows: soil is heated by sun rays.

This heat then moves in the atmospheric air upwards, and it gains potential energy. When it goes to the Poles with this potential energy, it can release this in terms of thermal energy: it falls down (at approximately 30N, 30S) in the regions of a zone called the *desert belt*. There are more deserts in those regions.

This is very important, because what the best modelers decided, when they started to design models a few decades ago, was that all the small scales would somehow serve the large scale. This hypothesis has been confirmed: small-scale processes do serve the Hadley cells functioning. The relationship between large and small scales in the atmosphere is extremely intricate.

The limit of Hadley cells comes from the angular momentum conservation law.

The angular momentum for an air parcel M_a, is a quantity that is conserved from an adiabatic motion (without any thermal transfer), which is the most frequent case, and it varies largely with height for a few days in the atmosphere.

Mathematically, we can write:

$$M_a = \rho \times (a \times \cos\Phi)\,(a\Omega\cos\Phi + u) \qquad [3.27]$$

with a being the radius of the Earth, and therefore, the distance to the Earth's rotation axis, Φ the latitude, $a\Omega\cos\Phi$ the absolute velocity of the solid Earth surface and u the West-East wind.

When we consider M_a in an absolute referential, the motion will be quicker. This is conserved for any kind of quick motion, and there is no viscosity involved and not much heat transfer. When the air parcel moves away from the equator, the quantity $a \times \cos\Phi$ diminishes with latitude. Only u moves up twice as fast from East–West, because it is nearer to the Earth's rotation axis, similarly to a situation where an ice skater brings their arms close to their body to turn faster.

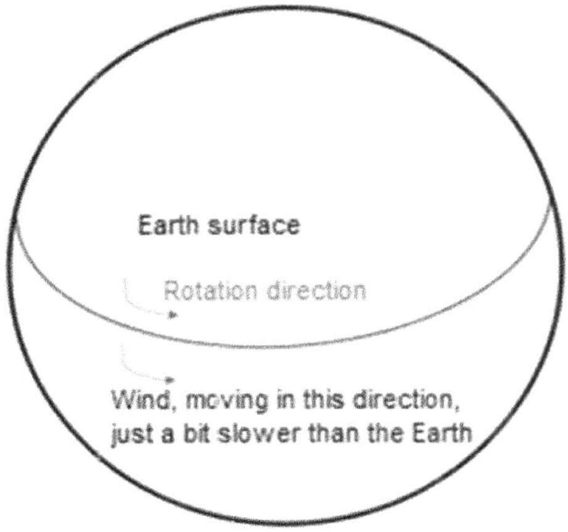

Figure 3.26. *Scheme of the Earth and wind speed compared to the Earth rotation*

When the motion is near to the Earth's rotation axis, it goes faster: strong winds develop like that from West to East, which is a result of the fact that the referential taken into account is an absolute referential.

There are two reasons why the air has a West–East motion: first, the way the energy is being transferred, and because of very strong West–East winds that are also the jets used by plane drivers when coming back to Europe from New York. For instance, a limit of the Hadley cell stops this circulation at the point where this jet is too quick, and therefore, the circulation becomes instable.

Another important question is whether the limit between the Hadley and the Ferrel cells is a limit that may change with climate change.

The answer is that it will not change very much because it is very linked with the speed of the Earth rotation that determines the strength of the jets/motions of air to the Poles: the acceleration from West to East primarily depends on the rotation of the Earth upon itself. The limit of the intertropical climate is set up by astronomical parameters (Earth rotation upon itself).

This has many consequences for the description of future climate. In fact, it is probably the only thing that we know for sure: there will be an intertropical climate that will stay in the same place, and an extratropical climate that will stay in the same place as well. But the climate will change differently according to their nature: more between 30N and 30S and something totally different at mid-latitudes.

When we speak about climate justice or injustice, the first reason is for astronomical motivations, and this is something humans cannot change. This feature will be important to keep in mind when we discuss the consequences of climate change associated with this transport of energy.

The Ferrel cells are the consequence of this: the 30S–30N limits are given by astronomical constraints and by a horizontal motion represented by the black lines in Figure 3.27, with a strong western wind component.

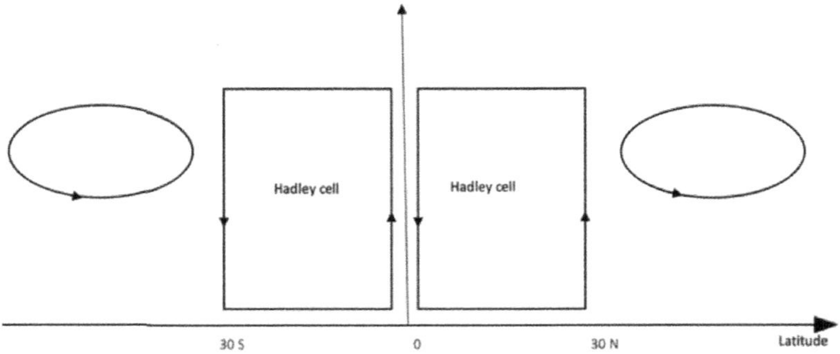

Figure 3.27. *Scheme showing the 30S–30N limits of Hadley cells*

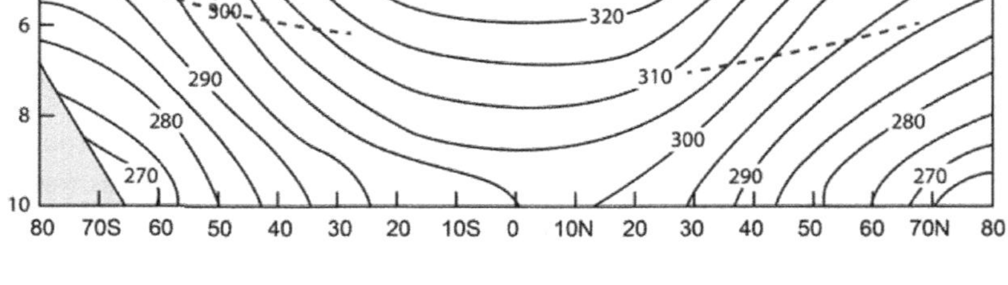

Figure 3.28. *Diagram showing the observed stratification of the atmosphere seen through the potential temperature field (Source: Oort and Peixoto (1992))*

Figure 3.28 shows a diagram illustrating potential temperature fields, which have a special role when it comes to mid-latitudes.

Entropy is a key notion to understand this figure. We have:

$$dS = \frac{Q^{\cdot}}{T} \quad [3.28]$$

If there is no heating, then:

$$dS = 0 \quad [3.29]$$

i.e. entropy is conserved in motion.

$$S = C_P \ln\theta \quad [3.30]$$

Stability in the system can be translated as follows:

$$\frac{\partial s}{\partial z} > 0 \Leftrightarrow \frac{\partial \theta}{\partial z} > 0 \quad [3.31]$$

When things are not stable, it means there will be motion in the atmosphere.

Let us now ask ourselves what may happen away from the equator.

Figure 3.28 shows a map of the potential temperatures θ averaged in the atmosphere: it is a dry one, as we do not consider latent heat. Latitude goes from 80S to 80N, and the vertical axis is pressure, expressed in db. We understand clearly that:

– θ increases with height.

– θ is larger near the equator.

This means that isolines have the kind of shape represented in Figure 3.28: in the regions of mid-latitudes, the lines of isoθ will be slightly translated upwards. Because of the relationship of equation [3.28], dS = $\frac{Q^{\cdot}}{T}$, and entropy increases with the heating.

Therefore, if we look at where the potential temperature is the largest, it is represented in the red zone in Figure 3.29. We found that the maximum is reached near the equator, at high altitudes of about 15 km above the surface.

Around the Poles, the coldest parts for potential temperature are represented with the zones cross-hatched in blue in Figure 3.29. For such small motions, air is going from higher θ to smaller θ (upwards): these are the conditions of instability. When θ is larger below, it means there is instability at mid-latitudes that are actually very specific: for all the motions that cross the isolines for the potential temperature that we can write as isoθ. This isoθ is almost horizontal, as represented in Figure 3.29. The atmosphere height is about 20 km, while the motions seen are over thousands of kilometers; therefore, an angle over the horizontal which is 1,000 of the radius represents a very small angle. The angle of the motion which is induced by that will be even smaller. Therefore, we have an effect directly linked with the equator-Pole gradient that will provoke almost horizontal motions that cross the isoθ at mid-latitudes.

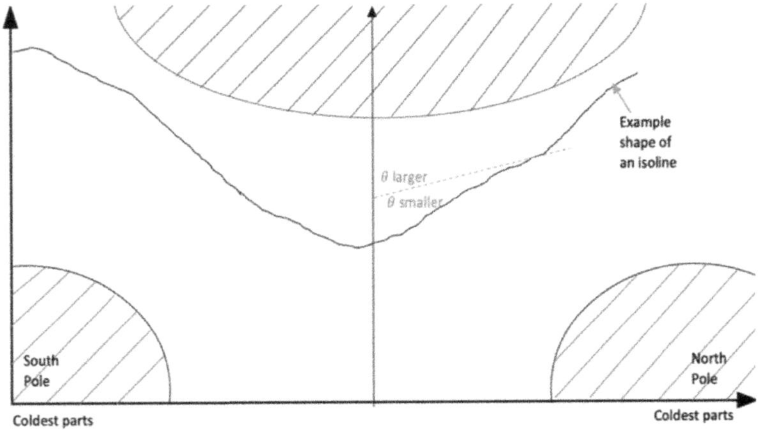

Figure 3.29. *Scheme detailing the previous diagram showing potential temperature fields*

The Earth engine functions in two ways to transport energy to the Poles:

1) The Hadley cells in the intertropical region transport energy within the limits of 30S–30N because of the Earth's rotation upon itself: latitude commands that phenomenon.

2) After that, some heat transfer is due to the horizontal mixing coming from the motions resulting from this kind of instability of the flow (almost horizontal). This dictates the consequences that we may derive from future climate change related to Hadley cells under global warming. We will have something in between 30N and 30S, with the increase in droughts and cyclones occurring in the intertropical regions. At mid-latitudes, we will have an increase in more powerful manifestations of very different phenomena. Cyclones in the intertropical regions and in the extratropical regions are very different, and there is a small extension of the intertropical regions, as we saw cyclones in Portugal, for instance.

In Figure 3.25, an interesting question to consider is why do we have a negative budget over the Sahara? Why is it sending radiation to space rather than getting and being warmed by solar radiation coming from space? In a region where there is a lot of sun, we could expect this to be heated, but in fact there is a negative budget.

The explanation for this phenomenon comes from albedo, with the reflectiveness of sand first, and also because the air above the ground is very dry, which means that terrestrial radiation crosses the atmosphere to space very quickly. There is another argument (chicken and egg solution) that is completely compatible if we consider circulation and the origins of the heat there. If we look at the cells in regions warmed by the Hadley cell, this region is warmed: heat moves near to the equator, brought up a bit more to the North, and moves down with a strong compression effect, heating near the ground. Therefore, Hadley cells are bringing heat there.

How does this heat escape? The answer is through radiation. There is no other way because it does not penetrate the ground very deeply: there is no storage of this heat very deeply in the ground because it takes a long time to heat the ground, so the radiation "has to escape". After solving a complex equation system fulfilling all those conditions, it appears that Hadley cells bring heat there, and the only way for the ground to get rid of it is to send it to space.

As a conclusion, a bit counter-intuitively, the Sahara is warmed by the atmospheric circulation and gets rid of this extra-energy through radiation. A bit of the same phenomenon occurs in Arabia. But the Sahara is a huge domain, remaining dry for the entire year, while the situation is a bit different in Arabia, which is more affected by seasonal cycles.

Figure 3.30 was created by Robert Sadourny, former director of LMD, more than 20 years ago, but it is valid. It illustrates transport by the atmospheric meridional cells in winter and in the Northern Hemisphere. The transport of internal and latent energy is represented with dashed lines. We note that the meteorological equator is located a bit to the South in this scheme; we see Hadley cells made of both the summer cells and the big winter cells.

The geopotential transport is represented in blue, and the total transport is represented in red, which goes to the North very strongly. The Ferrel cell has a negative value. The rationale of this diagram analysis is to show the very large amplitude of latent heat, which represents a huge part of heat transfer. It is the *geopotential transport* "that wins" in terms of dominating amplitude and that gives the direction of the energy flow to the Poles.

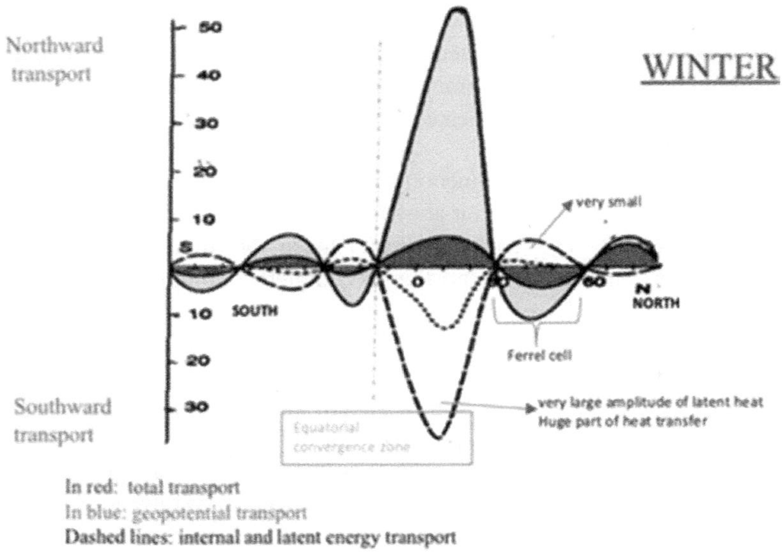

Figure 3.30. *Transport by the atmospheric meridional cells (Source: Sadourny (2001))*

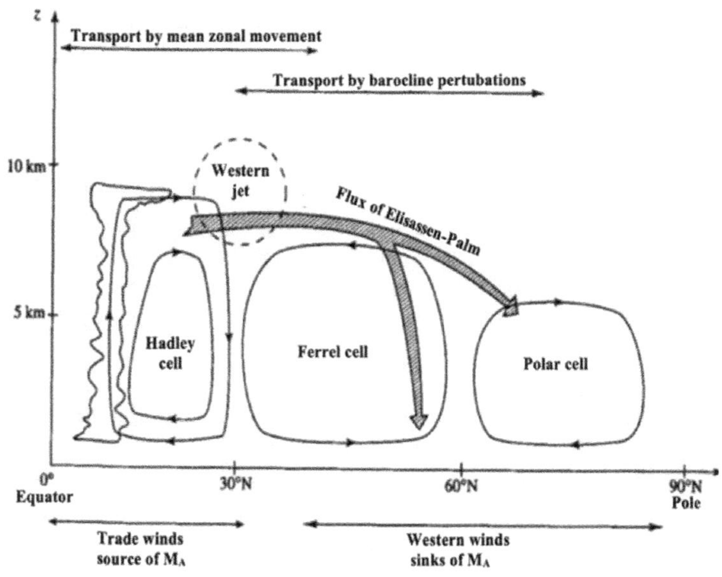

Figure 3.31. *Transport by atmospheric cells (Source: Sadourny (2001))*

Internal and latent heat represent a very small amplitude comparatively; this energy must be replaced by something that is not the energy transferred by the cell, but the energy transferred by all the motions that are non-stationary, non-axisymmetric, and that do transfer an important amount of energy to the North. Therefore, if we add up those terms, we obtain a continuity of the energy transfer. But there is another element: this important heat transfer brings a lot of energy from the Southern Hemisphere to the Northern Hemisphere, and it is actually specific to the seasonal cycle.

Another way to look at this is represented in Figure 3.31. A Hadley cell is represented on this diagram, as well as a Ferrel cell between 30N and 60S. What was missing in the previous description is something associated with the Ferrel cell that transfers energy to the North, as represented in Figure 3.31, with the Eliassen–Palm flux, and also momentum conservation back to the Earth's surface, which we have not described completely, which is also brought about by those motions in the Ferrel cell. They are not really the motions of the Ferrel cells that do not exist materially and are a mathematical construction, but in this area, many other mechanisms participate in the transfer of energy.

In Oort and Peixoto (1992), the representation of the change of latitude on average over the longitude from 80S to 80N for different seasons shows a gradient between the polar regions and intertropical regions. In the intertropical regions, there is stability of the temperatures: the sea surface temperature (SST) between 30S and 30N does not vary much, it is about 27°C over the ocean and some of the cyclones act to maintain the temperature at this level. This is linked with the humidity pattern over the same regions, as shown by satellite observations of the relative variations of water vapor to the saturation. Satellite observations show very humid conditions near the equator, and very dry conditions on each part of the equator. In fact, with the same temperatures, we have climates that are very different: there can be moist or dry climate for the same temperature.

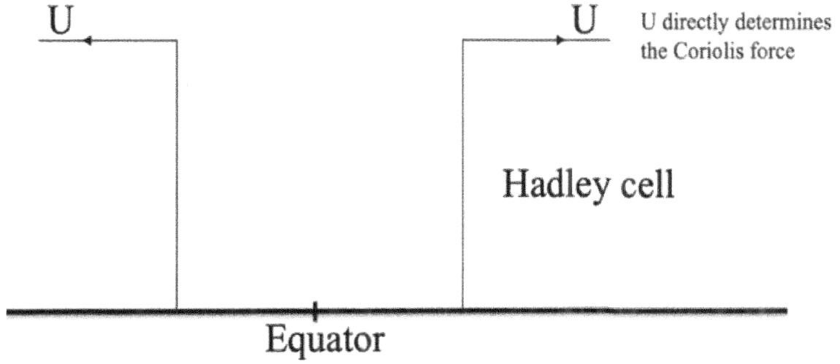

Figure 3.32. *Scheme of Hadley cell*

We have already seen a motion like that in the Hadley cells, and this determines the value of U, which is the wind blowing across the horizontal, which increases when moving away from the equator.

Furthermore, geostrophy, meaning an equilibrium between pressure forces and the Coriolis forces, depends on temperature: we have a motion that determines velocity by conservation of angular momentum M_a. This geostrophy, which depends on the heat gradient and which is also constrained, must be linked. Equations show that the conservation of angular momentum implies that the temperatures do not change very much; otherwise, geostrophic equilibrium cannot be achieved. Temperatures will therefore not change very much in the intertropical area. This has strong consequences: *near the equator*, we have *a temperature that is almost equal between 30S and 30N, especially around the oceans.*

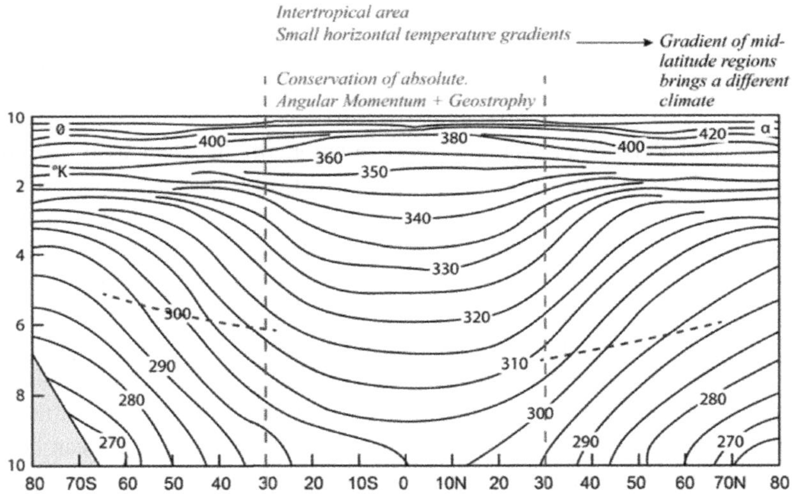

Figure 3.33. *Scheme of temperature gradient as a function of latitude and pressure/height (Source: Peixoto et al. (1992))*

Figure 3.33 was already discussed in Chapter 1. Let us consider just one climate of the Earth – the equatorial climate.

We have some temperatures, and above them, there is a lot of moisture, bringing much convection.

Above the ocean mostly or near continents, we have another climate that is very dry, and it is possible to show that we have the same equations for two stable equilibriums (a moist one and a dry one) close to each other: this is very specific to the climate of tropical areas.

There are, indeed, two different equilibriums: one stable and one unstable, which are close to each other and that constrain the climate of mid-latitudes. We can check the contribution of water vapor effects using satellite data, and the instability is restricted to equatorial areas, as shown in Figure 1.42, using the criterion of stability of equation [1.22].

Figure 3.34 shows the equations with two separate equilibriums, a dry one and a moist one, with the same equation and same temperature.

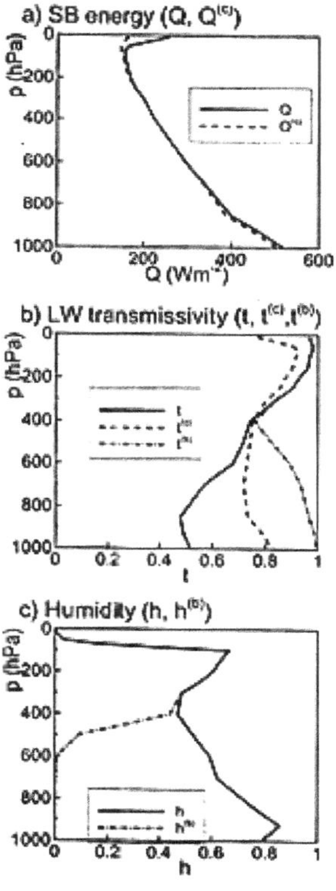

Figure 3.34. *Existence of two separate equilibriums (dry and moist) (Source: Ide et al. (2001))*

We use conservation equations for internal, geopotential and kinetic energies in equations [3.11] to [3.13], which we detailed in Box 3.2, already seen by summing them up together in equations [3.13] to [3.14] in Box 3.2, with another view presented in Box 3.3.

Another view of Potential to Kinetic energy conversion

$$\frac{Dc_vT}{Dt} + \frac{D\Phi}{Dt} = \mathbf{U}.(\nabla\Phi) - \frac{P}{\rho}\mathbf{U}.\nabla\mathbf{U} + \frac{1}{\rho}Fe \quad [3.32]$$

$$\frac{DK}{Dt} = -\mathbf{U}.(\nabla\Phi) - \frac{1}{\rho}U.\nabla P + U.F. \quad [3.33]$$

If we sum up over the total atmosphere, the two circled terms are opposite and equal, respectively, – Conversion and + Conversion.

$$\{P\} = \iiint_\Omega \rho\,(C_v T + \Phi)\delta\omega \quad [3.34]$$

$$\{K\} = \iiint_\Omega \rho\, K\, \delta\omega \quad [3.35]$$

Box 3.3. *Summing up the energy conservation equation over the total atmosphere*

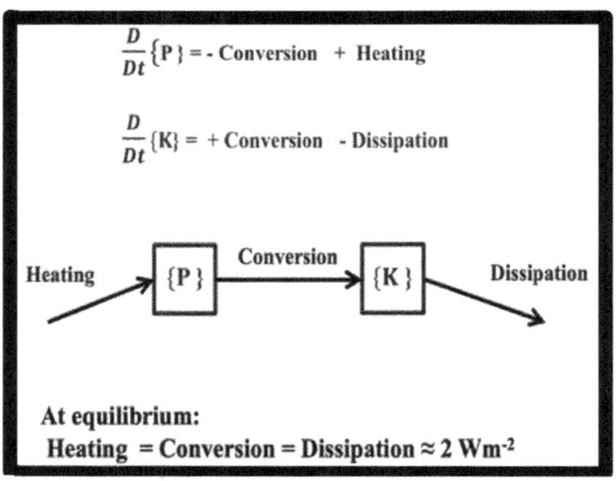

Figure 3.35. *Representation of the Lorenz cycle*

In fact, as shown in Figure 3.35, another way to look at this is to know how the geopotential energy forces the motion on the Earth. It is very frequent in the analysis of the motions of the atmosphere to study how the energy that gets into the system may be transferred into kinetic energy. It is not a huge quantity, but it is important because it creates the motion, and it is the motion that creates the climate. The steps consist of writing the equation in two parts:

1) the equation of the forcing: thermal + geopotential energy, as detailed in equation [3.32] in Box 3.3;

2) and for kinetic energy, the equation for $\frac{DK}{Dt}$, as detailed in equation [3.33] in Box 3.3.

Those terms are equal to each other. We can regroup the divergences into a common divergence term, and the sign of the resulting complete divergence after summing up over the total atmosphere is 0.

Therefore, we have the terms: $-U.(\nabla\Phi) - (\frac{P}{\rho}\nabla.U)$, which are opposite to one another: this shows the way energy is entering into the system.

Equation [3.32] in Box 3.3 is a warming. This warming changes not only this heat, but also a term for energy.

Equation [3.34] in Box 3.3 represents the total energy summed up over the whole atmosphere:

$$\{P\} = \iiint\Omega\rho(C_vT + \Phi)\delta\omega \qquad [3.34]$$

The kinetic energy summed up over the whole atmosphere is represented in equation [3.35] in Box 3.3:

$$\{K\} = \iiint\Omega\rho\, K\, \delta\omega \qquad [3.35]$$

The message of those equations is represented in Figure 3.35. We have already mentioned that the energy associated with kinetic energy is small and effectively, the energy that goes from the potential energy from the heating to the motion of the atmosphere is extremely small. As represented in Figure 3.35, at equilibrium, we have: heating = conversion = dissipation ≈ 2 $W.m^{-2}$.

The energy being used by the atmosphere from the heating is very small because of the stratification. If we go back to the notion of available potential energy, if the system is stabilized, there are energy levels represented by lines, and the energy only moves between two lines of isogeopotential. When the system is heated, it will try to relax: this relaxation of the system by heating is not conserving (the state of rest happens when there is no horizontal gradient, no motion induced by the energy): some unrest in the system is created. This unrest is measured by the difference between the energy and the "resting energy": this is called the *available potential energy*, and it is a very small part of the total energy. The energy in the climate system triggers many processes; the kinetic energy is not very large compared to the huge amount of heat stored stably, and that will not be used afterward, but it provides motions that will change the geography of the planet.

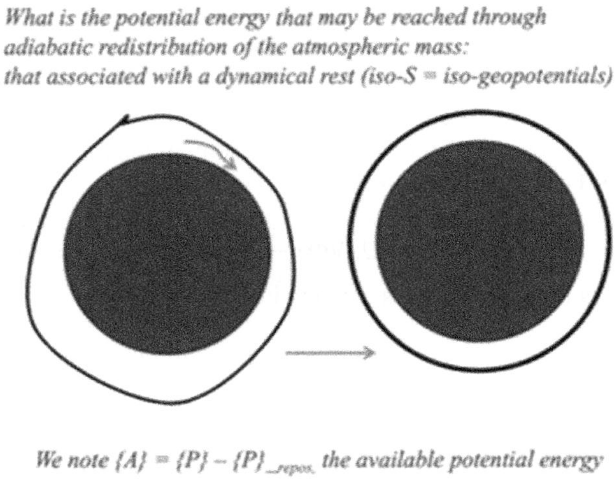

Figure 3.36. *Scheme for the potential energy that may be reached through adiabatic redistribution of the atmospheric mass*

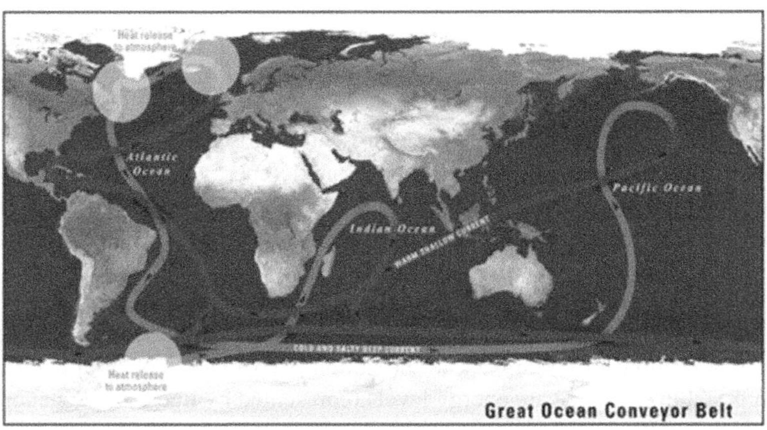

Figure 3.37. *The great ocean conveyor belt (Source: Broecker et al. (1991))*

The ocean may first appear as simpler than the atmosphere, as this fluid is not compressible; it has a fixed density, and its water travels over long distances.

In terms of energy, it is a mixing process that is rather simple, occurring almost horizontally or at least on one level near the surface.

The main difference between the atmosphere and the ocean is the fact that the latter is featured with the presence of salt, and the density of the ocean depends on salt, as shown in Figure 3.37 with the conveyor belt representation.

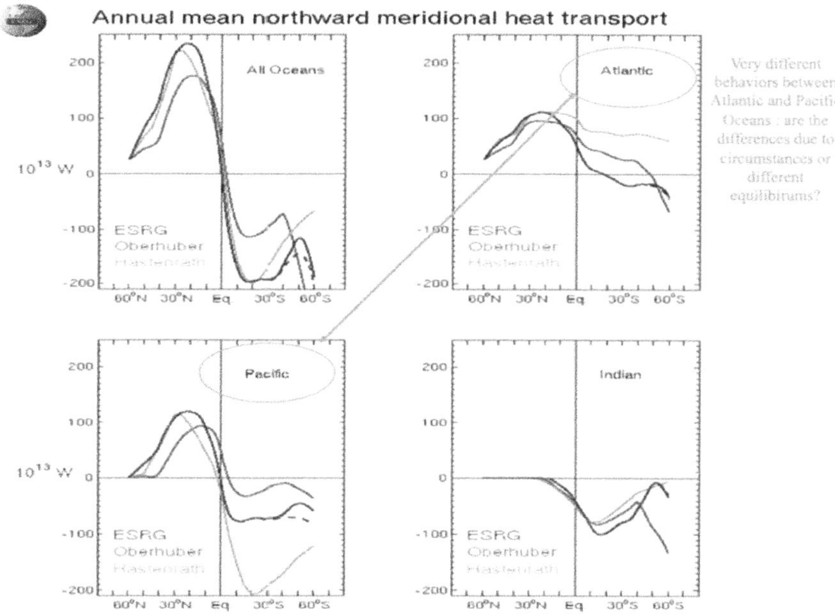

Figure 3.38. *Annual mean northward meridional heat transport in the Atlantic, Pacific and Indian Oceans (Source: satellite data (ESRG) and boat measurements (Oberhuber and Hastenrath))*

This is a feature that has no equivalent in the atmosphere, where everything that happens can be restored naturally. For example, if it is too hot, the atmosphere may send radiative energy to cool down. If the ocean is too salty, however, nothing can be done. The ocean is extremely different in the sense that this system has an extremely important parameter near the Poles, will contribute a lot to the motion/transport of energy and has no capacity to be self-regulated like all of the other terms in the atmosphere. Alternatively, when considering the atmosphere, we may believe that this system has only one kind of equilibrium. We may start a climate model with a completely isothermal atmosphere, let it run a little, and slowly it will create Hadley and Ferrel cells, mid-latitude separation, etc., "by itself" if it is well coded.

The oceanic circulation can therefore be very different among different basins, as shown by the contrast between the Atlantic and Pacific oceans in Figure 3.38, which presents the idea that we may have multiple oceanic equilibriums, due to salt, implying different capacities of the ocean to evolve.

Are those differences due to "circumstances" or are there possibly several equilibriums in the Atlantic and Pacific? The question is still open.

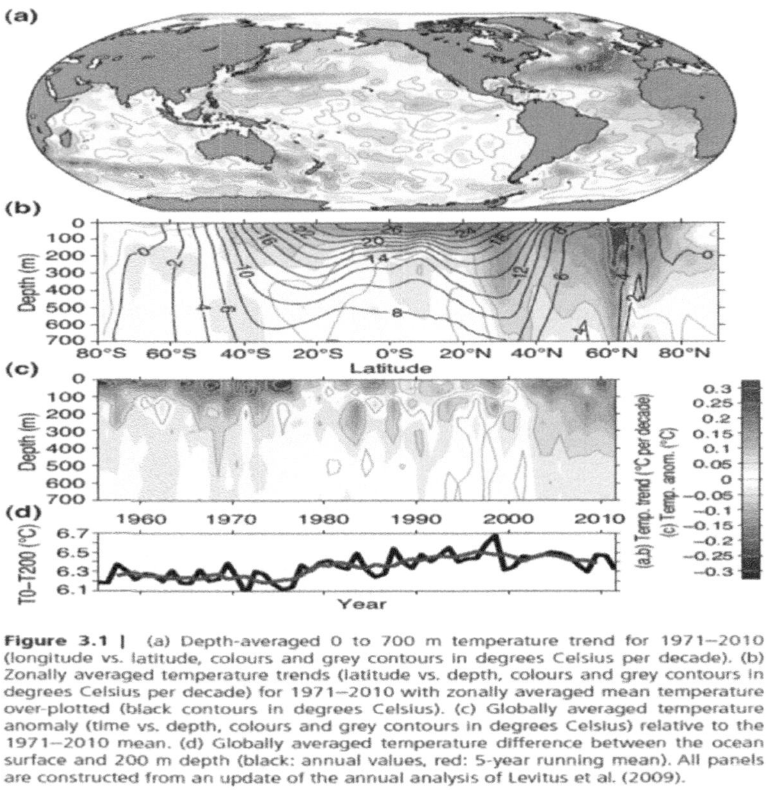

Figure 3.39. *Isolines of in-depth temperature evolution as functions of latitude and time over 1971–2010 (Source: Figure 3.1. of IPCC Fourth assessment Report AR4 (2013). Reproduced with permission from Rhein et al. 2013)*

The ocean is very important because it is capturing energy. Figure 3.39 shows the isolines of in-depth temperature evolution, with:

a) Depth-averaged 0 to 700 m temperature trend for 1971–2010 (longitude vs. latitude): we note that there is roughly a warming, not everywhere, however.

b) Zonally, averaged temperature trends: we see that there is mostly warming in the Northern Hemisphere, and this warming where waters are moving downwards more strongly is very efficient.

We are able to see this warming as a function of latitude and of time. There is clearly a progressive penetration of the depth of the ocean. Therefore, the ocean has a major role concerning heat: *it is the place where the energy is stored.* This feature is extremely important. We may have a decrease in the temperature near the ocean surface just because there is a certain quantity of heat-provoking evaporation. What matters here is the heat being stored that is not ready to escape. Another way to look at it is to analyze measurements over bigger depths, but it is not completely possible to do this in a routine manner, and the fact that there is a lot of noise in this assessment could considerably modify the diagnostic of future trends under climate change.

3.5. Conclusion

In Chapter 3, we discussed key mechanisms driving the changes of energy exchanges in the system naturally.

These first three chapters were dedicated to the description of natural energetic phenomena governing the Earth.

In the next two chapters, which are dedicated to the anthropogenic forcing of the Earth's energetic processes and its sociopolitical impacts, we will examine how the Earth's climate can be changed by human activities.

4

A Transition from Quasi-Equilibrium to Forcing from Anthropogenic Source: A Cartography of Predictable Phenomena, Uncertainties and Feedback Related to the Climate Response to Anthropogenic Disturbance

In this chapter, we will stay in the domain of the analyses of processes through the help of modeling and observational tools. General issues such as radiative forcing will be explained; methods to compare different greenhouse gases will also be discussed.

The focus of this chapter is the transition between an Earth climate that was at equilibrium, and a current phase that is getting out of equilibrium.

It is possible to scientifically describe a phenomenon that can be called the "way out of equilibrium". This is key about what can be said with some certainty: the scientific community is of course not able to describe every future climatic phenomenon, but some mechanisms are predictable, and it is linked with the idea

---

For a color version of all of the figures in this chapter, see www.iste.co.uk/letreut/energetics.zip.

Energetics of the Climate System,
by Hervé LE TREUT and Mounia MOSTEFAOUI.

that there was an Earth climate equilibrium in the past. In this chapter, we will describe what has been changing from the state of equilibrium. Associated with the description of feedbacks, we will map out what is predictable.

One key purpose of this chapter will be to draw a line between what we know for sure and what merely represents a risk, and its associated uncertainties. This distinction is important as certainties and risks cannot actually be managed similarly.

4.1. A transitory evolution

The idea that while climate alert has been existing for a long time, things have not changed much due to a lack of response from the political world is partly wrong.

Box 4.1 presents some key dates related to the climate alert by the scientific community.

Key dates regarding the climate alert

– *1886:* Svante Arrhenius was a Swedish Nobel Prize laureate of chemistry. His work included computations of the warming that could result from a doubling of CO_2. He obtained the result of a + 5°C warming. This result is often cited, and Arrhenius is sometimes referred to as the pioneer alert scientist regarding climate change that politicians did not listen to. However, this representation is actually wrong for at least two reasons: first, Arrhenius was already very famous at his time and had influence. Second, he actually thought that climate change was a positive phenomenon for human survival. He did not perceive the associated risk because at that time, the amount of CO_2 necessary to go to a CO_2 doubling compared to pre-industrial levels seemed huge and a matter of thousands of years, rather than a matter of a few decades. Given the large scale of time that he mistakenly anticipated, Arrhenius concluded that in thousands of years, global warming would represent a positive effect for humankind, because he thought that it would allow us to escape the next glacial era. That is the reason why he thought that using oil might help humanity to get rid of ice ages due to the scope of thousands of years that he wrongly foresaw. For the anecdote, Svante Arrhenius is also a direct ancestor of the Swedish activist Greta Thunberg.

– *1957:* this year was the International Polar Year and corresponded with the first time during which scientists spoke about climate change using a framing similar to terms used today, referring to the future risk associated with the scales of times of decades or centuries. The paper which first introduced those considerations of dangerous effects of a global warming was co-written by Roger Revelle, who was also American Vice-President Al Gore's mentor. Its title is: "Carbon Dioxide Exchange Between Atmosphere and Ocean and the Question of an Increase of Atmospheric CO_2 during the Past Decades". In this paper, the authors expose that they found that the exchange of carbon dioxide between the atmosphere and the ocean could be very slow compared to expectations. At that time, the ocean was seen as a sort of "garbage" and it was thought that it was possible to continue to use it the same way. The authors wrote that we could probably detect CO_2 increase with new devices to measure the CO_2 in the atmosphere. In 1957, there was the capacity to detect an increase of CO_2 with high quality measurements. However, this CO_2 increase was still weak at that time compared to now. As a consequence of the findings from this paper, a new observational device was created: the Mauna Loa Observatory (MLO), on the slopes of the dormant volcano in Hawaii, a place with particularly favorable conditions for sky observations and measurements as there were no disturbance sources around, such as industries or forests which have a very strong associated seasonal cycle. The MLO was first directed by Charles Keeling and then by his son Ralph Keeling. The outcomes obtained from this device represent the longest record in the atmosphere with direct measures of atmospheric carbon dioxide. The quality of the records coming from this station was good because the CO_2 goes in the atmosphere very quickly; therefore, these measures could be representative of the whole world. Moreover, other evidence of the reliability of the MLO's measures is that they are consistent with the measures that have been taken elsewhere in the world, such as in the Pacific or in Antarctica, showing similar trends. It is actually the MLO's measures that gave the alert: when the global CO_2 emissions passed over 400 ppm for instance, the measures were taken from there. The MLO belongs to the Earth System Research Laboratory within the National Oceanic and Atmospheric Administration (NOAA).

– *1967:* within the climate community, a research group in the space domain named the Global Atmospheric Research Project was run during that year, based on satellite reports.

– *1972:* this year, the first summit was organized with a focus on climate change in Stockholm, mostly linked with the development of developing countries.

– *1979:* Jule Charney published a historical report to the American National Academy of Sciences (NAS). Charney was an eminent and very influential scientist who notably developed, among others, the technique named the "quasi-geostrophic approximation". He worked on the creation of the MLO from the theory to its application. He also developed the theory of the droughts in Sahel as well as an explanation regarding why they were lasting so long. He died two years after the publication of his famous 1979 report to the American Academy of Science.

This important and fascinating report was actually quite small in terms of volume, with only 22 pages. The main results of Charney's report is future anticipations run by two climate models. This report contained everything needed to understand the problem of carbon dioxide emissions for climate change. Of course, Charney did not stop the whole fossil fuel industry with his small report; however, the necessity to decrease the exploitation of the oil industry came quickly after its publication. The main content of the report is synthetized in this sentence of the second page from the report: "we estimate the most probable global warming for a doubling of CO_2 to be near 3°C with a probable error of ± 1.5°C".

His estimate was roughly unchanged for more than 30 years, as shown by the estimate in the 2007 Intergovernmental Panel on Climate Change (IPCC) report (4th Assessment Report) pointing to "a best estimate value of about +3°C" of climate sensitivity and stating that it "is very unlikely to be less than +1.5°C".

– *1980:* the World Climate Research Program (WCRP) was created in 1980 mostly to include the ocean in the models as there were actually different traditions regarding the coupling or not of the ocean in climate models. The WCRP was sponsored by the World Meteorological Organization (WMO) together with the International Council for Science. Since 1993, the WCRP has also been sponsored by the Intergovernmental Oceanographic Commission (ICSU) of the UNESCO. In the early 1980s, the first Director of WCRP was the French scientist Pierre Morel, who was also the creator and the first director of the LMD (Laboratoire de Meteorologie Dynamique). This period also corresponded to the beginning of the Atmospheric Intercomparison Program and the Coupled Model Intercomparison Program.

– *1988:* the Intergovernmental Panel on Climate Change (IPCC) was created by a political decision from national government leaders at the occasion of a G7 under the pressure of the President of the United States of America, Ronald Reagan, and of the British Prime Minister, Margaret Thatcher. The IPCC was placed under the guidance of United Nations Program for the Environment (UNEP) and the World Meteorological Organization (WMO). However, the work done by the scientific community including Charney's 1979 report was the real trigger for this political decision of creating the IPCC and for the mediation of climate science by this entity in charge of regularly providing a synthetical bibliographical assessment of the best available international knowledge regarding the state of the art concerning climate change.

– *1990:* the first Assessment Report of the IPCC report was published.

– *1992:* the United Nations Framework Convention on Climate Change (UNFCCC) was adopted at the occasion of a major international meeting concerning climate change which took place in Rio de Janeiro and is referred to as the Rio de Janeiro Earth Summit. It was the Third Earth Summit (the first one took place in Stockholm in 1972 as already mentioned), meaning that it actually came quite early after Charney's report (three years later). However, after that, concrete actions to politically and globally translate the outcomes coming from the scientific community have come less quickly. The biggest difficulty was not to make the scientific diagnostic accepted but came from the fact that the decision-makers have been disagreeing for decades about the actions to be taken.

Box 4.1. *Key historical dates regarding the climate alert*

4.1.1. *Presentation of the Intergovernmental Panel on Climate Change (IPCC) and its missions*

The IPCC's missions are to provide regular updates on the state of the art of climate change science, as well as on the related risks, and suggest adaptation and mitigation solutions. The IPCC is an international structure composed of 195 member-states, who elect a bureau of 34 members, in charge of mobilizing thousands of scientists to write and review IPCC productions.

The different IPCC reports deliver a detailed bibliographical synthesis of the best available existing scientific knowledge about climate change, its impacts and its solutions, including both peer-reviewed articles, on the one hand, and papers coming from the "gray literature", on the other hand. Its purpose is to build a common and shared scientific knowledge recognized worldwide regarding climate change.

The reports are unanimously approved by all of the IPCC authors on a consensus basis. The summaries for policymakers must additionally be approved by all of the governments (195 countries in 2024) without deforming the scientific facts.

IPCC reports are primarily intended for country leaders as a neutral and non-prescriptive support for public policy decision-making and as a support for the implementation of national laws on climate change. They include factual uncertainty assessments as well as risks and benefit estimates.

The IPCC conducts no research on its own and only provides synthesis work on global scientific production. The vocation of IPCC reports is universal and to deliver an indisputable basis for international negotiations related to climate change.

One key feature of the IPCC work is its interdisciplinary aspect, both in terms of members and regarding the topics involved in the assessment, from the physics of climate change, to economics, sociology, public policy, national and international law, etc.

More specifically, the IPCC's missions are subdivided into three "working groups", plus one group focusing on GHG emissions and sink inventories.

The scientists involved in the IPCC report redaction represent 14 permanent staff members of the IPCC's Secretariat. Those permanent members and the IPCC Assessment Reports, published at every cycle of about 5–10 years, are organized following the three Working Groups:

– The Working Group I (WGI) focuses on the physical science basis of climate change. It typically delivers a state of the art regarding the knowledge related to the Earth system components (atmosphere, ocean, terrestrial surface, biosphere and cryosphere), the way they interact together and the impact of climate change on them.

– The Working Group II (WGII) is focused on the evaluation of the impacts, risks, adaptation and vulnerability aspects as well as action options in the face of climate change.

– The Working Group III (WGIII) is dedicated to the mitigation of climate change.

– A fourth group, named the Task Force on National Greenhouse Gas Inventories, works on methods and processes for measuring greenhouse gas emissions and their sinks.

The Sixth IPCC Assessment Reports were published between 2021 and 2023.

4.1.2. *A focus on IPCC productions*

4.1.2.1. IPCC Assessment Reports

Within one assessment cycle, the IPCC Assessment Reports are produced following different steps, including first- and second-order drafts, reviewed by hundreds of independent international scientists, on a voluntary basis. Expert reviewers are selected after an IPCC application process based on their international background, experience and previous publications. The chosen expert reviewers are hired based on the diversity of their background to ensure that as many ideas as possible are considered for a rich perspective on different topics. Each reviewer's comment is considered, ensuring that the reports are transparent and reliable.

For every IPCC report, an order of magnitude of thousands of articles from both peer-reviewed articles and from the "gray" literature is evaluated. This quasi-comprehensive review allows the widest possible range of scientific outputs to be considered.

4.1.2.2. IPCC Special Reports

The IPCC can also issue Special Reports, for instance at the request of a government for more scientific information on a particular matter. For example, the following Special Reports can be quoted:

– A Special Report on a Global Warming of + 1.5°C (2018).

– A Special Report on the Ocean and Cryosphere in a Changing Climate (2019).

– A Special Report on Climate Change and Land (2019).

– A Special Report on Cities and Climate Change (2027).

The Special Reports provide an interdisciplinary evaluation of those topics and involve more than one Working Group. The Special Reports are less dense in terms of volume and content than the regular IPCC Assessment Reports.

4.1.2.3. *IPCC Synthesis Reports*

The Synthesis Reports focus on a small number of important policy issues. They include the inputs coming from the Assessment Reports and the inputs from the Special Reports produced during a given 5–10-year assessment period. The Synthesis Reports contain several chapters, and may also include a Technical Summary.

4.1.2.4. *IPCC summary for policymakers (SPM)*

Within each assessment period, approximately 50 scientists are chosen and join the efforts of the Working Groups to deliver a first-order draft of another type of IPCC report called the summary for policymakers (SPM), mainly intended for government leaders. Such a report involves a separate development process and is based on a consensus for its framework.

First, each IPCC Working Group synthesizes its inputs coming from the Assessment Report in a first draft of the SPM. This draft must then be endorsed line by line by government officials through a dedicated discussion process to agree on a final version of the text, preserving the scientific factual results. Sentences are displayed on a screen visible to all participants, and as part of the rewriting process, paragraphs are added or deleted as required.

The new draft is then reviewed during a plenary session including both national government delegates and observers. The plenary sessions are placed under the complete supervision of the different IPCC Working Groups' Chairs to make sure that the scientific message is faithfully translated into a framing designated to the policymaking area.

This review process aims at clarifying the message of the SPMs while preserving the core of the scientific inputs. As for the policymakers' representation, each delegation sends about six representatives including their local experts that may belong to the IPCC, as well as other public officials or experts of diplomacy and negotiations.

In case of a major disagreement, a small group with the main actors involved is created by a Working Group Chair to expedite the process and negotiate a mutually satisfying framework.

4.1.2.5. *IPCC Methodology Reports*

Methodology Reports are another kind of document produced by the IPCC. Their purpose is to deliver operational advice regarding the elaboration of national greenhouse gas inventories. They include an "Overview Chapter" with the inputs of the SPM.

All of the reports described above may also contain a section named "Frequently Asked Questions" as well as a section named "Supplementary Material".

4.1.2.6. *Other diverse resources linked to IPCC productions*

Some other additional resources related to IPCC reports can be particularly useful, especially for pedagogical purposes, including:

– The main "locks" listed at the end of each IPCC report's chapters, and suggestions of further research investigation.

– The IPCC synthesis sheets by question, sector or by region.

– The IPCC summaries for all, to the destination of a high school level public.

– The frequently asked questions section.

– Global interactive atlas, freely available online: https://interactive-atlas.ipcc.ch/; https://atlas.climate.copernicus.eu/.

– An artificial intelligence searching engine based on IPCC reports, as well as based on its equivalent on biodiversity: the Intergovernmental science-policy Platform on Biodiversity and Ecosystem Services (IPBES) reports (see: https://www.climateqa.com/).

4.1.3. ***Examples of IPCC production impacts***

The direct impacts of the IPCC's extensive work are not necessarily easy to measure. However, in some cases, their consequences in terms of political decision are clearly identifiable. The most impactful reports regarding public and political awareness on climate change are undoubtedly the summaries for policymakers, with synthesized information on critical conclusions from the main assessment or Special Reports.

As an example, the European Green New Deal announced by Ursula von der Leyen during COP25 in 2020 directly quotes the scientific findings of the IPCC 1.5°C Special Report to substantiate the targets set, including a proposal to implement climate legislation to achieve climate neutrality by 2050.

The IPCC productions resulting from the different contributions from the three Working Groups plus the Task Force on GHG represent a major source of information for the international negotiations related to climate change, which take place every year at the occasion of the annual Conference of the Parties, as well as for decision-makers. IPCC reports largely contributed to the adoption of the United Nations Framework Convention on Climate Change (UNFCCC) in 1992, to the Kyoto Protocol in 1997 and to the Paris Agreement in 2015.

4.1.4. *Complexity, a main feature of the Earth climate system transitory evolution*

Complexity is certainly one of the main features of climate change. The difference between alerting and beginning to design processes, laws and agreements between numerous different nations is huge. Anyway, it is not possible to alert or manage the problem of climate change without going into the complexity of its physical processes.

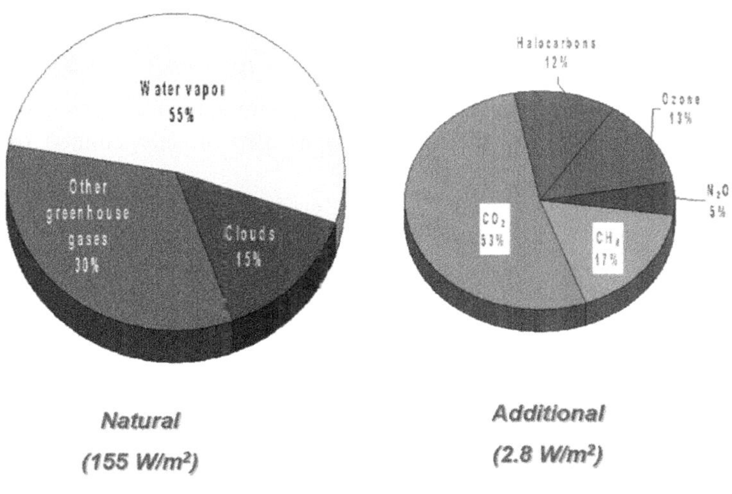

Figure 4.1. *Rough main contribution estimates to the greenhouse effect (natural and additional part as in 2016). Source: Le Treut (2016)*

The natural greenhouse gas effect is measurable by the difference between the radiation which is emitted by the ground and the emission of the terrestrial radiation that goes out to space; this difference is approximately 155 W.m^{-2} (see Figure 4.1).

This simple measure of the greenhouse gas effect can be represented in Figure 4.2.

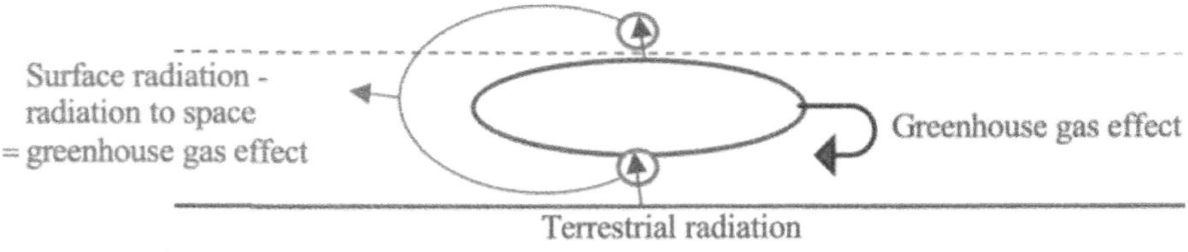

Figure 4.2. *Very simplified scheme of the greenhouse gas effect*

In terms of natural greenhouse gas effect, a huge contributor is the water cycle, as shown in Figure 4.1. In the same figure, if we consider the water vapor and the clouds, not as vapor or gases, but the particles that can reflect the solar light and also affect the terrestrial light, what is measured is therefore the terrestrial impact of clouds on the terrestrial radiation emitted to space: clouds and water vapor actually represent the largest component (~70%). All the other gases represent only ~30% of the proportion. Note that these are rough estimates, regularly subject to updates, but the order of magnitude is relatively stable.

When considering the changes, the gases representing the main greenhouse gases are CO_2, CH_4 and N_2O, i.e. they are the ones also directly emitted by human activities.

Ozone is complex and depends on stratospheric chemistry; it is a singular gas compared to the others. The additional greenhouse gas effect that those gases represent is ~2.8 W.m^{-2}.

The halocarbons are quite "new" in the atmosphere, as they are emitted by human activities. If we make an order of magnitude comparison between the additional greenhouse gas effect of ~2.8 W.m^{-2} and ~155 W.m^{-2} which represents the natural greenhouse gas, the additional radiative forcing due to human activities actually appears to represent a small change of less than 2%, but it actually creates a huge change in the climate system. The reason for the strong effect of a relatively small perturbation is that before humans started to act on disturbing the system, about 10,000 years ago, the system was more or less at equilibrium. Humans are disrupting a long equilibrium: that is a key point to attribute how much the climate system itself is reacting to a small change. When looking at changes in the temperatures during the Holocene, the scale is of about a few tenths of degrees (0.5°C) during the whole period: the Earth climate had therefore been then very stable at this scale of time.

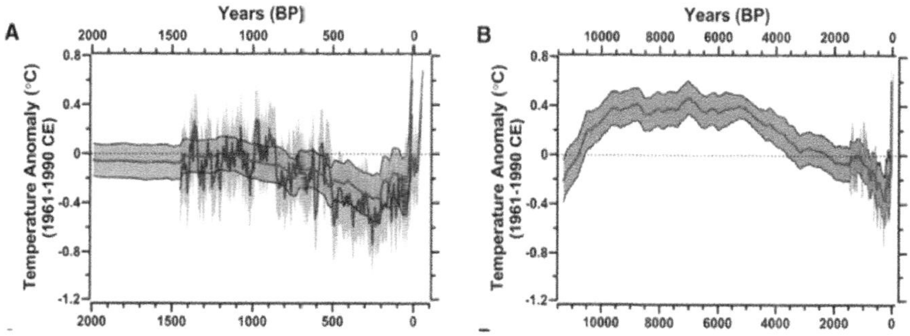

Figure 4.3. *A reconstruction of regional and global temperatures for the past 11,300 years by comparison of different methods. Globally stacked temperature anomalies for the 5° × 5° area-weighted mean calculation (purple line) with its 1σ uncertainty (blue band) and the global CRU-EIV composite mean temperature (dark gray line) with their uncertainty (light gray band). Source: Shaun et al. (2013)*

The little ice ages represent a cold climatic period that took place between the beginning of the 14th century and the end of the 19th century. It mainly affected the North Atlantic and it is featured by a series of long and cold winters. The temperature anomaly was then 0.4°C at the highest level, which is extremely stable.

What is emphasized in Figure 4.3 is that there has been stability in the Earth climate for several thousand years and human activities have been changing that. The temperature of the Earth is approximately 300 K. An important change such as the one from the glacial to the interglacial era also represents a change of a few percents in the radiative forcing: in the domain of climate, small changes actually matter.

A theory developed by climatoskeptical groups was the idea that the existence of many feedbacks was responsible for the stability of the mean annual temperature of our planet for such a long time. Their point was also that those feedbacks would show up again and therefore the warming would not happen. Actually, the warming is already occurring now, and we have to take the figures where they are: it does not mean that all processes are fully understood by the scientific community, and the fact that uncertainties exist is a right source of concern. However, what climatoskeptical groups have absolutely no excuse for, is the risk denial.

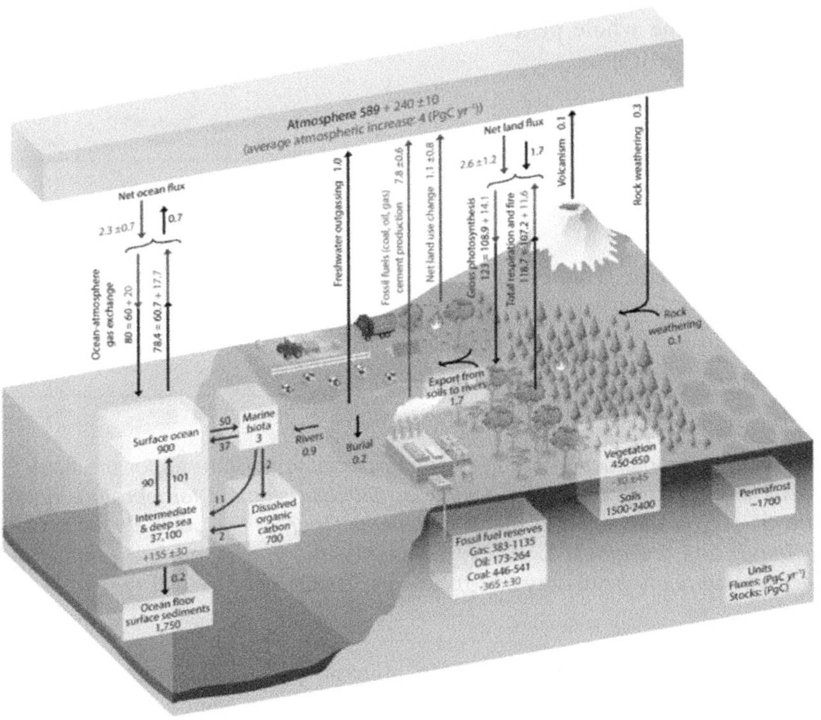

Figure 4.4. *A simplified schematic of the global carbon cycle. Source: IPCC, WGI, AR5 (2013). Reproduced with permission from Rhein et al. (2013)*

A simplified scheme of the global complex carbon cycle, taken from the fifth IPCC report, with figures expressed in Petagrams of Carbon per year (PgC. yr^{-1}) is delivered in Figure 4.4. In the boxes of Figure 4.4, quick descriptions of the stocks are given, with the related amounts of CO_2 contained in the atmosphere and in the soil. These are expressed in mass of carbon rather than in CO_2. Carbon takes the form of other molecules that are of organic nature in the soil, which represents a large part of the cycle, more than the atmosphere. CO_2 itself has almost no internal chemistry: one thing making the reasoning very simple is the consideration of the exchanges with the surface. The CO_2 appears and disappears in the atmosphere through specific types of emissions and removals, and various exchanges with the surface. A main process is the biochemical process of dissolution by the ocean and the role of biochemistry a bit deeper in the ocean. After the direct role of the physics of exchanges at the surface, the marine biota dissolves organic carbon, etc., down to potential sediments.

The fossil fuel production of CO_2 is about 7.8 $PgC.yr^{-1}$ (see Figure 4.4). This looks small in comparison to the amount of what is exchanged. It is extremely important to realize that what makes the difference is that with anthropogenic greenhouse gas emissions, there is currently a rupture in a system which was more or less at equilibrium during the Holocene. In fact, in the climate models, modeling teams take as an assumption that the model was at equilibrium previously, which is a strong assumption.

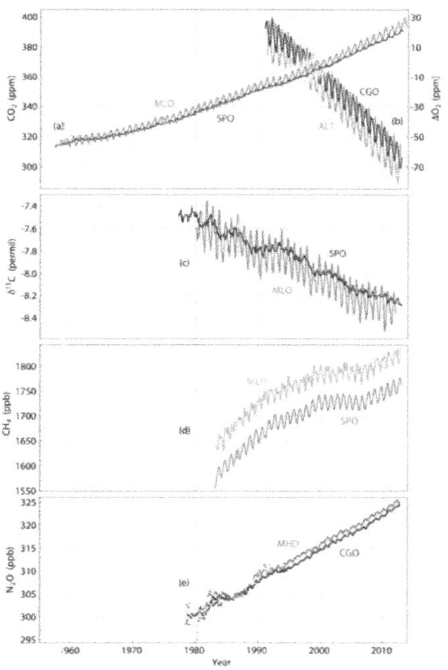

Figure 4.5. *Atmospheric concentration of CO_2, oxygen, carbon stable isotope ratio $\delta^{13}C$ in CO_2, CH_4 and N_2O recorded over the last several decades at representative stations (a) CO_2 from Mauna Loa (MLO) Northern Hemisphere and South Pole Southern Hemisphere (SPO) atmospheric stations (Keeling et al. 2005). (b) O_2 from Alert Northern Hemisphere (ALT) and Cape Grim Southern Hemisphere (CGO) stations (see: http://scrippso2.ucsd.edu/right axes, expressed relative to a reference standard value). (c) $\delta^{13}C$: Mauna Loa, South Pole (Keeling et al. 2005). (d) CH_4 from Mauna Loa and South Pole stations (Dlugokencky et al. 2011). (e) N_2O from Mace-Head Northern Hemisphere (MHD) and Cape Grim stations (Prinn et al. 2000). Source: IPCC, WGI, AR5 (2013). Reproduced with permission from Ciais et al. (2013)*

The Mauna Loa observations started in 1958 (see Figure 4.5). The oscillations of the seasonal cycle slightly delayed the perception of the greenhouse gas increase. The rate of increasing CO_2 concentration increases with time because it is cumulative in the atmosphere. Atmospheric CO_2 concentration from MLO is represented in panel (a), while atmospheric O_2 concentration is represented in panel (b). Panel (c) shows the concentration of the carbon stable isotope ratio $\delta^{13}C$, and in panel (d), the N_2O atmospheric concentration is displayed.

As for the lifetime of the gases in the atmosphere, for CH_4, it is about 11.8 years (according to IPCC 2021b, c). CH_4 is mixed between the two hemispheres, but with some lags. This is also mixed because CH_4 does not stay as long in the atmosphere as CO_2. There is a period of time where CH_4 was leveled out and then began again to increase in 2000. Around 2010, it began to increase again: the scientific community is not really sure why. One hypothesis is the effect of climate change on dry regions: there is less CO_2 in swamps due to for example, human activity (because of limitations of humid zones). Shell gas exploitation may also result in emissions of CH_4 in the atmosphere, following a process which is not completely understood. N_2O emissions are largely linked with agriculture and fertilizers. Its features are close to those of CO_2, with a stronger impact with the same order of magnitude for N_2O lifetime in the atmosphere (about 109 years according to IPCC 2021b, c). Even the carbon cycle is complex; atmospheric CO_2 is mainly absorbed after about 100 years (IPCC 2022a, b), even though part of it may subsist for more than 1,000 years.

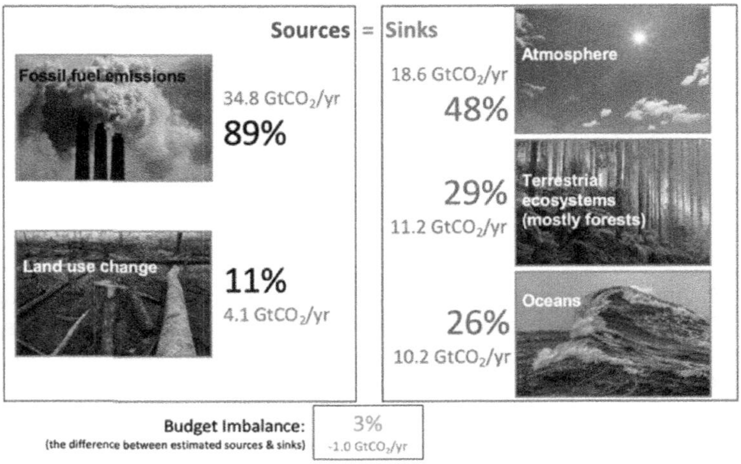

Figure 4.6. *Anthropogenic perturbation of the CO_2 budget (2011–2020). Source: Friedlingstein et al. (2021)*

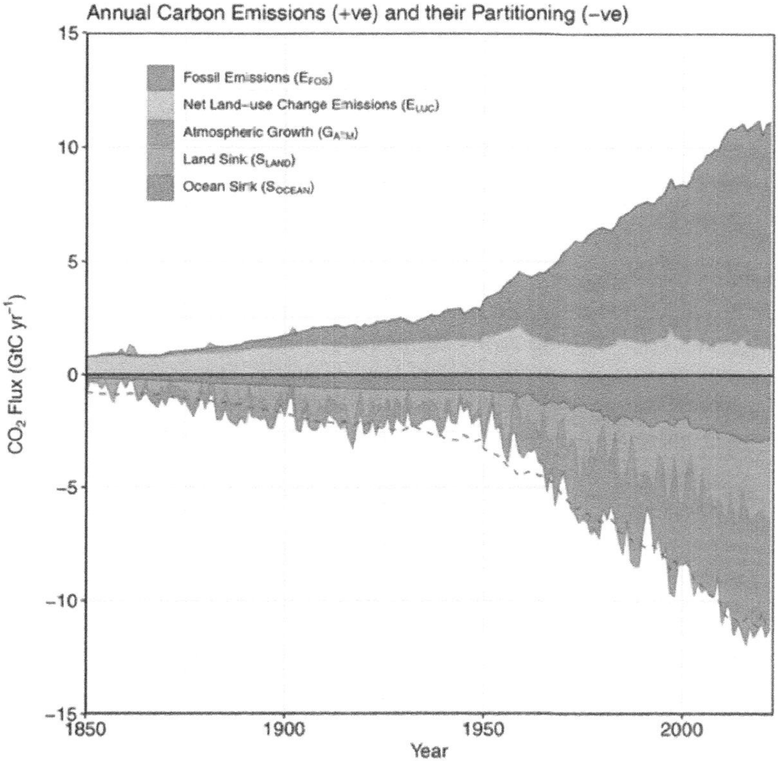

Figure 4.7. *Global carbon budget as a function of time since the year 1850 for emissions and sinks. Source: Friedlingstein et al. (2023)*

The amount of CO_2 emitted through the use of fossil fuels (about 89% of anthropogenic emissions) and through land-use change, including deforestation (about 11% of emissions) is regularly assessed by a research group named the Global Carbon Budget (see Figure 4.6). A communication problem for the scientific community is the difficulty of making pictures with greenhouse gases which are by the way fully transparent contrary to a common false belief that they would be white or apparent, and they have no smell either. It is not innocent practice to usually represent GHG with smoke as they indeed act silently. For instance, the smoke visible in Figure 4.6 on the upper left image representing fossil fuel emissions is actually made of droplets and dust: of course, the emission locations are about the same, but the dust is mostly reflecting radiation to space; therefore, contrary to what a non-expert may think, their effect is actually a cooling. The effect of this kind of smoke is mostly a problem of pollution and health (causing respiratory diseases/cancers), and it stays very little time in the atmosphere. In that sense, the use of electric cars for instance is a resolution in terms of particle pollution and health locally, but from the global point of view of climate, it is a matter of fossil fuels used and how much is emitted for the whole process of building electric cars and their carbon footprint, including for their transportation and the material they require. The problem of climate is different from the problem of air quality, even though they are often confused with each other.

The right side of Figure 4.6 represents the GHG sinks: the atmosphere takes up about 48% of GHG emissions, while terrestrial ecosystems take up 29% of emissions, and oceans take up 26% of emissions.

In the representation of the global carbon budget as a function of time since 1850 for emissions and sinks (see Figure 4.7), we can note a remarkable linearity between the sink response and emission forcing, suggesting that land sinks are sensitive to climate, at least on an interannual time scale.

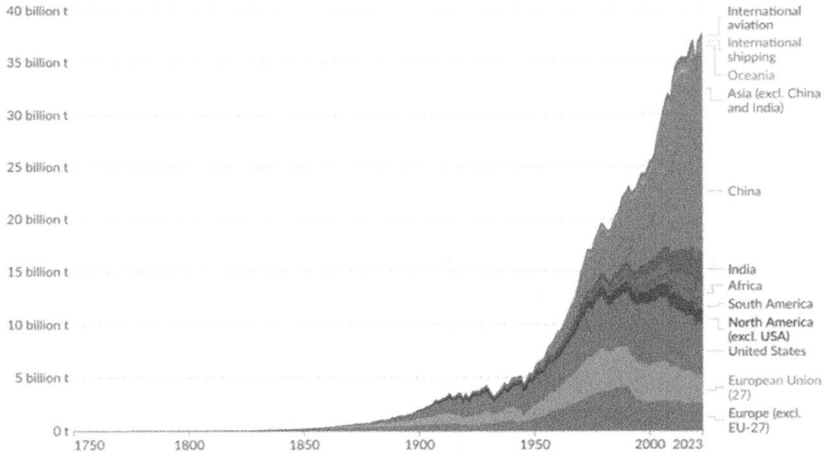

Figure 4.8. *Annual CO_2 emissions from the use of fossil fuels and industry (no land-use change emissions). International aviation and shipping are included as separate entities. Source: Global Carbon Budget (n.d)*

The diagram representing annual CO_2 emissions from the use of fossil fuels and industry over time (see Figure 4.8) shows the speed at which the emission increase is happening, and there is something very structural in that phenomenon.

The emissions of the countries of Western Europe and North America plus the addition of emissions coming from new countries have been rapidly increasing since 1950. In fact, since 1870, the CO_2 emissions from the use of fossil fuels has always been increasing quickly, but the acceleration in 1950 is striking.

An emphasis must be put on the fact that the situation is always changing and the margin of what can be done to reverse the curve is getting smaller and smaller with time. Many climatologists who started their career in the 1970s, including one of the authors, did not think that they would witness the consequences of climate change during their lifetime, but it is clearly already happening.

There are three experiments corresponding to three main generations of model assessments directed toward climate change:

1) The sensitivity to CO_2 doubling (from one equilibrium to the other): the beginning of this field of investigation started with personalities such as Svante Arrhenius, Jule Charney, etc. In the 1970s, there was of course no ocean or swamp ocean model (the ocean was first represented in climate models as not moving, and able to take the heat with a small layer of 50 m.) At that time, there were no capabilities to have an ocean in the climate models where it was merely introduced as a very simplified component. At the beginning, the models coming from this first phase could not be used to make comparisons with the real world either.

2) During a second phase starting from the third IPCC report, people could use coupled atmosphere/ocean models and introduce a 1% yearly CO_2 increase while the first two IPCC reports were using models without ocean (IPCC (2001a, b) and IPCC (2007a, b)). The oceans appeared in the third IPCC report, giving a time transition that was more realistic, as one of their key roles is actually to be a "timekeeper" in the climate system.

3) During a more recent phase, the CO_2 doubling was prescribed through "RCP scenarios" that are policy-oriented in the IPCC (2013) fifth Assessment Report (AR5).

The first IPCC scenarios based on futurology were named "SRES" scenarios, referring to the Special Report on Emissions Scenarios (SRES) produced in 2000 by the IPCC. Those scenarios were used in IPCC reports in 2001 and 2007, with the aim to deliver a synthesis concerning the future for political evolution (including demographical considerations), with a sensitivity to environmental issues; however, it was policy-free which means that in those scenarios, no policy parameters were accounted for. These were replaced in 2014 by Representative Concentration Pathways (RCP) for the fifth IPCC Assessment Report (AR5).

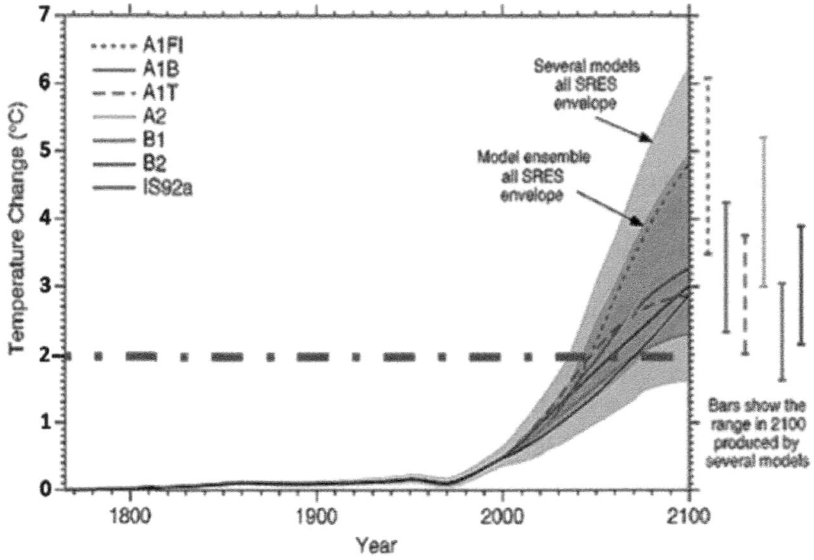

Figure 4.9. *Global temperature change as a function of time in the case of "policy-free" scenarios. Source: IPCC/SRES, 1999–2001–2007. Reproduced with permission from IPCC (2001b)*

The RCP have themselves been replaced by more refined scenarios for the IPCC Sixth Assessment Report (AR6) (IPCC 2021a, b, c), named shared socio-economic pathways (SSP).

In the following paragraphs, we describe the evolution from SRES to SSP scenarios.

For SRES scenarios displayed in Figure 4.9, different energy mixes were used, and scenarios were added, mixing information among consistent stories. Gathering information from social and political science derived some conclusions from reference scenarios in the diagram (e.g. B1 corresponds to a scenario with less warming). For each scenario, there is an associated range of uncertainty corresponding to the results of the experiments' models. The models had to reproduce what happened until 2000, and then climate modelers let the models evolve toward the corresponding future following those scenarios. The B1 SRES scenario is no longer possible with the rationale of reaching 10 million tons of carbon per year in 2050, as we have already reached it: we have about 40 years in advance on that scenario. We take as a reference the +2°C warming, which has progressively been jointly chosen by the scientific and the political community as a simple threshold where danger may lie. This figure was first mentioned in a technical note of the European Commission, mentioning the need to maintain the CO_2 emissions below a level compatible with a limited climate impact. The +2°C figure has been much mentioned since the Copenhagen conference in 2009. At that summit, the participants prepared a draft document with open sentences before being signed (with different possibilities to be decided later on). At the end of the two-week duration conference, the leaders took the floor and the heads of states arrived, but the document was not ready for signature on time; only half a page was ready, stating that they should stay under +2°C. After that, in the following COPs, the negotiators did not reconsider the 2°C figure before the Paris Agreement, where the willingness to make efforts to stay under 1.5°C was added. The topic is still a burning issue: do we need to follow what we already stated should be done, even if we know that it is almost impossible to reach this aim? “Let’s be reasonable and ask for the impossible” Che Guevara used to say, this could also be a motto for the climate ambition of staying below +2°C warming in 2100 as compared to the preindustrial era. If the +1.5°C aim has already been unachieved according to 2024 Copernicus data, the +2°C may still be met with a short window for action, however, the probability to reach it given the geopolitical distribution of climate policies by the main emitters seems low. The +2°C limit is supposed to represent an infinite equilibrium: this threshold is linked to the level of temperature that corresponds to the moment we get to it.

Several intellectuals and activists have argued that there is a right for the people to stay under +1.5°C, and this right is not fulfilled: to them, it represents a debt toward future generations. It would be highly unrealistic to pretend that we will reach the +2°C aim for sure, especially as carbon storage solutions for instance are currently not deployable at large scale, nor in a cost-effective manner. Some scientists even defend the idea that no technology such as carbon storage should be used for mitigation. There are also more and more voices defending ideas of radical adaptation, meaning that people should change their life habits including their life environment, for example, with the idea that inhabitants of zones threatened with submersion should merely move from those shores.

The model using the SRES B1 scenario to stay under 2°C actually does not do it the right way, as it is not stabilized. Some scenarios in the models bring the annual mean temperature to +5°C/6°C if bio-feedbacks are included. According to the projections, it can also be much more than that for the highest possible values.

The fact that the range of those results (2–5°C) is similar to the results in Charney’s 1979 report may be disturbing. But this similarity comes from different reasons: in IPCC reports, this range comes from physical uncertainties and feedback about future climate policies. In the discussions among scientists, a figure that is often quoted is a +3°C warming "in a business as usual case".

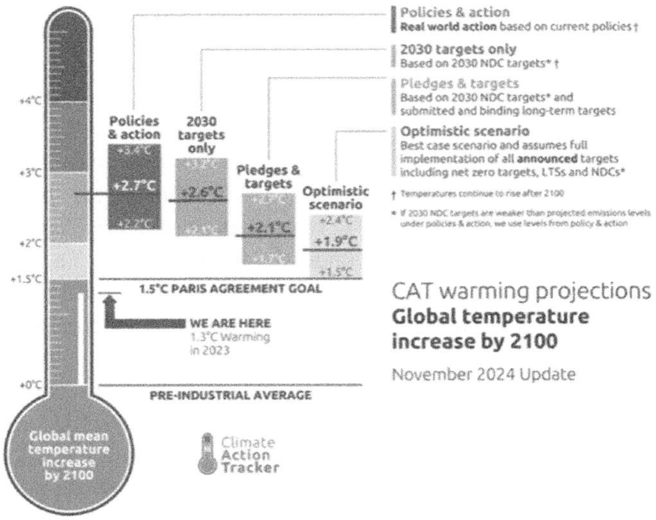

Figure 4.10. *The climate action tracker (CAT) thermometer. Source: Climate Action Tracker (2024b). See: https://climateactiontracker.org/global/cat-thermometer/*

The climate action tracker (CAT) research group worked on a simple representation of various warming projections in terms of global temperature increase by 2100, including an estimate of the global situation in terms of temperature increase in 2023 (see Figure 4.10). The median values are represented with a straight line in the boxes representing the numerical magnitude of temperature elevation, with the minimum and maximum values accounting for uncertainties. The climate model used for the computation of probability distribution is named MAGICC 7.

Four scenarios are examined:

– A scenario named "policies and actions", based on current policies and that could therefore also be named "business as usual scenario".

– A scenario named "2030 targets only", which is based on 2030 NDC goals.

– A scenario named "pledges and targets", which is based both on 2030 NDC goals and on legally binding pledges on long-term climate goals.

– Finally, an "optimistic scenario" which is actually a "best-case scenario", assuming a full implementation of all the targets announced in countries pledges, including net zero targets.

The CAT late 2024 analysis shows that current NDC trajectories have a 99.8% chance to stay below +2.9°C, but above 99.8% to exceed +1.5°C.

Indeed, if we stay at the current level of emissions, according to the analysis made by the CAT in 2024, as shown in Figure 4.10, with rough estimate, the pledges including binding long-term climate targets made voluntarily are not enough to stay below +2°C and they point instead to a median warming value of +2.1°C in 2100, with the result that the actual value is likely to stay under +2.3°C, with a 94.3% chance to exceed +1.5°C; however, even if the pledges were fully respected, they could bring the Earth's mean annual temperature up to +2.7°C in the worst-case scenario, considering the tail risks related to uncertainty estimates. But in case some main emitters do not respect their initial promises – for instance, the USA withdrew once again from the Paris Agreement in 2025 – then we can go up to much higher values for the mean annual global temperature, and this has been a bit forgotten in the communications toward the general public. +2.7°C is indeed likely to happen if all the countries stick to what they promised and did that for the whole century, which is unpredictable at this stage due to policy choice uncertainties. This is absolutely not granted, and the range (2–5°C) is still valid.

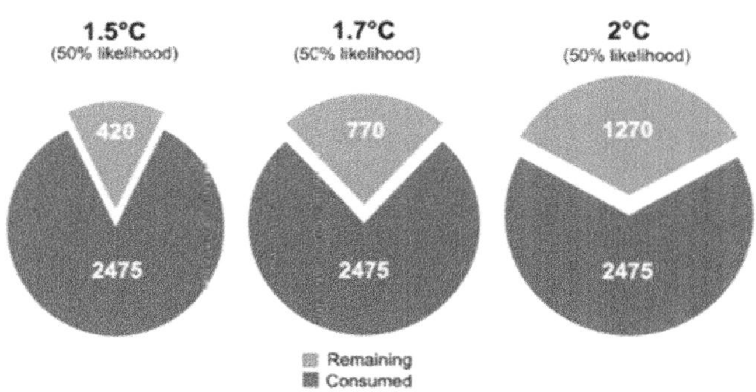

Figure 4.11. *Total carbon budgets to stay below +1.5°C, +1.7°C and +2°C in the 2100 global surface mean temperature starting from 1850 up to 2021, with the partition of the historical carbon budget and the remaining carbon budget. The remaining carbon budget since January 2022 is represented in orange. Total CO_2 emissions include emissions from the combustion of fossil fuels, land-use change and cement production. Budgets are based on IPCC methods (Canadell et al. 2021) and on updated data from Friedlingstein et al. (2021). Unit is $GtCO_2e$. Source: Friedlingstein et al. (2022)*

Some research groups such as the Global Carbon Budget also use the reverse process, asking themselves: If we want to stay under +2°C, what are the emissions compatible with this aim? That approach allows us to map out the necessities in terms of emissions if we want to cope with a + 2°C warming in 2100.

This remaining budget approach is represented in Figure 4.11.

The Representative Concentration of GHG scenarios (RCP) of GHG Pathways (RCP) are based on the idea of federating consistent scenarios with a given goal in terms of radiative forcing to be reached out at the 2100 horizon. A difficult question is: What is the amount of CO_2 that we can allow in the atmosphere to stay below +2°C? Models do not give the same results because they do not react exactly the same in terms of temperature to the amount of CO_2 added in the atmosphere. Therefore, it becomes important to find a less uncertain unit. In fact, models show less differences when we look at the way they translate the greenhouse gas level into radiative forcing.

The principle was to choose radiative forcing at some point as a way to describe the scenario in which we hope to stay under a certain level of global temperature elevation, and this is not something necessarily easy to expose to the general public.

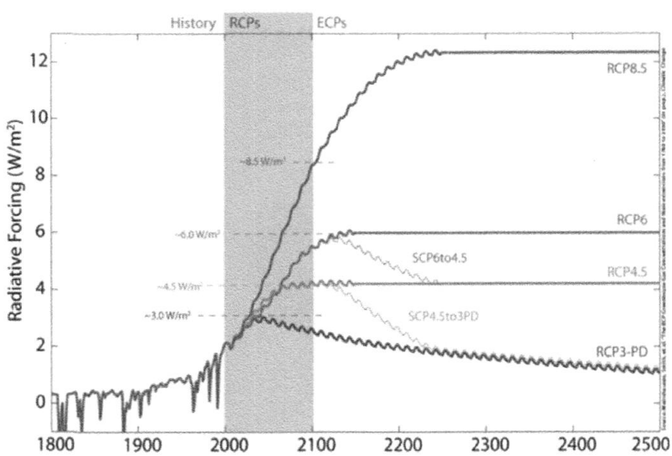

Figure 4.12. *Radiative forcing past and future projected evolution according to different Representative Concentration Pathways (RCP). Source: http://www.pik-potsdam.de/~mmalte/rcps/*

The RCPs are classified in terms of $W.m^{-2}$ and an RCP is actually the pathway in terms of concentration of GHG that we want to allow in order to match a certain level of $W.m^{-2}$, which itself is associated with a certain level of risk.

The way it was designed can be described as follows: as shown in Figure 4.12, the studied parameter is radiative forcing, expressed in $W.m^{-2}$. CO_2 is progressively added in the model, and the radiative forcing increases (it is something prescribed in the model). The scenarios help understand what should be done in terms of emissions for a given goal in terms of radiative forcing and therefore of temperature (expressed through a given prescription).

Radiative forcing is actually not an obvious notion to measure or use, but it is absolutely central to the debate about climate change.

RCP scenarios aim at correcting the previous situation with SRES scenarios which roughly functioned as follows:

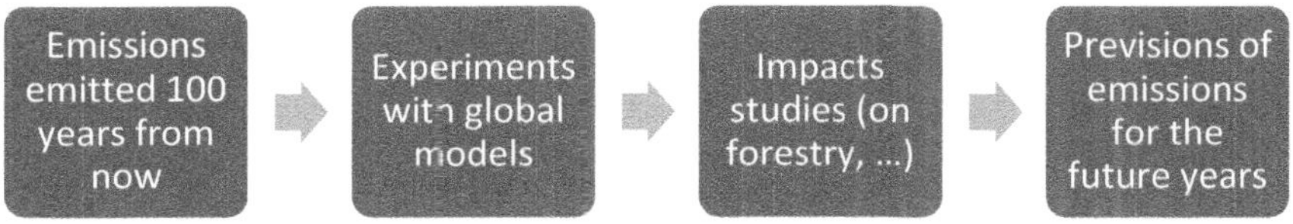

The impact studies associated with RCP scenarios are diverse and interdisciplinary, studying biodiversity, social impacts, etc. The next step is to explore scenarios with fewer emissions. In this case, the difficulty comes from the work on the social aspects.

With RCP, two things are done in parallel: the construction of the scenario required to fulfill them, and the investigation of the impact at the same time.

This means that for RCPs, two different scientific communities work in parallel:

– the community investigating what are the appropriate energy mixes in order to stay in the domain of emission scenarios and for this same domain of emissions;

– the community investigating the impact in terms of climate.

This work in parallel functions nicely. For instance, one question to investigate was framed as follows to the IPCC by decision-makers: "We would like to know what is possible in terms of staying under +1.5 °C versus +2°C?". Addressing this complex question would not have been easy without this kind of research setting up in parallel for RCPs.

Once again, the main problem is that models are very different. The common feature for the different models is the use of the radiative forcing. This metric does not differ much while there is a large range of results in terms of temperatures for the same forcing.

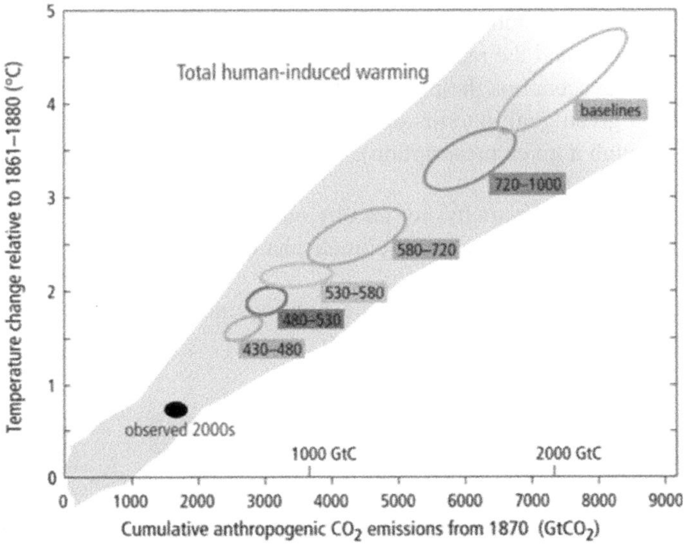

Figure 4.13. *Warming versus cumulative CO_2 emissions. Source: IPCC Summary for Policymakers, WGI, AR5 (2014). Reproduced with permission from IPCC (2013)*

Another feature putting the emphasis on radiative forcing is the fact that as greenhouse gases tend to store in the atmosphere, it is their accumulated effect which matters the most.

Figure 4.13 was made using models: it shows the temperature change relative to 1861–1880 (in °C) as a function of cumulative anthropogenic CO_2 emissions from 1870, the year corresponding to the beginning of the industrial era.

When considering the temperature change relative to the beginning of the industrial era, the cumulative emissions provide some agreement and a linear response, which means that we should focus on an accumulated effect in terms of CO_2 diagnosis.

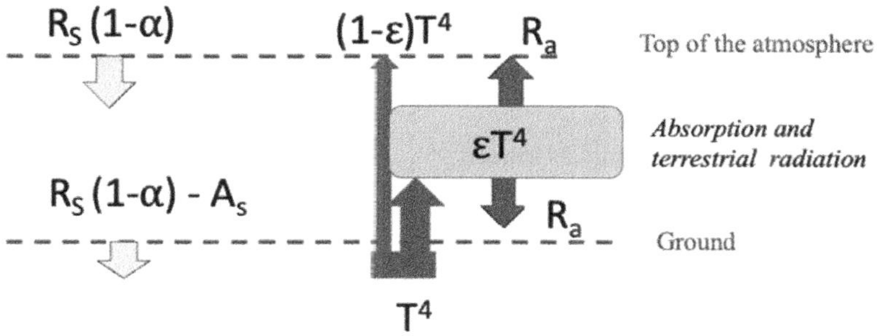

Figure 4.14. *A scheme showing the concept of radiative forcing*

4.1.4.1. *What do we call radiative forcing?*

In order to deliver a definition of radiative forcing, let us first consider the solar radiation coming at the top of the atmosphere, as described in the scheme in Figure 4.14. One part is reflected, the other part gets down to the Earth. A part of the solar radiation is absorbed, noted A_s in the scheme: this corresponds to the warming. There is no clear indication that it will change with temperature, so this can be left a bit apart. Then, there is a phenomenon much more linked with temperature, which is the emission absorption. A part of the absorption is re-emitted at the top to the bottom and the major part of the radiation goes back to space. If we increase the part which is absorbed, we then also get a decrease of the part of emissions which is emitted back to space: this means that in such a case, there is a disequilibrium in terms of radiation emitted to space.

Generally, what modelers do is to let the system warm very rapidly, with an increase of the amount of greenhouse gases, and an increase of the absorption. This leads to a warming because some equilibrium must be re-established at the top of the atmosphere. We can compute the kind of warming needed to maintain the equilibrium with an additional amount of greenhouse gases. As for the general value corresponding to the greenhouse effect, we obtain the value of the temperature near the ground as follows, when neglecting the absorption of solar radiation A_s, as a function of the temperature T_0 without greenhouse effect:

$$T = \frac{T_0}{\sqrt[4]{(1-\frac{\varepsilon}{2})}} \qquad [4.1]$$

and in order to specify how it will change, ε becomes ε+Δε, which merely means that when we add a slight amount of greenhouse gases in the atmosphere, the equilibrium is slightly changed too.

Let us imagine first that the ground temperature of the Earth does not adjust and remains equal to an "initial" temperature, which we can write as T_{init}: then in that case, the radiative equilibrium is disrupted.

This radiative equilibrium was indeed:

$$R_S(1-\alpha)-(1-\varepsilon)\,\sigma\,T_{init}^4 - R_a = R_S(1-\alpha)-(1-\tfrac{\varepsilon}{2})\,\sigma\,T_{init}^4 = 0 \qquad [4.2]$$

The radiative forcing RF becomes:

$$RF = (\tfrac{\Delta\varepsilon}{2})\,\sigma\,T_{init}^4 \qquad [4.3]$$

When it is relaxed:

$$T_{adjustment} = T_0\,\sqrt[4]{(1-\tfrac{\varepsilon+\Delta\varepsilon}{2})} \qquad [4.4]$$

The integrated value of the radiative forcing over 100 years provides the way to define "equivalent CO_2" value for all other greenhouse gases.

Equation [4.2] represents the equilibrium at the top of the atmosphere; $R_s - 1$ represents the net budget of the solar input at the top of the atmosphere.

The amount of energy that gets through the atmosphere at the ground is: αR_s.

The emission by greenhouse gases that go back to space at the top of the atmosphere can be written as follows: $(1-\varepsilon)\sigma\,T_{init}^4$ if we suppose that there is no solar absorption.

The radiative forcing corresponds to the amount of disequilibrium at the top of the atmosphere, and this disequilibrium corresponds to the fact that the radiative energy going out to space is limited.

Before we include some additional greenhouse gases, this very simple model in which we do not consider the solar absorption, delivers the result that the radiative forcing is the same at the bottom and at the top of the atmosphere with the value: $\frac{\Delta\varepsilon}{2}\sigma\,T_{init}^4$.

This simple computation consists of showing that if we add GHG, there will be a radiative forcing corresponding to what was described above.

An important question is: How does the radiative convective equilibrium (RCE) theory affect this equilibrium at the top of the atmosphere?

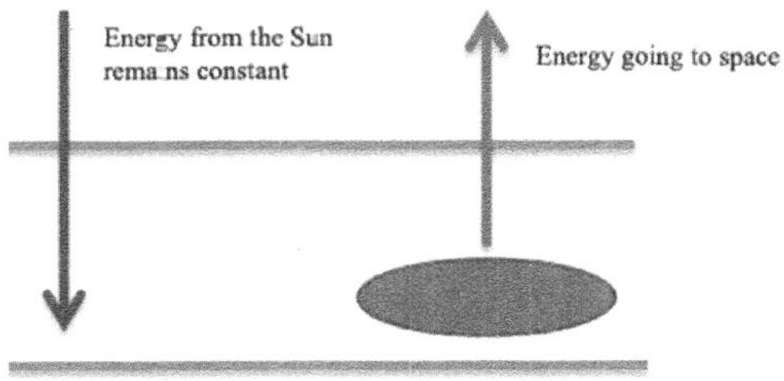

Figure 4.15. *A very simplified scheme of the incoming and outgoing fluxes at the top of the Earth's atmosphere*

If we consider the scheme in Figure 4.15 which is a very simplified representation of the incoming and outgoing fluxes at the top of the Earth's atmosphere, if we introduce the notion of RCE in our simplified model, the value of $R_S(1 - \alpha)$ in equation [4.2] does not change. However, the energy going to space, represented with the blue arrow in Figure 4.15, diminishes in a radiative convective model. Therefore, a warming in the area represented in blue is needed in order to balance this disequilibrium: the fact that the energy going to space is diminishing translates into a warming in order to re-establish the equilibrium, and so that more energy is sent to space. In terms of radiative forcing, this is equivalent to freezing the warming and preventing the system from warming, by just taking the disequilibrium at the top of the atmosphere.

The radiative forcing expressed in $W.m^{-2}$ basically measures a disequilibrium in the radiative balance of the atmosphere at the top of the atmosphere. This disequilibrium is meant to be followed very quickly by a warming or by a cooling: there is indeed an adaptation of the system to re-establish equilibrium quickly at the top of the atmosphere. This notion of radiative forcing is complex because it represents a few percent compared to the system's energetic budget, and therefore only a few $W.m^{-2}$. Studying the way it may change is complex because it cannot really be measured from space: the diagnosis of the radiation changes at the top of the atmosphere can be done with an accuracy of only 4–5 $W.m^{-2}$. In fact, locally, the accuracy is much less, because a measurement is made at one point, at one moment: it is actually very difficult to rely on measurements to know how emissions of a given greenhouse gas do affect temperature, and modelers face important uncertainties in the interpretation of small radiative forcing changes. Despite this complexity, the steps followed by climate models to assess the effect of a given radiative forcing as a function of temperature can be simplified as the following: they let the system accumulate greenhouse gases, and they do not allow the surface temperature to change. The modelers let this state run on over many decades in the models, and at the end of this long enough time period, the system is then authorized to warm up.

This temperature measure issued from the radiative forcing perturbation is not easy to assess for the different gases. The comparison of the different radiative forcing associated with different gases is not easy neither. It is generally done only for the troposphere because the stratosphere would let the system equilibrate quickly. Generally, there is a dial system:

– In the stratosphere, when modelers want to diagnose radiative forcing, they let the system equilibrate quickly.

– In the troposphere, modelers freeze it and keep this warming to maintain the disequilibrium at the top of the atmosphere: this then causes a radiative forcing which then manifests itself as a real warming when releasing the condition.

There are different ways to compute those steps numerically, from simple to complex models. Complex models are generally used by climatologists in order to estimate the radiative forcing of CO_2. Those complex models look line by line at all the CO_2 absorption rays. The radiative–convective equilibrium theory cannot currently be integrated, but it is described as part of the feedback.

Radiative forcing estimates in 2011 relative to 1750 and aggregated uncertainties for the main drivers of climate change have been assessed in the IPCC 5th Assessment Report, and presented in a pedagogical way (see Figure 4.16), which we will discuss here, even if an update for numbers exist in the AR6 that we, however, find less complete (see Figure 4.17). The nature and names of the gases are reported in the first columns on the left part of Figure 4.16, and the impact of their concentrations increase relative to 1750 is reported in terms of radiative forcing, expressed in $W.m^{-2}$ by emissions and drivers. Anthropogenic and natural causes of the forcing are distinguished.

Note that the radiation forcing is computed relatively to 1750; therefore, modelers act to freeze the system for a large amount of time, during which the gases accumulate and the history of when the warming occurs is not very important in that regard, and just an assumption.

When modelers release the system, then a warming corresponding to the amount of greenhouse gases put in the atmosphere is obtained: this is how the greenhouse forcing is accumulated.

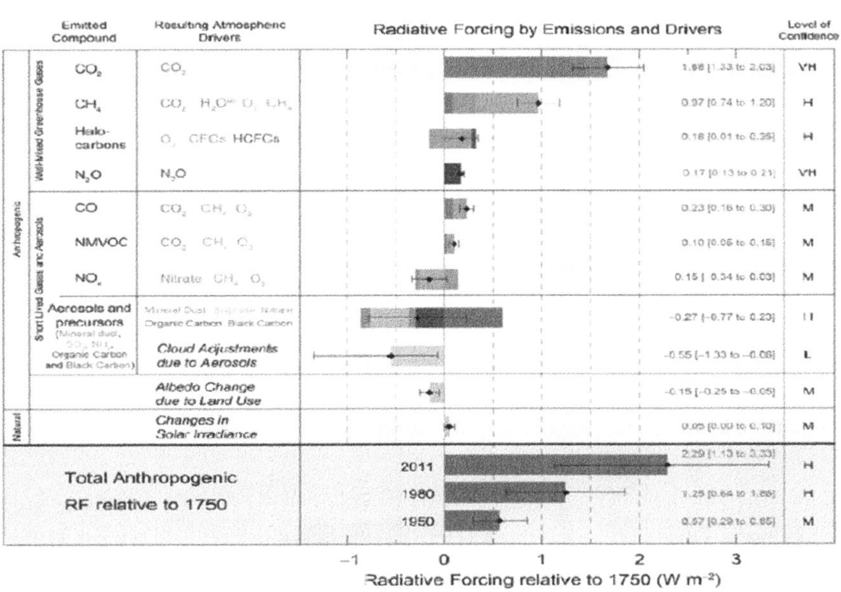

Figure 4.16. *Radiative forcing estimates in 2011 relative to 1750 and aggregated uncertainties for the main drivers of climate change. Values are global average radiative forcing (RF 14), partitioned according to the emitted compounds or processes that result in a combination of drivers. The best estimates of the net radiative forcing are shown as black diamonds with corresponding uncertainty intervals; the numerical values are provided on the right of the figure, together with the confidence level in the net forcing (VH – very high, H – high, M – medium, L – low, VL – very low). Albedo forcing due to black carbon on snow and ice is included in the black carbon aerosol bar. Small forcings due to contrails (0.05 W. m^{-2}, including contrail induced cirrus), and HFCs, PFCs and SF 6 (total 0.03 W. m^{-2}) are not shown. Concentration-based RFs for gases can be obtained by summing the like-colored bars. Volcanic forcing is not included as its episodic nature makes is difficult to compare to other forcing mechanisms. Total anthropogenic radiative forcing is provided for three different years relative to 1750. Source: IPCC Summary for Policymakers, WGI, AR5 (2013). Reproduced with permission from Stocker et al. (2013)*

Gases with different levels of forcing shown on a numerical scale can be represented (see Figure 4.16). On the x-axis representing the radiative forcing, on the left side of the 0 line, a cooling is shown, while on the right side there is a warming.

A unique color in the column named “resulting atmospheric drivers” means that the considered gases experience (almost) no chemistry in the atmosphere: therefore, there is no change in the related exchanges with the surface. If we consider CH_4 for instance, there are actually changes occurring because there is some chemistry in the atmosphere, linking CH_4 with other species from the nitrogen family and O_2, creating secondary gases: CO_2, H_2O (water vapor) and O_3 (with a more complex behavior), through chemical equations. Therefore, the disappearance of CH_4 in the atmosphere is linked with other gases. The radiative forcing of CO_2 represents ~1.66 W.m^{-2}, with a range related to uncertainties. For CH_4 with additional gases, it represents a radiative forcing of ~ 0.97 W.m^{-2}. This point was largely debated during the release of the 2013 IPCC report, because it means that CH_4 is far from being negligible. Even if it acts for a smaller time scale than CO_2 when looking at how CH_4 has currently increased in the atmosphere, if methane was not constantly re-emitted, then it would not retain much of what has been emitted a long time ago: this is not the case due to constant re-emissions.

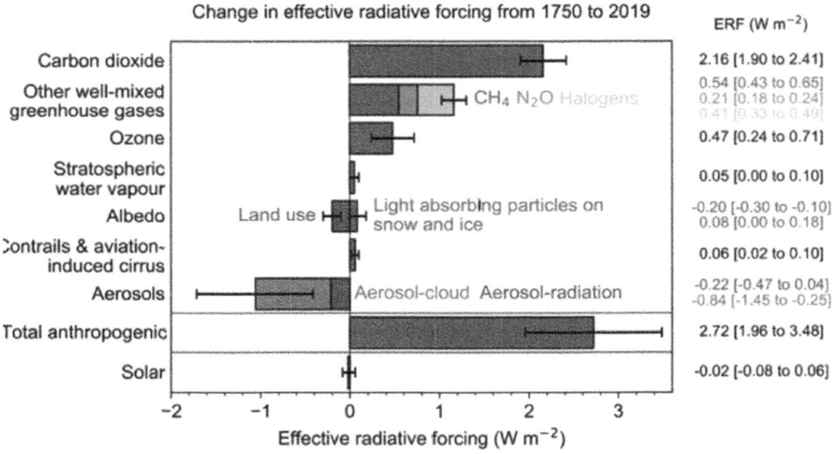

Figure 4.17. *Change in effective radiative forcing (ERF) from 1750 to 2019 by contributing forcing agents (carbon dioxide, other well-mixed greenhouse gases (WMGHGs), ozone, stratospheric water vapor, surface albedo, contrails and aviation-induced cirrus, aerosols, anthropogenic total and solar). Solid bars represent best estimates, and very likely (5–95%) ranges are given by error bars. Non-CO_2 WMGHGs are further broken down into contributions from methane (CH_4), nitrous oxide (N_2O) and halogenated compounds. Surface albedo is broken down into land-use changes and light-absorbing particles on snow and ice. Aerosols are broken down into contributions from aerosol–cloud interactions (ERFaci) and aerosol–radiation interactions (ERFari). Volcanic forcing is not shown due to the episodic nature of volcanic eruptions. Source: IPCC, AR6, WGI (2021). Reproduced with permission from Chapter 7 of IPCC (2021a)*

As for halocarbons, some of them, the CFCs, attack the Ozone hole and we can see in Figure 4.16 the disappearance associated with the CFCs. The HCFCs as well as the CFCs are active greenhouse and they tend to have a warming effect.

N_2O is a gas mainly associated with agriculture. It is called nitrous oxide, and it is a long-lived greenhouse gas that stays long in the atmosphere (it has very little chemistry), about 109 years according to the IPCC AR6 report (see Figure 4.17).

NOTE.– establishing the simple diagrams of Figures 4.16 and 4.17 required a huge amount of work for analyzing all those gases and chemical reactions between themselves. Those gases have different bounds of absorption, which are complex to study, including computations from quantum mechanics with which scientists assess the absorption in a given bound of absorption.

Among the short-lived gases, there are also species that are not really gases but rather particles, which reflect solar light. Generally, when there is a total radiative forcing in a diagram, it is not only the terrestrial part but the total that is considered (because there is less GHG for the reflection).

There is negative radiative forcing because there is reflection.

All the gases in Figures 4.16 and 4.17 that are very short-lived represent a source of concern for health but they do participate rather strongly to the GHG effect as well.

There are also nitrates and particles that reflect/diffuse light, as well as aerosols and precursors. They tend to reflect the suit (black carbon) which also has a greenhouse effect.

The way the IPCC reports functions is that for all of the figures, there is an array giving the range of uncertainty bar: it is based on the kind of computations made by different labs and teams. There is also a column named the "level of confidence" which was once called the level of scientific understanding (but the name changed in the course of the various IPCC reports), and it reflects the feeling of security that scientists have in understanding a given process. Among the different results obtained by scientists, this indicator is about assessing how strong the physics are that they are based on. There are currently difficulties to set the last digits of the results, but we know that the science is self-consistent through the given level of confidence. Of course, there are some matching between the length of uncertainty and the level of confidence: when there is a large uncertainty range, there is a related very high level of confidence, but in some situations, there might be some nuances too.

The method has always been done this two-way by the IPCC, and the final result is the total anthropogenic radiative forcing: there are large uncertainties but this has to be put in perspective with the debate “can we stay under +1,5°C?”: only, for the radiative forcing, the climatologists already have estimates with large error bars. When translating that into temperatures, we similarly obtain large error bars too. It does not mean that a world warming under the given conditions is not dangerous but that we should be ready to consider all of the correlated possible resulting options in a situation of unknowns. Precisely because many unknowns exist, we have to elaborate different scenarios of strategies to protect the Earth from what may be dangerous in the future.

When considering greenhouse gases radiative forcing results with large error bars, all show that the change will be large compared to 1750, even if the situation is difficult to fully diagnose. Behind those considerations is one important question: How much we can compare the “different gases”. The metric mainly used is the CO_2 equivalent, notably for the comparison of various greenhouse gas effects. In fact, there are four main different units which are: tons of carbon, tons of CO_2, tons of carbon equivalent and tons of CO_2 equivalent.

Through the use of those different metrics, there are differences in terms of working methodology between the domain of physics of climate change and the domain of economy.

In the case of climate science, climatologists tend to mix only with care the different gases using the CO_2 equivalent metric which is convenient to obtain orders of magnitude, but is nevertheless less precise for computations when the aim is to compare the role of different gases at different scales.

Both for carbon budgets and carbon markets, a single metric is necessary to assess the effect of the main greenhouse gases. It is largely used by the Global Carbon Budget (Friedlingstein et al. 2023), which establishes a global assessment of the main greenhouse gases worldwide, and by various studies focusing on emissions trends and the possibility of the verification of greenhouse gas trends regionally, such as the RECCAP 1 and 2 projects led by the member of the French Academy of Science, Philippe Ciais, and by recent related papers (Petrescu et al. 2023; Mostefaoui et al. 2024). It has been defined by an equivalent unit between the different greenhouse gases which have appeared since the 1990 IPCC report, using the Global Warming Potential (GWP) tool to convert all greenhouse gases into the same CO_2 equivalent unit. The GWP principle is quite simple: it conventionally takes CO_2 as a reference because it has historically been the most discussed greenhouse gas. The GWP equation for a given greenhouse gas X is a dimensionless parameter that can be written as follows, with RF (X) being the radiative forcing of a given greenhouse gas X, and RF (CO_2) being the radiative forcing of CO_2:

$$GWP_{(X)} = \frac{\int_0^T RF(X)dt}{\int_0^T RF(CO_2)dt} \quad [4.5]$$

The principle of equation [4.5] for the GWP computation is to compare a situation referring to initial emissions of 1 kg of CO_2 and another one for 1 kg of a given greenhouse gas X. CH_4 will be taken here as an example, because it is one of the main GHG, but other gases may be considered, such as N_2O.

At a given emission time t, $GWP_{(X)} = 0$.

Let us take $GWP_{(X)}$ as a reference and consider now the relationship between CO_2 and CH_4 on a logarithmic scale as represented in Figure 4.18.

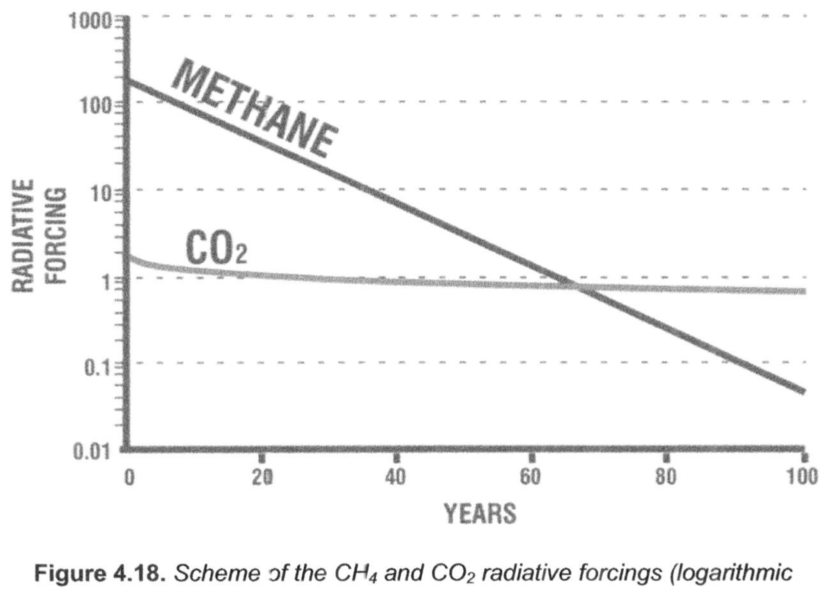

Figure 4.18. *Scheme of the CH_4 and CO_2 radiative forcings (logarithmic scale) as a function of time. Source: Edwards and Trancik (2014)*

When starting at a certain point of concentration for CO_2 and CH_4 with 1 kg of each, the CH_4 absorption at the beginning of the considered scale of time is about 100 times stronger than CO_2 (for the same amount of gas injected in the atmosphere), as represented in Figure 4.18. This scheme shows that CO_2 radiative forcing is more continuous over time than CH_4 and smaller, but at some decisive point, approximately T = 67 years, the two lines cross. A given GWP assessment is always done using a given fixed value of time T, which really matters for the analysis due to the lifetimes of the greenhouse gases, being smaller or larger than T.

Most of the time in IPCC reports, the chosen value is T = 100 years. When doing that, we actually lower the GWP for CH_4, which varies between 27.2 and 29.8 according to IPCC AR6 estimates; this computation has refined with time in the successive IPCC reports as shown in Table 4.1.

Greenhouse gas	100-year time period				20-year time period			
	AR4 2007	AR5 2014		AR6 2021	AR4 2007	AR5 2014		AR6 2021
	Feedback not included		*Feedback included*		*Feedback not included*		*Feedback included*	
CO_2	1	1	1	1	1	1	1	1
CH_4 fossil origin	25	28	34	29.8	72	84	86	82.5
CH_4 non-fossil origin				27.2				80.8
N_2O	298	265	298	273	289	264	268	273

Table 4.1. *GWP evolution over time during the various IPCC Assessment Reports*

The GWP metric is therefore an imperfect but useful tool to obtain a common constructed metric to measure the relative importance of the different greenhouse gases based on a common standard. It is important to note that the use of the constructed GWP metric is not completely realistic in terms of application: for the estimates related to the GWP metric, we indeed suppose a device in which we add 1 kg of a given greenhouse gas in its atmosphere and let it go. In fact, in the real world, the mechanisms at stakes are more complex for greenhouse gas production over time than the simplifications made for GWP computation. For CH_4, for example, removal occurs over many years as well; that is the reason why it does not correspond fully to what we can model. However, once again, GWP is useful as an index, even imperfect, for conversion in order to deal with greenhouse gas budget and also for carbon pricing which also uses the GWP converting parameter. It is important to remember how the convention for the GWP metric was built in order not to forget the various simplifying assumptions behind the obtained figure in the CO_2e unit, and to avoid being completely dogmatic about it.

The GWP figure should be considered with care as for instance, 1 kg of CH_4 is not literally worth 27.2 or 29.8 kg of CO_2, and must always also consider the gas lifetime and the considered scale of time. If, at a point in time, we want to avoid high values of warming by focusing on a gas with a relatively quick reaction, methane would be a good candidate. It is also important to keep in mind that the way the carbon markets are designed with the current definition of GWP can be discussed and even questioned. Maintaining the expertise about such a metric is important, with clear ideas about the science behind the figures, even for policymakers and journalists.

In the domain of processes that can be better understood regarding climate change, the sensitivity to radiative forcing is also important.

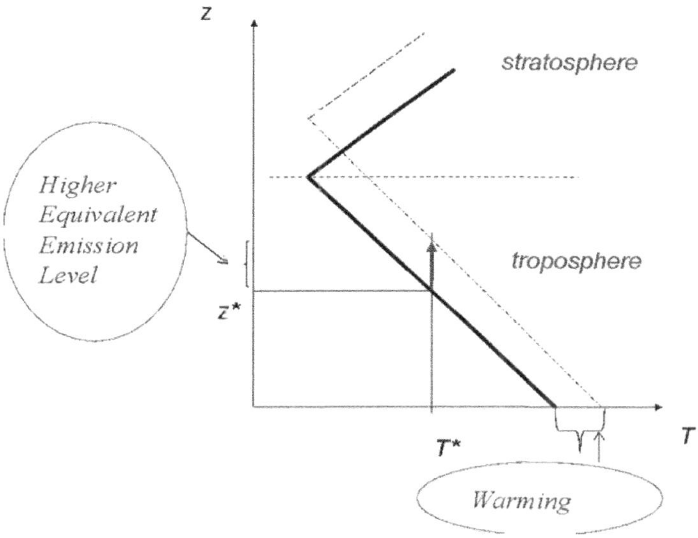

Figure 4.19. *A simplified scheme representing emission level z as a function of the temperature T to discuss the shift from sensitivity to radiative forcing*

The simplified scheme of the atmospheric temperature decreasing from the surface, not considering the convective effect, based on a mean trend from observational data, shows that the limit of the increases and decreases of the temperatures is drawn at a level corresponding to the tropopause, at a height of 10 km (see Figure 4.19).

The definition of the equivalent emission level is necessary for modeling the behavior of the planet, in an atmosphere with an increase of greenhouse gases emissions.

A planet with only O_2 and NO_2 would follow the rule of equilibrium represented by the following simple equation:

$$\sigma T_S^4 = R_S(1 - \alpha) \quad [4.6]$$

with T_S corresponding to the surface temperature, and R_S to the solar radiation.

4.2. Global temperature change and the role of feedbacks

Now, if we add some CO_2 and CH_4 to a planet with atmosphere (at first range), we can write the following equation:

$$\sigma T_e^4 = R_S(1 - \alpha) \quad [4.7]$$

The term $R_S(1 - \alpha)$ on the right side of equation [4.7] corresponds to the absorbed part of the solar radiation energy.

T_e is the equivalent temperature, representing a temperature equivalent to a situation in which there would not be any greenhouse effect, seen from the planet. It therefore represents a situation with the radiation emitted by a planet at the temperature T_e when seen from space, because of the equilibrium between what is emitted to space and what is absorbed of the solar energy.

The point at which $T^* = T_e$ means that this temperature crosses the temperature profile at a height z^*, which is the equivalent emission level. For someone who looks to our planet from space and sees our emissions as the emission of a black body at this temperature, whatever is done in terms of greenhouse effect, we can always see the temperature externally as being this value, except that the height at which this occurs may change.

In fact, when greenhouse gases are added, climate modelers do not change T^*: it is bound to be providing the equilibrium with a solar radiation that does not change. What changes is the level of emissions. The level of emissions means that we are emitting from a similar temperature but at a higher level, if the lapse rate of the temperature remains more or less the same, and this higher level will provide the surface warming.

The way the warming occurs is a little more complex than what non-experts may think with the wrong idea that there is just some disequilibrium at some point. This disequilibrium occurs in an atmosphere where there is a gradient of temperature which is hugely important to determine the warming.

A point usually raised by climato-skeptical groups is that if we look at the rays of absorption of CO_2, they are all saturated. If we take an emission that comes from the bottom, almost nothing goes out to space. It is absorbed by the rays of CO_2 at some point when it goes through the atmosphere; this is fine, except that then it does not mean anything for the capacity to impede the warming. The warming is the result of something subtler: Where is the current equivalent temperature located, and how does it change? More complex computations show that in fact this equivalent emission level is 1 optical thickness away from what we may call the top of the atmosphere. If we increase the optical thickness, we get the amount of greenhouse gases in the atmosphere – we will meet this level of 1 optical thickness a bit higher, consistently with the results, but it does change the surface temperature. This permanent gradient of temperature when we want to describe the system is partly based on the convective instability. We follow more or less the level of adiabatic transformation, i.e. the change with height of a parcel that would flow from year to year along an adiabatic transformation. The shape represented in Figure 4.19 also has an impact when studying the different responses of the atmosphere.

Let us consider diverse atmospheric thermal responses for three different modeled experiments referring to three different cases of forcing, and first a warming due to solar radiation increase: the warming is diffused, near the surface where there is more absorption (see Figure 4.20(a)). We consider no ocean, with a simple equilibrium only. In the experiment with results presented in Figure 4.20(b), the warming is due to increasing greenhouse gases: the warming is everywhere down to below the tropopause, and then there is a cooling above it.

Near the equator, the maximum warming is a bit more in altitude as a result of convection: it is a very structured phenomenon in altitude. Figure 4.20(c) shows that the aerosols' increase experiment has a result which is opposite to the previous experiment with greenhouse gases: the effect is a warming in the stratosphere, and a cooling below. Further, this cooling is occurring in both hemispheres with some differentiation, but not much in terms of amplitude. In fact, it looks more or less as the reverse effect of Figure 4.20(b), except the fact that as aerosols are mainly emitted in the northern hemisphere in largely industrialized countries, they do not travel very far away; that is the reason why the larger cooling is located there. The fact that there is such a symmetry is a bit strange because a much more local behavior for aerosols would be expectable. The only reason explaining this phenomenon is that the cooling due to aerosols is more or less opposite to the warming due to greenhouse gases. This calls our attention and it cannot be explained without feedbacks.

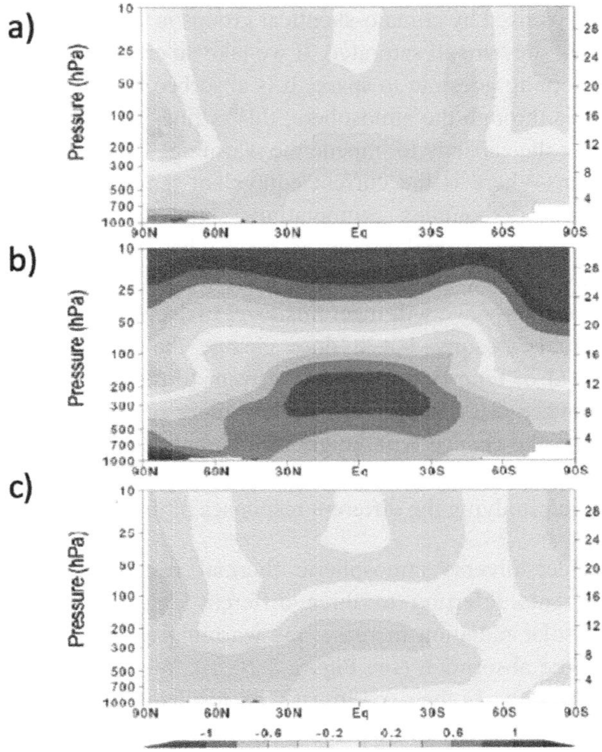

Figure 4.20. *Representation of the atmospheric thermal response to a) solar radiation increase; b) greenhouse gas increase; c) aerosol increase, with the x-axis representing latitude and the y-axis pressure in hectopascal/altitude in kilometers. Source: IPSL*

What could be the common feedback between Figure 4.20(b) and Figure 4.20(c)? The answer is that there is a given structure for the greenhouse gas increase, and there is a similar phenomenon explaining the aerosol increase. What could be similar between the two?

This similar phenomenon is the water vapor feedback, which functions as follows: the system is warmed in different places in the world; we may expect that it was warming through aerosols which are produced over regions of high pollutions, but knowing that those aerosols would not go very far away, it could be local. However, if the cooling due to aerosols produces water vapor, this water vapor goes further away and there can be a feedback associated with water vapor that is more global.

There is a very strong difference between the way the climate system reacts with a solar increase and to greenhouse gases. This is one of the reasons why the warming can be attributed to the greenhouse effect.

The interpretations must be cautious as many phenomena are complex. The geography of the phenomena occurring also depends on many feedback.

Our purpose in this chapter is to understand how temperatures could be affected in different scenarios of climate evolution. We will try to relate that a bit more with the energy processes previously mentioned, our red thread being the description of the climate system for past and future issues.

When stating that it is obvious that we are facing climate change, we have to be very cautious. Climatologists believe that climate change is happening for scientific reasons based on robust evidence, that are not necessarily obviously true but are complex.

The solar forcing considered in the experiment shown in Figure 4.20(a) is very often advocated as the main cause of warming by climato-skeptical groups (this is mostly wrong, but it may be right in some cases). There are some correlations between solar fluctuations that actually exist and have been the cause for cool episodes in the last century, but generally speaking it is difficult to find a real correlation between the solar cycle and the present situation.

When using a model, modelers are therefore not considering a completely true world, but a world of model equations, with a warming present all over the space represented in Figure 4.20(a): it goes to very high altitudes (28 km) and it is also very uniform on the horizontal scale, which is the scale of latitudes (90 North to 90 South). Further, there is a larger warming in Figure 4.20(a): on the yellow band, in the middle band and also in Figure 4.20(b), showing the results when submitting the model to greenhouse gas increase. It is a phenomenon which seems to be occurring in all cases. Figure 4.20(c) shows the role of aerosols that tend to reflect the solar light and therefore aerosols tend to cool down the system and the pattern of cooling is reversely similar again to the pattern of warming.

This tends to indicate a common feature to all those diagrams as, for example, the role of aerosols is much stronger in the Northern hemisphere (where there are much more emissions of aerosols, from anthropogenic actions). Those human-related emissions are mostly in the Northern hemisphere but they are extending in both hemispheres with some patterns, such as warming, which may be intriguing at first sight; this common feedback is actually the water vapor feedback, distributed according to the dynamics of the climate system, obeying differentiated rules depending on the considered zone.

There is a difference between the dynamics in the intertropical and the extratropical regions, and those dynamics represent a strong regulation. This strong regulation of the dynamics may mean that the water vapor in a warming world will be distributed more independently than the warming currently occurring. As water vapor is a greenhouse gas, it is therefore a strong behavior with increased greenhouse gases: in Figure 4.20(b), there is a warming which is displaced in altitudes due to the fact that there is a convection occurring which displaces the warming a bit upward and the part of atmosphere which really emits to space is above; there is a part in which the terrestrial radiation is trapped, and a part in which the terrestrial radiation emits strongly to space, and a consistent behavior associated with greenhouse gases.

A phenomenon linked with aerosols is happening with always the same kind of patterns; even the big differences are subject to similar patterns that may be due to one kind of greenhouse gas which is modified in response to the warming or the cooling in all cases. It means that when saying that the nature of the warming is clearly due to anthropogenic effect rather than to natural solar fluctuations, this is a robust result, but modelers do have to make the computation to take out the part of the feedback which may be responsible for the final signal.

A large part of the debate initially between climatoskepticals and climate scientists came from a number of theoreticians thinking that modelers were going too quickly in their interpretations. This view was not completely wrong. After some time, the climate science community went into a more detailed analysis that largely confirmed model results, which are now even doubled-confirmed by the warming currently happening.

It is still a pertinent issue to know, at present, the exact speed the climate will be warming. The amplitude due to climate change will depend quite a lot on feedbacks and on the complexity of the response to those feedbacks. The +1.5°C Special IPCC Report was produced on a governmental demand, asking what should be done to remain under +1.5°C. Of course, it is obvious that maximum efforts must be undertaken, but politicians generally like figures in order to show that they are "monitoring the process". If figures are useful to see how a process functions and what kind of scenario it supports, we have to be cautious about the way it is described, and to avoid giving and using figures without their associated uncertainties.

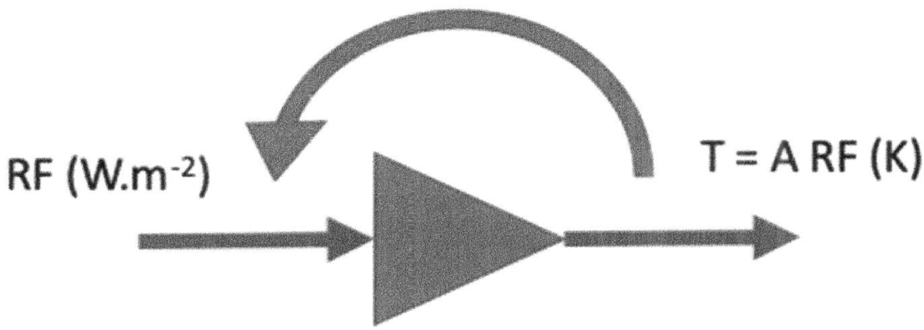

Figure 4.21. *A simplified scheme for explaining the equation representing a feedback*

Let us now propose a definition of what is feedback, schemed in Figure 4.21. The principles involve basic concepts, but not completely understood in their mechanisms. The idea of feedback has been duplicated from the domain of electronics: there is an operational amplificatory device, generally an electronic circuit materialized, as in Figure 4.21.

There is a forcing in $W.m^{-2}$ and a response of the system to this forcing, which we would like to measure in terms of the change of surface temperature, which is not the easiest way to study the phenomenon as it needs a conversion from radiative forcing with the dimension of an energy per unit of surface in $W.m^{-2}$. A is a parameter allowing us to go from the forcing to the answer of this forcing in terms of temperature, which can be written as ΔT.

Without allowing feedbacks, the models generally tend to agree that it is exclusively the radiative part which is playing a role. Also, this radiative part is, for example, the part associated with a CO_2 doubling in the atmosphere; this is how the forcing is established by line-by-line models where the quantity of absorbed radiation is studied. The result for the radiative forcing without feedback is $RF = 4\ W.m^{-2}$ (all models are very near this value).

Then, to convert this number and metric into a warming with a simple atmospheric model, without feedback, may mean basically freezing the water vapor content between the clouds' distribution: when stating that we do not touch that, we actually just consider the thermo-dynamical change/energy exchanges, and get $\Delta T \sim 1°C$.

$$\Delta T = A\ RF \cong \sim 1°C \qquad [4.8]$$

This result is important as there may be models for which doubling CO_2 induces a temperature change that can go up to 4–5°C of warming, meaning that the largest part of the answer is due to feedback so far.

Therefore, the idea that greenhouse gases induce a warming and that there is some additional answer of the system to this warming is actually not that simple: the largest part of the forcing actually comes from feedbacks which are complex to precisely determine.

Let us now introduce the contribution of the forcing in the equation: ΔT changes the radiative forcing, and the number defining the ratio at which this change occurs is λ, defined as the feedback parameter, expressed in $W.m^{-2}.K^{-1}$. We therefore obtain a simple equation with ΔT applied not only to the radiative forcing noted RF, but to the sum or subtraction $RF \pm \lambda\Delta T$, which corresponds to the way we define a positive or negative feedback. We can therefore write:

$$\Delta T = \frac{A}{1-A\lambda} RF \qquad [4.9]$$

This is how a feedback operates. It changes a multiplicative term to the forcing.

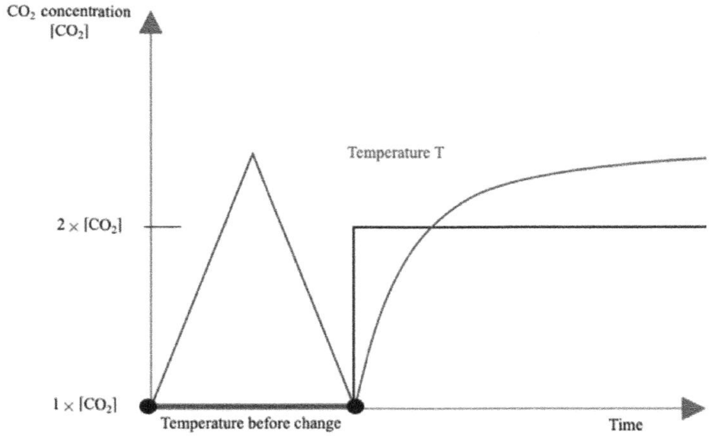

Figure 4.22. *A very simplified schemed of the Earth CO_2 concentration (in black) and temperature (in red) as a function of time*

Besides the radiative forcing FR, there is modulating term $\frac{1}{1-A\lambda}$, meaning that we are not merely adding a feedback in this equation, but rather multiplying the results by a feedback term. All of those multiplicative feedbacks are linked to each other, and it is not completely easy to solve them.

Despite the underlying complexity of the phenomena at stake, climatologists try to roughly model the mechanisms involved. There is a long history of CO2 doubling models. The way feedbacks are designed in CO_2 doubling experiments can be (very much) simplified as follows:

If we want to draw a very simplified schemed of the Earth CO_2 concentration and temperature as a function of time, the assumption is that the system is at equilibrium with $1\times CO_2$ and there is a ΔT with the temperature T before change represented (see Figure 4.22). This means that:

1) The temperature T is stabilized.

2) The temperature starts to change (increases) after a certain amount of time, which depends on the coupling with the ocean as represented in single models, generally with an ocean slab of 50 m.

Modelers then "wait" for the system to run for a time period representing ten years, until it gets to equilibrium, allowing then to compare the climate in those two different configurations.

The CO_2 concentration doubling experiment started by the time of Charney's 1979 report to the American National Academy of Science, and there has therefore been reports of model experiments of the climate response to those feedbacks for a long time.

In the previous paragraph, the intention was to deliver a grasp on admitted figures and trends regarding how the Earth climate responds to feedbacks, especially when considering a given forcing with a CO_2 doubling, corresponding more precisely to a 4 $W.m^{-2}$ forcing and to a warming of approximately +1°C with a simplified model (not considering all uncertainties).

4.2.1. *About models' uncertainties and radiative forcing associated with CO_2*

There is uncertainty associated with how much a given model evaluates a radiative forcing, and additionally to this uncertainty, there is the uncertainty related to the translation into temperatures that is superimposed (see Figure 4.23).

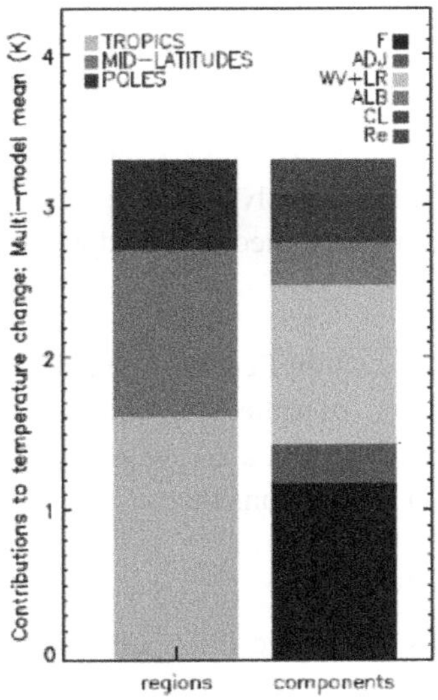

Figure 4.23. *The contribution to different feedback processes and different regions to the response in terms of temperature to a radiative forcing associated with a CO_2 doubling, from CMIP3 simulations, with models' uncertainties. Source: Vial et al. (2013)*

If we study the response in terms of temperature without the feedbacks, i.e. when we just consider one gas and observe how the different lines of absorption of the gas are modified by an increase of this gas, we obtain results relatively consistent between the models, up to ±10% of agreement, an order of magnitude which is not negligible.

However, climate modelers know that it is possible to decrease those uncertainties by introducing more physics in the models, which are getting increasingly complete with time and the development of IT and computing capacities. The uncertainty on the measure of the radiative forcing under a CO_2 doubling experiment, for example, has been clearly diminishing through time.

As for the feedbacks, it is possible to model an atmosphere with no change of water vapor, clouds or albedo, even if this is not an easy task. A warming of approximately +1°C is then obtained and when climate modelers let the water vapor evolve, this warming is multiplied by two, leading to a +2°C temperature increase. This is some kind of cumulative effect obtained by considering a given feedback by successive multiplications, from +1°C to +2°C, to a range of several degrees when considering the clouds' effect.

Efforts were made to mimic what would happen by just abruptly increasing the CO_2 and considering the change of temperatures with a certain temporal delay. After a while, it also gets to some equilibrium. A way to make the computation is to compare the difference between initial equilibrium and the equilibrium reached after the CO_2 concentration increases. In recent models, the accent is put on the moment at which phenomena are happening. The behavior in the ocean is different and complex as it is not completely at equilibrium, and that is the reason why it has been a bit forgotten intentionally for simplification.

The risks associated with an increase of aerosols in the future may seem counter-intuitive: in fact, our planet's mean annual surface temperature is already warming now; modelers are not sure exactly for the quantitative attribution of the phenomena causing this warming, partly involving greenhouse gases, partly aerosols. If the scientific community was able to determine what is the exact part due to aerosols, then the part due to greenhouse gases would be known as well. The aerosols' contribution is very difficult to assess because they are continuously being renewed; they do not spread out over the Earth, such as greenhouse gases. Aerosols also create some clouds with sometimes complex shapes to be represented in the models, such as large brown clouds over Asia. It is difficult to measure in detail the effect of additional aerosols due to human activities as the nature of the aerosols and the size of the droplet count for much in this assessment and are complex to study. Another phenomenon which may seem counterintuitive is that the small particles may have a strong effect, which is, however, very difficult to measure.

Let us consider the hypothesis that there is a large effect due to aerosols, that would be hiding the greenhouse gas effect: if that hypothesis is true, it would mean that the greenhouse gas effect could be even stronger than estimated, but it has been diminished by the aerosol effect. However, the aerosol effect will not increase as much in the future as the greenhouse gas effect because aerosols do not stay long in the atmosphere (they do not accumulate contrary to greenhouse gases). A stronger warming in the future can therefore occur, in the cases of a strong aerosols' effect comparatively, and that would become smaller over time in proportion.

NOTE.– The problem of global warming hiatus, also named global warming pause, referring to a period of relative stability in the global annual mean Earth surface temperature, is due to the natural variability of the Earth system. The North Atlantic fluctuation El Niño is the best known, but several other modes of fluctuations of the Earth also exist, and sometimes can even happen in phase. In the case of global warming hiatus, there are at least three modes more or less in phase, and it has already been required to wait for about 15 years to move away from when the natural fluctuations were all in phase around 1998–2012, with the effect of "dumping" the warming a little bit. Conversely, the warming is larger just after that.

If we use a simplified scheme, Figure 4.24 describes a radiative forcing.

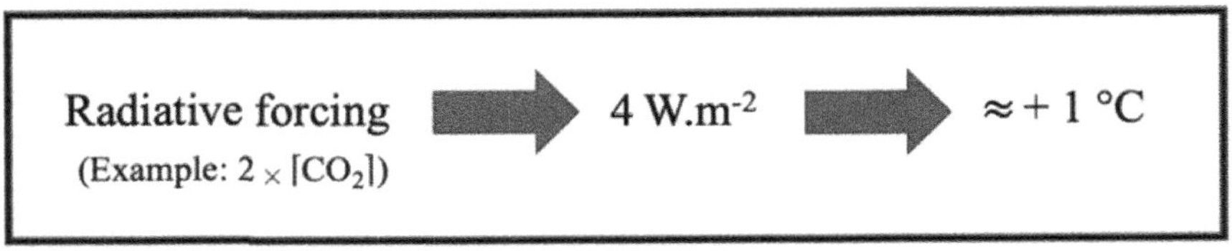

Figure 4.24. *Simplified scheme showing the sequence associated with a radiative forcing*

Let us now focus on the different feedbacks, serving as a multiplier of the system answer in terms of temperature to a perturbation.

4.2.2. *The water vapor feedback*

The water vapor feedback is the most important. It multiplies by an order of magnitude of about two an answer of +1°C, getting to a warming of +2°C. As this factor two, which is associated with the role of the water vapor feedback in the atmosphere is a multiplier, it also multiplies the other feedbacks, and even by more than two in fact.

Indeed, the water vapor dependency on temperatures goes through saturation value of water vapor, decreasing very quickly: at a temperature of 30°C, for instance, there is a certain amount of water vapor in the atmosphere. At –40°C, there is about 10,000 times less water in the atmosphere. Therefore, there is a huge decrease of water vapor between the bottom and the top of the atmosphere.

An interesting question is: While the air is ascending in a Hadley cell, do we have a strong feedback or not? The answer is that there is a feedback if water reacts strongly to temperature. This is the case near the equator, where this dependence is clear, because the water is always at saturation in this location. When air is descending, there is less and less water: precipitation occurs, taking out part of the water, and we are always at a saturation value of water near the equator. This saturation value is a direct function of temperature following the Clausius–Clapeyron formula; therefore, there is a phenomenon extremely clear in terms of feedback.

The concept of water vapor feedback in the first IPCC report is described as an extremely simple phenomenon because it depends on the saturation value which itself depends on temperature: there is a very strong embedded feedback. Climato-skeptical authors published a paper named "Some cooling about global warming" in the 1990s, stating that they agree with the description of the ascending air. But there was more doubt for them regarding the case of descending air: there is indeed dry air all along the path. What defines the amount of water vapor in the air is the level of drought present at the top of the atmosphere. What is obtained when the water is descending is the water vapor which is low and does not depend directly on temperature. This dependence is indirect because at the top of the atmosphere, the amount of water will depend on the coldest place that the particle of water vapor is meeting. If this is a very cold place, it will let down part of its water, and retain this value of water vapor amount when it gets down; there is an increase of temperature, but it is not a direct and simple link.

There were many complementary works for IPCC in the second and third reports, which generally concluded that there is effectively a water vapor feedback and that its order of magnitude given in the previous IPCC report seemed correct. It has been a long process of about ten years before the scientific community agreed on the fact that the water vapor feedback is a rather safely understood phenomenon on which models can rely on.

4.2.3. *The albedo feedback*

The albedo feedback is the second one in terms of conceptual simplicity; however, its detailed mechanisms are actually not simple at all. The albedo feedback depends mostly on snow at short time scales (time scales of future decades).

At longer time scales, the albedo feedback mainly depends on glaciers, on how they melt, but there is a question mark as the experts do not know exactly at what speed glaciers will melt, even if there is some indication through the way the sea level is rising; however, in terms of quick feedbacks associated with snow and sea ice, it is a much more complex assessment.

For instance, "young" snow has a certain grain; its crystals reflect very strongly, while for "old snow", the crystals are bigger and also with dirt: to simplify the problem of the snow albedo assessment, snow is basically not reflecting as much when it gets older. This principle is simple. We know that snow and ice will disappear with climate change and there will be less reflection. We know what the reflection was before through satellite measurements, but what will happen in the future depends on the physics of snow and on the way snow may be, for example, covered by vegetation. If there is snow in a region where forest is growing (for example, in Russia), it can make a big difference.

The figure taken for many years for the albedo feedback is a multiplication factor of 1.25. Therefore, it means that if we have a +2°C warming, the albedo feedback induces an increase from +2°C to +2.5°C, which is considerable. Indeed, the 2019 IPCC Special Report about the importance regarding staying below +1.5°C or +2°C, showed the importance of every half degree, and if we want to limit global warming to those levels, those kinds of orders of magnitude are very significant.

4.2.4. *The cloud feedback*

Cloud feedbacks generally bring results in the domain range between +1.5°C and +4.5°C warming. As the cloud feedback is multiplicative, it can also strongly change the answer. Oddly enough, this figure has not changed since the Charney report in 1979 (while at that time, it was based on two models only).

They are different explanations to the fact that this range has not varied since Charney's report, despite the refinement of scientific methods and data. As a scientist, we have to ask themselves if there is a kind of social constraint related to that: a hypothesis could be that a given group of climatologists obtain a result coming from their model; if the result is not in the domain forecasted, they may think that their model is wrong and make another computation until they get a result within the expected range. They may be satisfied with a model in line with an accepted range of "good models". For the anecdote, decades ago, the American scientist Robert Donald Cess, professor of atmospheric sciences at the Stony Brook University in New York, made comparisons of models based on satellite measures, with results of global feedback estimates. During the first phase, there was a huge spread of results among the models. All of the models then converged with the NASA observed results. There was a third experiment in which Cess separated solar and terrestrial radiations and found the same initial spread. The teams then managed to change parameters in their models to obtain results perfectly aligned with the results they were expected to find. However, when looking separately, it was actually wrong twice: for the solar radiation and for the terrestrial radiation. This kind of social constraint effect of course has to be considered when assessing the value of a model's inter-comparisons; as a scientist, we must be alert that people tend to align with what is considered as accepted.

Something not completely resolved with the feedback issues is also that there is a range in terms of global change which is still very large, even after decades of research.

There is also the idea that maybe we will get to "something better" to validate the model: that "better something" is the observed warming in the atmosphere.

NOTE.– 30 years ago, Hervé Le Treut thought that there would be no difficulty, when the warming would effectively occur, to determine the climate sensitivity. In fact, it is not that simple.

As a reminder, when knowing T, ΔT is obtained, and after computing the change in the forcing F that can be written as ΔF, we get the warming that the Earth is currently experiencing, as well as the possibility to make predictions for the future. When people ask what we should do to stay under +2°C for instance, a sharp answer is usually expected. However, it turns out that the answer is not that simple. A lot of processes are interacting with radiation and a challenge for the assessment comes from the fact that we mix two kinds of investigation: the study of greenhouse gases and their consequences; and the anthropogenic emission of aerosols. Also, the sensitivity of the warming due to GHG is mixed with the sensitivity of the climate to aerosols which has a cooling effect. Therefore, there is globally a warming effect from the greenhouse gases as well as a cooling effect from the aerosols. The attribution of the net effect either to the greenhouse gases or to aerosols is not obvious at all.

Even if there is a strong aerosol effect, aerosols stay only a few weeks in the atmosphere, as they are constantly renewed. On the contrary, greenhouse gases stay decades/centuries in the atmosphere for long-lived ones.

Therefore, a critical question is: What happens if the aerosols are particularly numerous? That would mean that global warming has not been larger because it has been "hidden" by aerosols. But this effect may not last for a long time because aerosols are not accumulating in the atmosphere, whereas GHG is accumulating.

If the warming is partially hidden, then what will happen in the future? This would mean that a larger increase of temperature will happen, and aerosols cannot increase as much as greenhouse gases; therefore, that would create a larger increase, in this hypothetical case of a strong aerosol effect. Also, that would be the reverse in the assumption that there has been a small number of aerosols until now.

Such a study is not obvious at all; it happens little by little with a large effort to better assess the cloud effect which is superposed on that. We definitely lack a solid point to grasp on that problem; aerosols also tend to make clouds more reflective: with more droplets, there is more reflection, and therefore a larger cooling.

About the study of water vapor and cloud feedbacks

The water vapor feedback can be studied by considering other climatic events: of course, we cannot observe future climate change, but there are many phenomena that can be predictable through model results. For instance, the Pinatubo volcano explosion was used as a way to check the water vapor feedback with air ascending/descending. The effect of this volcanic phenomenon was an order of magnitude of 0.5°C cooling for one year. This result can be used to check the feedback representation. Of course, there are also some seasonal cycles, with the need to always check if they are correctly represented in the models, including, for example, El Niño events.

It is also necessary to check if the feedbacks are the same under different circumstances or if they may differ. Many efforts are made by the scientific community. Regarding the cloud feedback, it is difficult to look only at the cloud processes because the cloud radiative effect depends on several parameters such as the size of the droplets, the height of the cloud droplets, the shape of the clouds themselves, etc. When adding all of those parameters, they represent a lot of uncertainties. Complementary tools are needed to assess the trust which climate scientists can put in the models representing those phenomena. Also, the level of trust concerning the water vapor feedback is absolutely key: if we are wrong on this, there is pretty much a mistake on everything. We know "for sure" only what is happening some time away from us, but the way global warming is already appearing is proof that we are not completely wrong, even if we are not absolutely sure of the amplitude of the future warming.

Box 4.2. *How can we study the water vapor and cloud feedbacks?*

4.2.5. *The lapse-rate feedback*

The lapse-rate feedback refers to the way temperature changes with height, which is represented in Figure 4.19. The effect of greenhouse gas on the vertical stratification of temperatures is to shift it upward; therefore, when we look at the bottom; a shift upward also means a shift of temperature at the surface. This depends on whether the line which is reflecting the variations of temperature with height is more or less equal to what it was before the greenhouse gases were inserted. But the lapse rate may change a little, and this will have an influence on the surface temperature. Therefore, this surface temperature change is something to additionally consider. In fact, this is a forced feedback, and generally we accumulate it with the effect of water vapor because when we study the lapse-rate feedback effect on water vapor, we are not interested in how water vapor is changing, but in how emission to space may change. This is a combination of the amount of water vapor at each level in the atmosphere; also, the way each level in the atmosphere is emitting in response to water vapor changes. Both effects are working together and tend to act to stabilize each other. The amplitude of the water vapor feedback is slightly stabilized by the feedback of the temperature lapse rate.

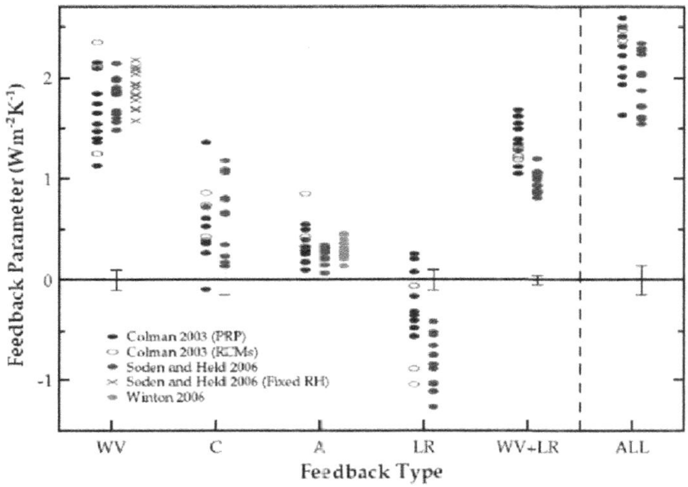

Feedbacks from different models

(WV= water vapour, C = Clouds, A= Albedo, LR= Lapse-Rate)

Figure 4.25. *Feedback parameter comparison from five different models for water vapor, clouds, albedo and lapse rate. Source: IPCC, WGI, AR4 (2007). Reproduced with permission from Randall et al. (2007)*

A summary of the feedback and the way they have been studied by models since the fourth IPCC Assessment Report, without many references to observations, is represented in Figure 4.25. We go from processes that people knew at that time (which did not change much) and that were inserted into the models, to model results giving the feedback type for those feedback parameters. We observe a large feedback parameter for water vapor with a multiplication factor of roughly two. The lapse-rate feedback tends to cool the system a little bit, because the change of stratification in the system tends to cool it down. When adding those parameters, we actually multiply one by the other, with also a ponderation parameter slightly reducing the difference between the models. Even if the processes are different, in terms of global result in the end, it does not matter that much. However, this compensation parameter to diminish the kind of ranges for the models is also an indication of some fragility in the climate models. Each point in Figure 4.25 is a study for different models. The colors represent different types of studies, all based on theoretical methods, not directly on observed changes in the climate system.

When observing clouds and albedo feedback, they are always increasing the elevation of temperatures (some uncertainties are mapped out from zero to approximately 0.5 W. $m^{-2}.K^{-1}$); the effect is a bit more diverse because of the multiplication factor. The clouds effect in terms of feedback is subject to high uncertainties and can be either positive or negative: clouds feedback is the most uncertain among all feedback parameters. It is the better understanding and representation of physical processes that will help to solve that problem and to diminish the uncertainties which are due to the numerous interfering parameters. Therefore, many efforts have been made to go through observations in order to find observed processes in which all of the feedbacks go together similarly.

The contribution to different feedback processes and different regions to the temperature response to a CO_2 doubling is shown in Figure 4.23. This figure contains the separations of feedbacks distinguishing the tropics, midlatitudes and Poles. The tropical regions are actually key as they represent the locations where the feedbacks are the most powerful. When studying the contribution of different feedbacks to a temperature doubling, instead of accumulation, climate researchers try to attribute this in terms of feedbacks; there are some residues, but again, the factor water vapor plus lapse rate are the main forcings, as the water vapor feedback is very strong, plus the effect of the lapse-rate feedback complete uncertainties in the current climate models.

A pertinent point is that the greenhouse gases are not part of the feedbacks, but of the forcing. The forcing shown in Figure 4.23 is calculated as the initial effect of greenhouse gases which is multiplied by the feedback. In fact, this added part is actually larger than the forcing itself. We need to know the forcing to get access to the value at rest.

The future behavior strongly depends on the complex physics of the feedbacks: the more research time dedicated to these problems, the more we can limit those uncertainties and provide relevant figures to the destination of decision-makers. Also, we only mention what has been modified by human activities. Figure 4.23 is not a diagram of the entire energy budget but a diagram of the way the energy budget is affected by human activities. We have an increase in greenhouse gases which is a forcing, and this increase in greenhouse gases leads to an increase in other different terms. The science of this modification is much more difficult than the global understanding. When observing the budget at the beginning with the entrance/output of solar radiation, all are big figures which equilibrate correctly. What we depend on are actually small modifications of this big picture, and this is very important for human life on the Earth because the human civilization has been developing under a short range of modifications only, compared to this big picture.

Note that the considered changes are occurring at equilibrium. It is a bit different in that case: we would have to wait until the energy stored in the ocean is given back in some way in order to be at equilibrium; therefore, presently, all those initial methods to consider the system are a bit wrong as we are actually not at equilibrium. Researchers study how we go from one equilibrium to another and we are in the middle of that. Moreover, this analysis is done without considering the ocean. As the heating is largely stored in the ocean, this represents another large source of unassessed uncertainty in this simplified scheme.

Figure 4.26. *The six satellites composing the A-train. From left to right: Aura, Parasol, Calipso, Cloudsat, Aqua, OCO. Source: CNES Illustration P. Carril (October 2004)*

A method to diminish the source of uncertainties mentioned before is to use observation data through the use of satellites, with several measurements. We already mentioned that the radiation budget at the top of the atmosphere depends on greenhouse gases and also on the other gases of the atmosphere, where everything is linked with the atmospheric water cycle (water vapor, clouds). The radiation budget at the top of the atmosphere also depends on aerosols. Therefore, all of those parameters must be measured. Aerosols (and clouds) depend on many detailed processes which include diffusion, absorption, height, cloud shape, droplet geometry, etc.

The scientific community realized that there was a real need for a consistent measurement of those parameters, obtained by satellites. The "constellation of satellites" in Figure 4.26 for instance shows lidars measuring the height of the cloud with a very good precision. Other French satellites are looking at the radiation diffused back from the clouds, considering different directions. From that, they diagnose the kind of crystals or droplets diffusing the light for a clear diagnosis of the nature of the studied clouds. All of those satellites forming the “A-train” are on the same orbits, separated only by a few seconds/minutes. They are mostly looking at the same things, one after the other. The information they deliver is not easy to use for models because many satellites are going through the Earth on different points.

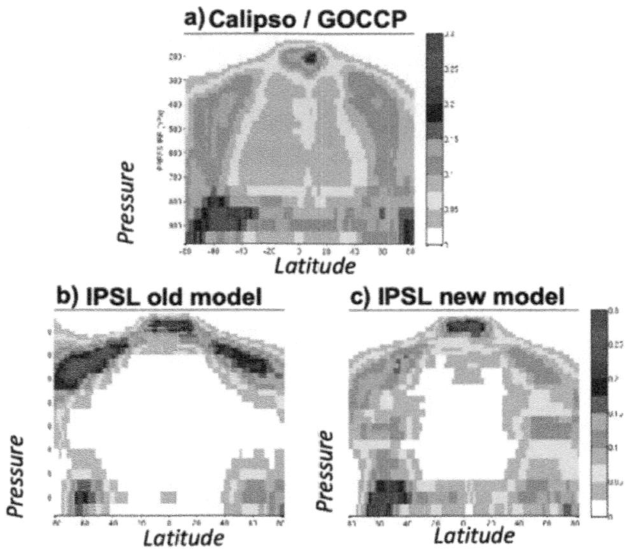

Figure 4.27. *Cloud fields: comparison between a) observation data from Calipso; b) model data from IPSL old model; and c) model data from IPSL new model. Source: Kontsa et al. (2016)*

A French expert on clouds, Hélène Chepfer has been trying to design a process to go from satellite images to models and backward consistently. This type of research results is shown in Figure 4.27 with a comparison between satellite observations and model results regarding cloud fields. When analyzing what satellites observe, there is a view which goes from South to North, and from the bottom to the top of the atmosphere. We place ourselves for observations at 900–800–700 mbars, and at 500 mbars, which also represents the middle part of the atmosphere. 200 mbars correspond to the tropopause. The green spots in Figure 4.27 represent the part of the atmosphere where there are less clouds/particles. In the purple zone, we see the radiation from the top; it is absorbed in some areas, and we are not completely able to detect the cloud in the white zones.

We can imagine that there is a satellite in our model, and that we analyze how observations and models compare. There are two IPSL versions of the model, an old one and a new one, in which the physics of the clouds has been changed; it is satisfying as they appear to be in good agreement.

As for the satellite radiation penetrating in a model, there are regions where satellite data are available over the models, and other regions where we do not have any signal. The new version of the IPSL model is actually satisfying when observing the satellite image.

The increasing use and research with satellite data also illustrates the necessity of looking seriously at those feedbacks.

4.3. Measured and anticipated impacts: what is predictable?

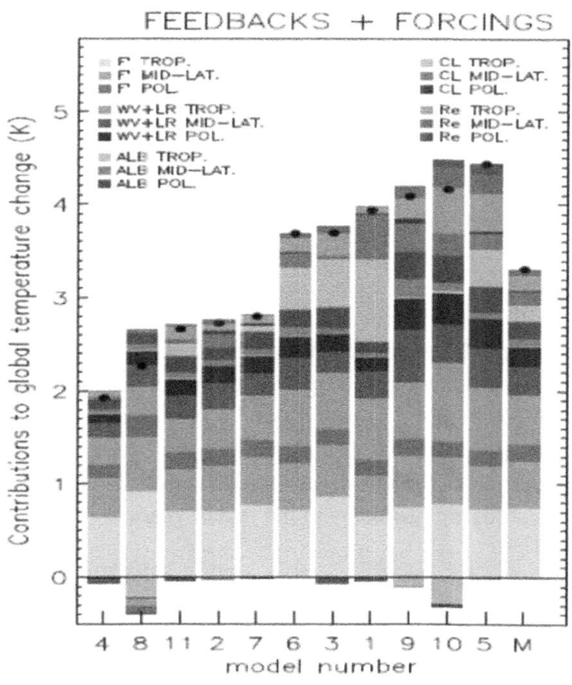

Figure 4.28. *Assessment of the contribution to global temperature change in Kelvin (K) for different models representing diverse feedbacks and forcings (albedo, clouds, lapse rate, etc.), relative to different Earth areas (tropical, mid-latitude, polar area) (Vial et al. 2013)*

The contribution to global temperature change in Kelvin (K) for different models representing diverse feedbacks and forcings (albedo, clouds, lapse rate, etc.), relative to different Earth areas (tropical, mid-latitude, polar area), can be assessed by models (see Figure 4.28). The effect of feedbacks that increase the direct role of a given forcing (such as a greenhouse gas increase) shows a differentiation by models. All of the models represented in Figure 4.28 correspond to the best available science at their publication time, and they show a reasonably large range. For this exercise, the problem was analyzed backwards: the initial hypothesis is that we want to stay under a +1.5°C warming in 2100, and modelers translate this aim into an amount of GHG that we are allowed to emit; this task is complex, and there is a range of uncertainties associated with the result.

4.3.1. *Three generations of model assessments directed toward climate change*

We already discussed sensitivity experiments with CO_2 doubling from one equilibrium to the other.

We also studied specific experiments where scientists studied scenarios, such as the SRES (Special Report on Emissions Scenarios with 1% yearly CO_2 increase (SRES) scenarios (IPCC 2001a, b, 2007a, b, c, d) in the IPCC Third Assessment Report (TAR) and in the IPCC Fourth Assessment Report (AR4) that we mentioned previously, with policy-oriented scenarios including transitory aspects such as the oceans.

We also mentioned RCP scenarios in the IPCC Fifth Assessment Report (IPCC 2013) that are policy-oriented, meaning that the scientists working on it wish to understand what should be done to stay around a certain amount of climate change, as agreed by climate policy decision-makers. We will not go into too much detail regarding those scenarios as the aim is just to deliver the essence regarding the way scientists have been studying problems in the recent years.

As a historical reminder shown in Figure 4.29, for the IPCC SREX Scenarios, it started with considering a real societal scenario and its climate policy aims. The greenhouse gases were then included, with the work done by the Working Group I of the IPCC in charge of the "hard science" in the scenarios. All of the gases were not mixed in the same single scenario with the use of equivalent CO_2 emissions; there were separate assessments for CO_2, CH_4, N_2O, CFs, etc. All greenhouse gases were considered with associated scenarios for each one, imposed to the models. Another level of work is to go from annual emissions to what has been stored in the atmosphere, represented in the right part of Figure 4.29 in terms of concentration. Let us consider the emission scenarios represented in Figure 4.29; the green one is the most favorable, which we are already not complying with. It is a scenario where anthropogenic emissions increase until 2050 and then emissions decrease, starting in 2050. When considering the concentrations in the atmosphere (right diagram of Figure 4.29), the trend is an increase even if the increase is supposed to get slower with time: there is no decrease of CO_2 concentrations if we continue to add CO_2 in the atmosphere; the most favorable outcome is merely a slower increase. We may at some point reach a place where the sinks of CO_2 are a bit stronger than the emissions, but it is very near to the time of zero net emission around 2050 according to the related emission scenario.

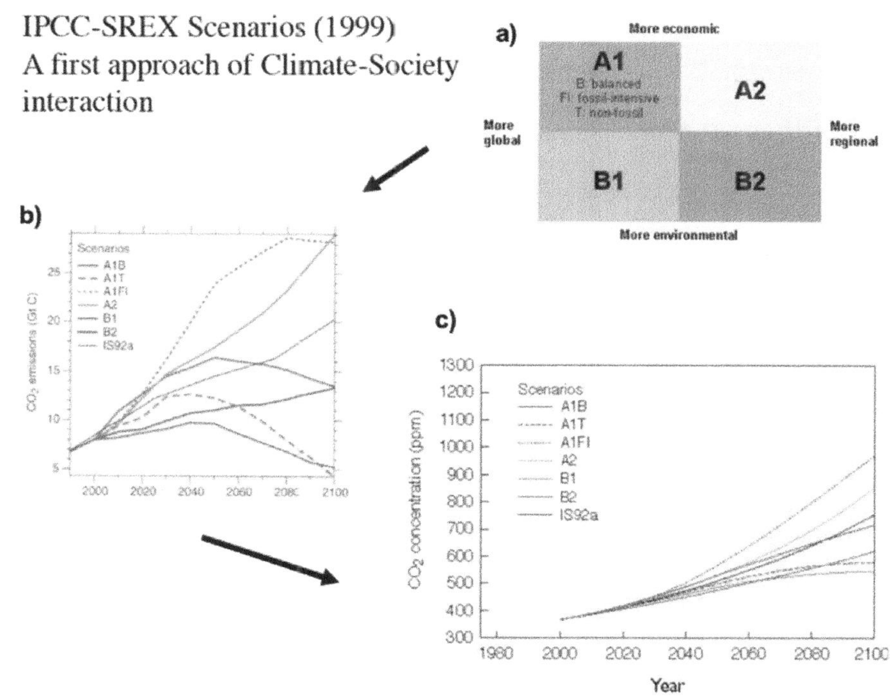

Figure 4.29. *IPCC-SREX Scenarios. A first approach of Climate–Society interaction. a) Matrix presenting the main orientation of four types of scenarios: A1, A2, B1 and B2; b) CO_2 emissions in GtC as a function of time according to various types of SREX scenarios; c) CO_2 concentration (in ppm) as a function of time. Source: adapted from Figure 17 in, and with the permission of, IPCC (2001b)*

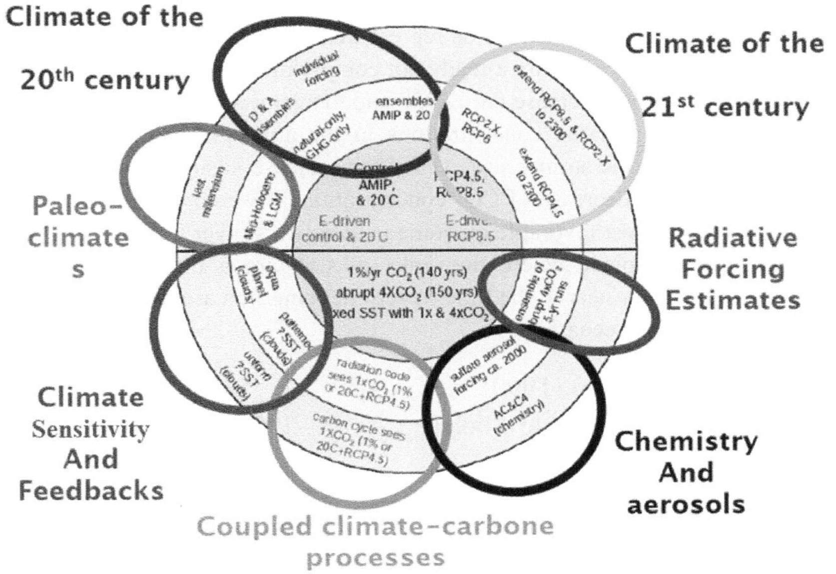

Figure 4.30. *Scheme of the main principles of simulations proposed by CMIP-5 model*

As for the simulations proposed by the Coupled Model Intercomparison Project (CMIP-5) model, the main principles are shown in Figure 4.30. Just a few main steps are necessary, which in the general publications have been left quite open to the different climate modeling teams. The National Center for Atmospheric Research provides access to CMIP data, which largely serves as a reference in terms of CO_2 concentrations in models.

CMIP is a program grouping all modeling centers producing simulations for the analysis of past, present and future climate. CMIP is a program part of the World Climate Research Program (WRCP), ensuring the coordination of scientific results and knowledge, as well as of infrastructures, as the supercomputer outputs need data homogenization and structuration, in order to share the processed data to the destination of states, researchers, students, etc. The simulations furnished for the sixth edition of CMIP have been evaluated in the IPCC report.

About 50 climatic centers participated for CMIP6, either for data furniture or for the protocol building in order to manage all of those simulations (see Figure 4.31).

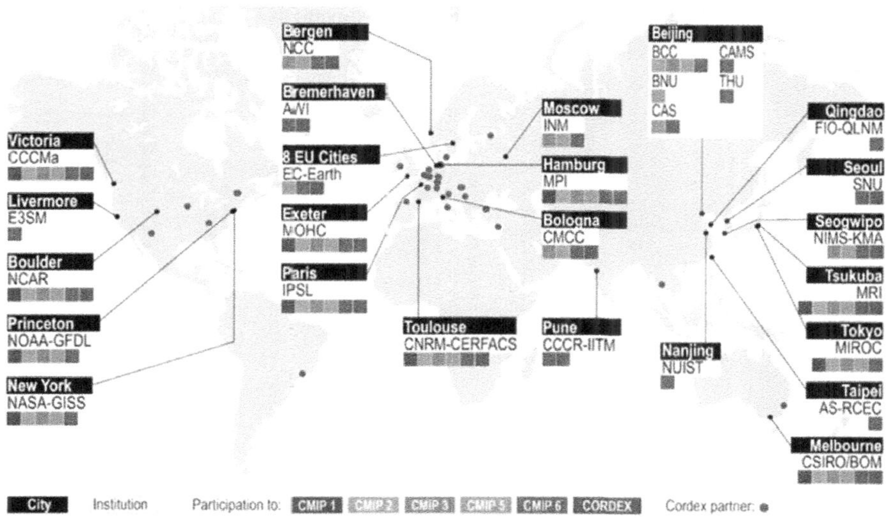

Figure 4.31. *Map of the participating climatic centers worldwide to the CMIP exercise. Source: IPCC, WGI, AR6 (2021). Reproduced with permission from Chapter 1 of IPCC (2021a)*

A main feature of the CMIP6 exercise is that they have a small number of numerical experiments of reference, that have not evolved for about 20 years, when those programs started. Those experiments consist of increasing the atmospheric CO_2 concentration by 1%, or in fixing CO_2 at pre-industrial values to assess how the climate would evolve without anthropogenic action. These are reference experiments which were conducted for all of the successive CMIP editions.

Therefore, we can intercompare the evolution of results for those core experiments: CMIP6 delivers much better performance for surface temperature simulations for instance as compared to the previous versions, CMIP3 and CMIP5, both over continents and over oceans, at the global and at the regional scales (see Figure 4.32).

We also have a better representation of precipitation, and an improvement in the representation of the atmospheric circulation, i.e. winds, large flowing streams, heat transport, and transport of moisture. Errors and differences among models remain nevertheless; they represent what are called model or structural uncertainties; those uncertainties are accounted for in the IPCC report evaluation of CMIP in order to evaluate the robustness of conclusions and analyses that remain at the IPCC experts' disposal.

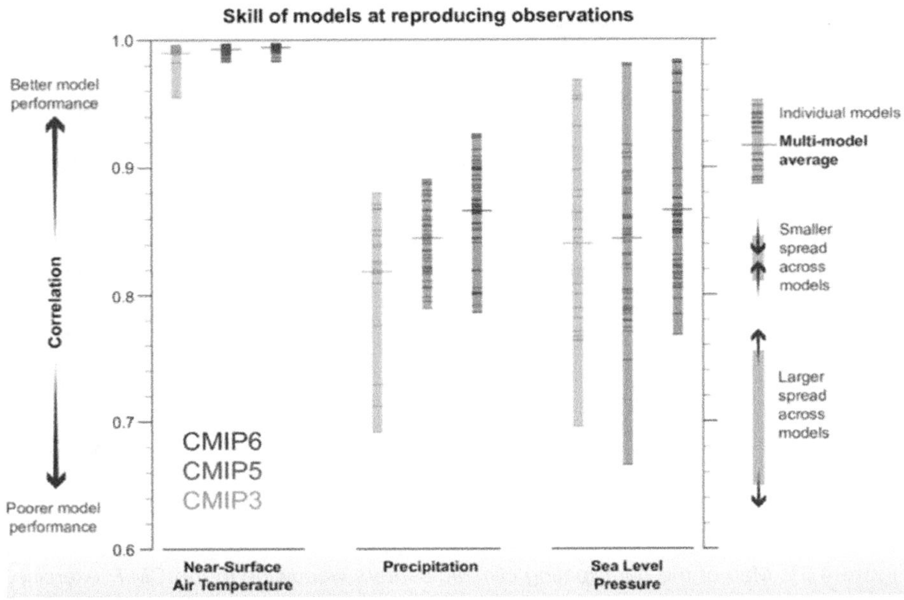

Figure 4.32. *The improvement of climate models' skills at reproducing observations through successive CMIP versions thanks to the increasing computer power and a better understanding of climate processes. Source: IPCC, WGI, AR6 (2021). Reproduced with permission from Chapter 3 in IPCC (2021a)*

Another important point is that for what we call scenarios, a recommended and necessary attitude for comparability purpose is to use the same model in the course of different exercises. At Institute Pierre Simon Laplace (IPSL) for instance, it was decided to make all of the experiments with the same model; this was the last CMIP5 experiment.

Five years ago, from CMIP5, for instance, there was a core mandatory experiment in order to be part of the IPCC as a laboratory. There were also two experiments that were not compulsory, but were useful for delivering more information for studying different purposes: the climate of the 20th century, past climates, climate sensitivity and feedback processes, coupled climate-carbon processes, radiative forcing estimates, etc., all within the same model. This type of huge computation lasted for one year, using a supercomputer in each laboratory. The diagram displayed in Figure 4.9 shows the result of this huge exercise, but some commentators have stated that climate models showed only the results in terms of the global annual mean temperature for an atmosphere with a doubled CO_2 concentration. However at the Laboratoire de Météorologie Dynamique (a laboratory member of the IPSL), this is not the case; there have been petabytes of various results, not just in terms of the global temperature. Thousands of scientists have been mobilized to analyze them.

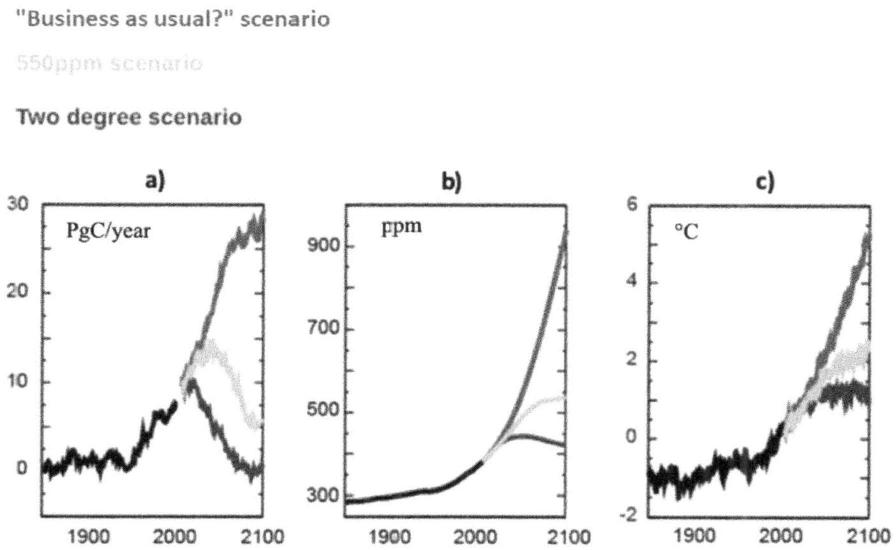

Figure 4.33. *Diagrams of a) global carbon emissions in PgC/year as a function of time until 2100; b) global CO_2 emissions in ppm as a function of time until 2100; and c) global mean temperature in °C as a function of time until 2100. Source: simplified scheme adapted from data from WGI, AR4 (2007)*

In IPCC reports since AR4, scientists have been looking at scenarios in some reverse ways, as shown in Figure 4.33 with a series of diagrams representing historical and projected emissions for three representative concentration pathways: "business as usual scenario" in red, "550 ppm scenario" in light blue, "two degrees scenario" in dark blue.

Each RCP may give rise to different temperatures in 2100 as shown in Figure 4.33(c). Figure 4.33(a) shows emissions expressed in $PgC.yr^{-1}$, and these emissions are translated into a greenhouse gas amount (diagram of the center shown in Figure 4.33(b)) expressed in ppm, and then in terms of temperature raise (in °C) until 2100, as shown in Figure 4.33(c).

Global Warming Between 1850–1900 and 2010–2019 (°C)		Historical Cumulative CO_2 Emissions from 1850 to 2019 ($GtCO_2$)					
1.07 (0.8–1.3; likely range)		2390 (± 240; likely range)					
Approximate global warming relative to 1850–1900 until temperature limit (°C)[a]	Additional global warming relative to 2010–2019 until temperature limit (°C)	Estimated remaining carbon budgets from the beginning of 2020 ($GtCO_2$) *Likelihood of limiting global warming to temperature limit*[b]					Variations in reductions in non-CO_2 emissions[c]
		17%	*33%*	*50%*	*67%*	*83%*	
1.5	0.43	900	650	500	400	300	Higher or lower reductions in accompanying non-CO_2 emissions can increase or decrease the values on the left by 220 $GtCO_2$ or more
1.7	0.63	1450	1050	850	700	550	
2.0	0.93	2300	1700	1350	1150	900	

[a] Values at each 0.1°C increment of warming are available in Tables TS.3 and 5.8.

[b] This likelihood is based on the uncertainty in transient climate response to cumulative CO_2 emissions (TCRE) and additional Earth system feedbacks and provides the probability that global warming will not exceed the temperature levels provided in the two left columns. Uncertainties related to historical warming (±550 $GtCO_2$) and non-CO_2 forcing and response (±220 $GtCO_2$) are partially addressed by the assessed uncertainty in TCRE, but uncertainties in recent emissions since 2015 (±20 $GtCO_2$) and the climate response after net zero CO_2 emissions are reached (±420 $GtCO_2$) are separate.

[c] Remaining carbon budget estimates consider the warming from non-CO_2 drivers as implied by the scenarios assessed in SR1.5. The Working Group III Contribution to AR6 will assess mitigation of non-CO_2 emissions.

Table 4.2. *Estimated remaining carbon budgets are calculated from the beginning of 2020 and extend until global net zero CO_2 emissions are reached. They refer to CO_2 emissions, while accounting for the global warming effect of non-CO_2 emissions. Global warming in this table refers to a human-induced global surface temperature increase, which excludes the impact of natural variability on global temperatures in individual years. Source: IPCC SPM, WGI, AR6 (2021). Reproduced with permission from IPCC (2021b).*

It is also possible to work in the reverse way with a type of model that now integrates a full cycle of greenhouse gases: in fact, the two methods are present in the current research teams. For the three temperature evolution scenarios for the model represented in Figure 4.33, the main question that scientists are investigating is: "What are the allowed emissions per year needed to stay under +2°C of warming?", referring to the aim of the Paris Agreement (in the case of the dark blue scenario). The dark blue line evolution for this +2°C scenario in Figure 4.33(c) representing the temperature fluctuations, shows stabilized temperatures in the pre-industrialized era, with some small fluctuations due to the little ice age, then a temperature increases with a stabilization to stay under the stated aim of +2°C in 2100. The exercise consists of reversely going back to the kind of CO_2 emissions needed to stabilize the temperatures at +2°C in 2100. When considering the trend related to the study of this metric, it appears that a progressive decrease of CO_2 emissions is needed in order to reach zero net emissions by 2050 for that aim. This is an integrated computation, considering all of the processes and sinks. It shows the real nature of the effort needed when stating the necessity to have a net zero carbon emission level. The effect if the reference level is set at +1.5°C warming is that it just shortens the time which is required to reach the net zero emission level. This net zero emission purpose would require us to have no cars using fossil fuels anywhere else in the world, and before 2050. Of course, this is physically possible. However, we let the reader conclude if this seems socially and politically feasible. The authors believe that it seems reasonable to state that under the current geopolitical and social global climate/economic system, this is impossible unless major changes occur very quickly.

The IPCC Sixth Assessment Report delivers detailed estimates of historical carbon dioxide (CO_2) emissions and remaining carbon budgets (Table 4.2).

4.3.2. *Socio-economic pathways (SSP) in AR6*

Figure 4.34. *A graphical representation of the five families of SSP scenarios in AR6 and their associated socio-economic challenges for mitigation (y-axis) and for adaptation (x-axis). Source: O'Neill et al. (2014)*

There are different socio-economical possible evolution pathways, tracing various trajectories for GHG emissions or land-use evolutions. The AR6 is more specifically based on five families of scenario, corresponding to five different ways of "being a society" at the global scale. Those five types of scenarios are based on different demographical, economical, technological, geopolitical hypotheses and lifestyle change assumptions: various values of those various criteria as a whole form a given family of scenario (see Figure 4.34). SSP1 is the "sustainability" scenario, taking the "green road"; SSP2 is the "middle of the road scenario"; SSP3 refers to a "regional rivalry, with a rocky road" scenario; SSP4 refers to an "inequality" scenario with a "road divided"; and SSP5 refers to "fossil-fueled development" scenario "taking the highway".

For instance, the SSP3 scenario corresponds to a narrative that the world is close to in 2024–2025. In the SSP 3 family, the societal developments are featured with the resurgence of regional rivalries/conflicts, nationalism, by a low sharing of technological improvements, by energetical and agricultural crises, and by food and energetical insecurities.

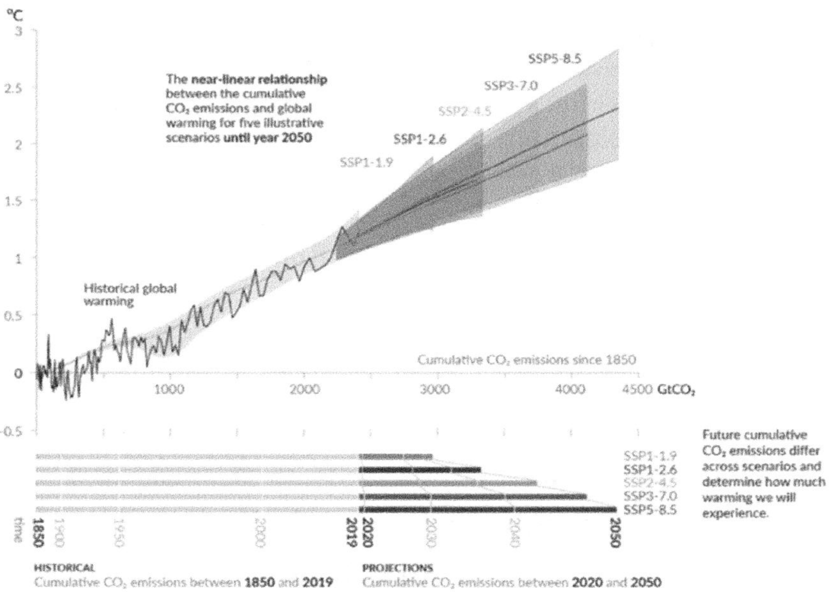

Figure 4.35. *Global surface temperature increase since 1850–1900 (°C) as a function of accumulative CO_2 emissions ($GtCO_2$) showing a near-linear relationship between accumulative CO_2 emissions and the increase in global surface temperature. Source: IPCC, WGI, AR6 (2021). Reproduced with permission from IPCC (2021b)*

	Near term 2021–2040		Mid-term 2041–2060		Long term 2081–2100	
Scenario	Best estimate (°C)	*Very likely* range (°C)	Best estimate (°C)	*Very likely* range (°C)	Best estimate (°C)	*Very likely* range (°C)
SSP1-1.9	1.5	1.2–1.7	1.6	1.2–2.0	1.4	1.0–1.8
SSP1-2.6	1.5	1.2–1.8	1.7	1.3–2.2	1.8	1.3–2.4
SSP2-4.5	1.5	1.2–1.8	2.0	1.6–2.5	2.7	2.1–3.5
SSP3-7.0	1.5	1.2–1.7	2.1	1.7–2.6	3.6	2.8–4.6
SSP5-8.5	1.6	1.3–1.9	2.4	1.9–3.0	4.4	3.3–5.7

Table 4.3. *Warming best estimate and very likely range for different SSSP scenarios, for near-term (2021–2040), mid–term (2041–2060) and long-term (2081–2100). Source: IPCC, WGI, AR6 (2021). Reproduced with permission from IPCC (2023b)*

This SSP3 scenario leads to an increase of GHG emissions until the 1970s, which goes to a level of temperature increase that ranges very likely between +2.8°C and +4.6°C and 3.6°C as the best estimate, as compared to the preindustrial era (see Figure 4.35 and Table 4.3).

The SSP3 scenario family is very contrasted with another SSP type named "sustainable development scenario". Sustainable development family scenario included the SSP1-1.9 and SSP1-2.6 scenarios, which are compatible with the aim of the Paris Agreement. It is based on ambitious climatic policies, allowing us to massively and quickly reduce GHG emissions to reach out the carbon neutrality around 2050s to 2060s.

4.3.3. *Articulation between SSP scenarios from AR6 and RCP scenarios from AR5*

In the AR6, SSP scenarios were used, but there must be a continuity among the different IPCC evaluations in terms of scenarios, especially with AR4 and AR5 which served as a basis for the Paris Agreement in terms of climate policies, based on RCP scenarios. RCP scenarios are featured with a level of radiative forcing corresponding to a trajectory of GHG emissions since the beginning of the century until 2100; the hypothesis being of a continuity with a future increase of GHG emissions, with a high radiative forcing of 8.5 $W.m^{-2}$ for instance at the end of the century for the scenario RCP 8.5 used in AR3 and AR4 (see Figure 4.36).

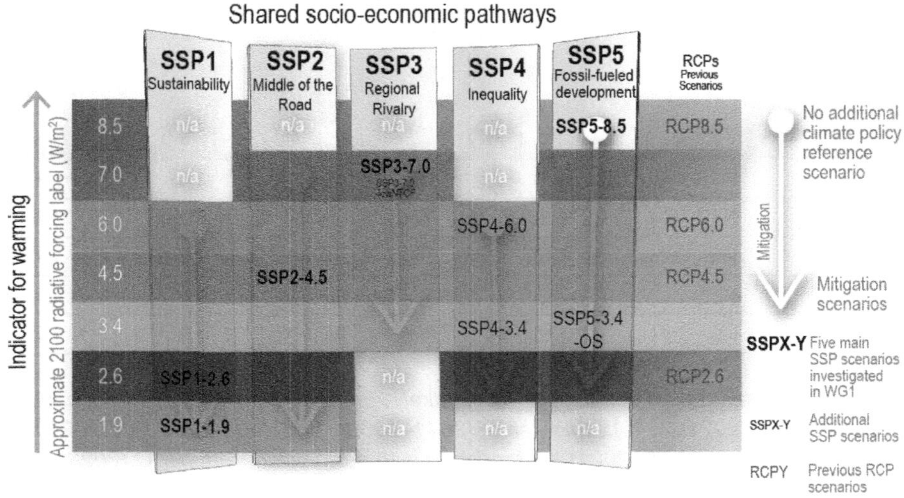

Figure 4.36. *Five families of shared socio-economic pathways (SSP) in AR6 findings and their impact in terms of radiative forcing as a result of mitigation action, and their correspondences with RCP scenarios, with emphasis underlined in a black box on the correspondence between SSP5-8.5 and RCP8.5. Source: IPCC, WGI, AR6 (2021). Reproduced with permission from Chapter 1 of IPCC (2021a)*

In the instance of a drastic reduction of GHG, the energetical disequilibrium used to a radiative forcing of 2.6 W.m^{-2}, as referred to in the RCP2.6 scenario, corresponds to a stabilization of the Earth temperatures below +2°C at the end of the century (see Figure 4.36).

It was therefore necessary to ensure a continuity between the old RCP scenarios and the new SSP scenarios integrating more hypotheses than RCPs, in particular for socio-economical ones.

Therefore, illustrative scenarios of the five families of SSP were chosen in order to match with RCP to generally get the same radiative forcing at the end of the century in order to ensure this continuity and obtain a sort of equivalence table (see Figure 4.37).

In each scenario family, there is a reference scenario/baseline (see Figure 4.38) that is the scenario without public policy. In addition to this baseline, there are other trajectories corresponding to other hypotheses, especially to technical hypotheses; for example, for the SSP5 scenario family, featured with increasing GHG emissions, and a stabilization around 2070s at a level that is three times that of today: in this case, there is a very high radiative forcing of 8.5 W.m^{-2}, leading to a global warming of +4.4°C as the best estimate (with a very likely range of between +3.3°C and +5.7°C) (Table 4.3).

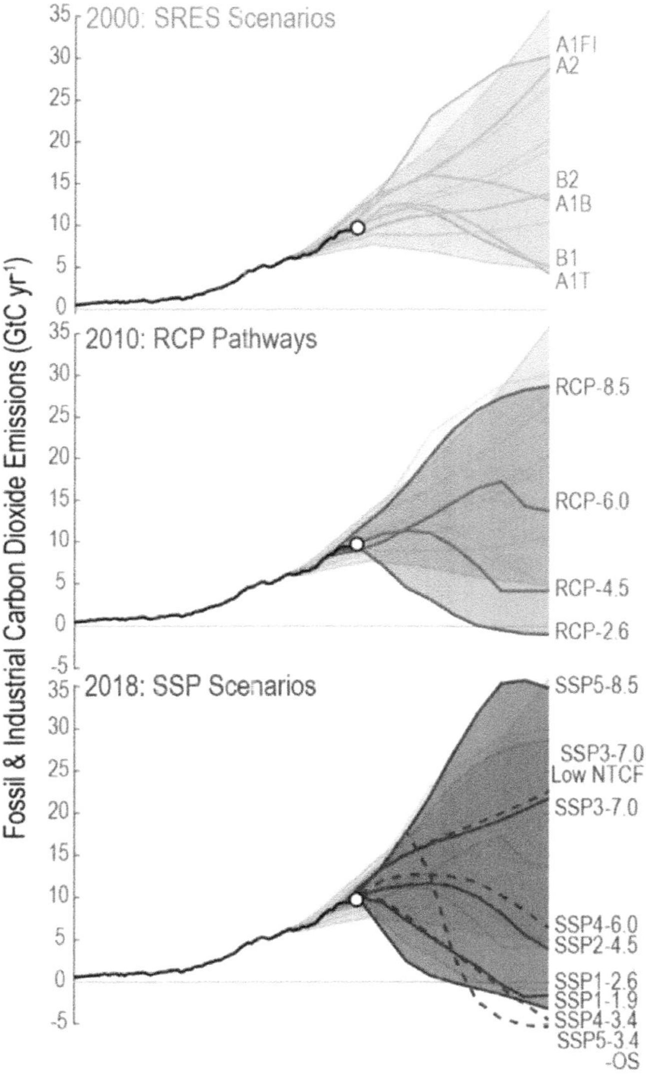

Figure 4.37. *Comparison of the range of fossil fuel and industrial CO_2 emissions from scenarios used in 2000 SRES scenarios, 2010 RCP scenarios and 2018 SSP scenarios. Source: IPCC, WGI, AR6 (2021). Reproduced with permission from Chapter 1 of IPCC (2021a)*

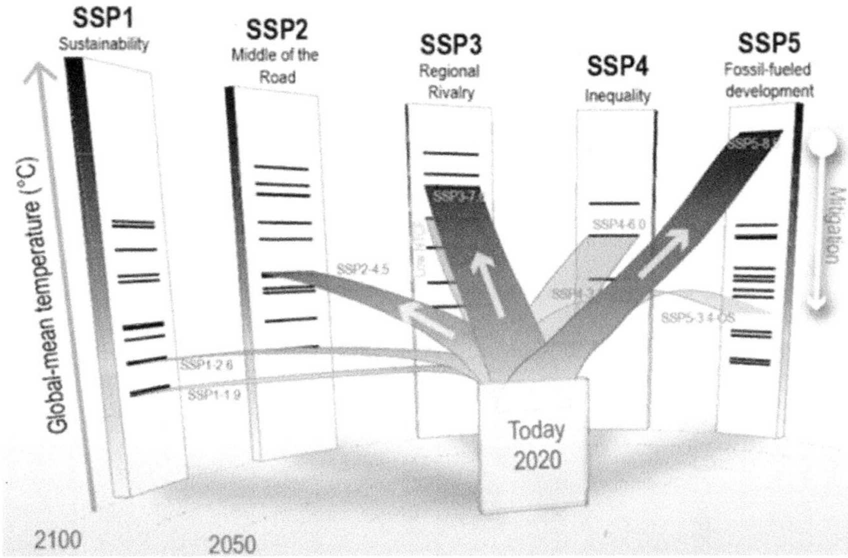

Figure 4.38. *Five families of shared socio-economic pathways in AR6 findings and their impact in terms of the global-mean temperature in degrees Celsius as a result of mitigation action. Source: IPCC, WGI, AR6 (2021). Reproduced with permission from Chapter 1 of IPCC (2021a)*

In the case of this SSP scenario family, with the hypothesis of the use of carbon and storage GHG in the atmosphere because the use of fossil fuel energies is not stopped in this pathway, a reduced radiative forcing is reached out, which may, with extreme hypotheses, maintain the global warming below +2°C. However, there are no probabilities associated with those different scenarios. It is just possible to say that some scenarios have rather unrealistic hypotheses, for example, in the SSP5 family, in the case of a continued increase of GHG emissions, the carbon and storage should be very efficient and deployed at very large scale in order to counterbalance the atmospheric GHG anthropogenic emissions.

4.3.4. *As for the most probable trajectory currently followed*

SSP scenarios were first created at the beginning of the 2010s: in 2025, there is therefore a 15-year hindsight to assess the angle of the GHG emission trajectory that the world is following, and it seems that in 2025, this trajectory has detached from the most emitting scenarios (see Figure 4.39), and the current trajectory is closer to

SSP2-4.5 and SSP4-6.0, corresponding to a stabilization of GHG emissions until 2050.

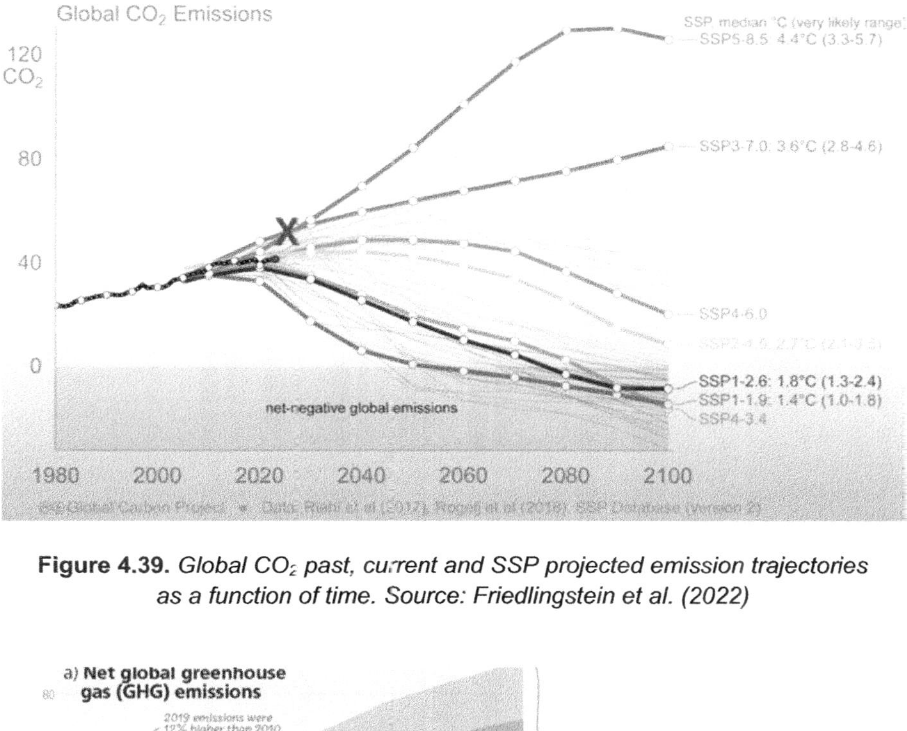

Figure 4.39. *Global CO_2 past, current and SSP projected emission trajectories as a function of time. Source: Friedlingstein et al. (2022)*

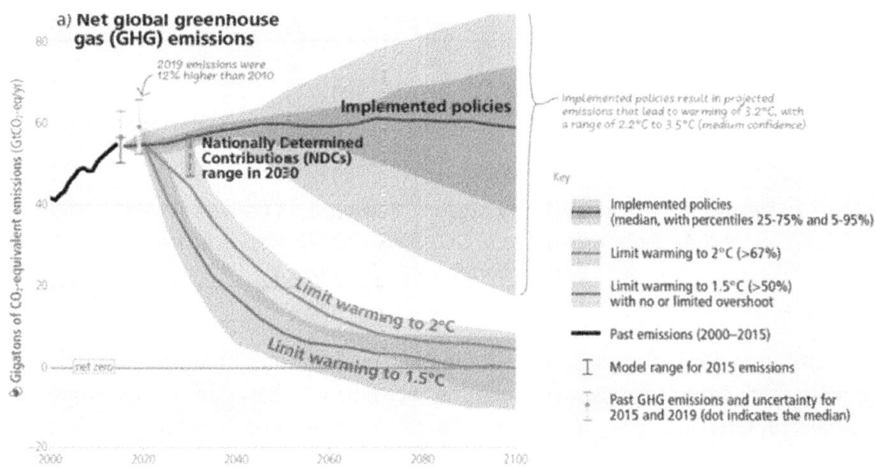

Figure 4.40. *Net global greenhouse gas (GHG) past, current and SSP projected emission trajectories as a function of time. Source: IPCC, SYR, SPM, AR6 (2023). Reproduced with permission from IPCC (2023b)*

The world is currently not following scenarios limiting the global warming to less than +2°C in 2100, as in SSP4-3.4, SSP1-1.9 or SSP1-2.6 for the three scenarios that could reach the Paris Agreement aim: humanity is clearly away from that (see Figure 4.40).

Even if the world seems safe from the most very impactful scenarios, we must consider the risks anyhow, especially in the case of the highest warming, as the Earth is not protected from a potential global warming that could be much stronger than what is expected on average for a given level of emissions. There could be climatic surprises.

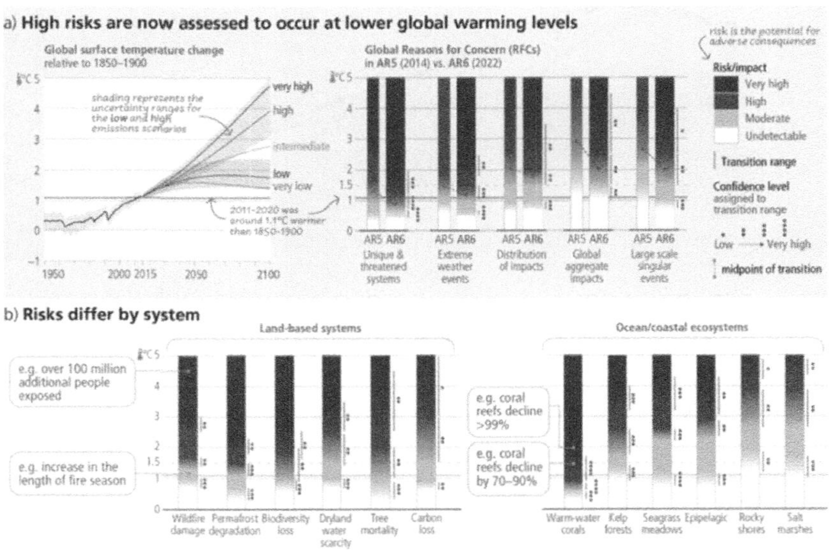

Figure 4.41. *Risk differ by system, comparison between AR5 and AR6 data. Source: IPCC, SYR (2023). Reproduced with permission from IPCC (2022a)*

For instance, through processes inducing an acceleration of temperature increase, especially for natural processes such as permafrost that can degas important quantities of CH_4 and CO_2 not accounted for in projections, inducing a level of anthropogenic emissions not necessarily high to be very impactful, as the quantity of accumulated CH_4 and CO_2 in the atmosphere will be higher because of natural wetlands that may also emit a large amount of GHG (see Figure 4.41). In terms of risk assessment, these scenarios with the highest warming must be considered too in case climate surprises occur.

4.3.5. *Main novelties for the seventh IPCC Assessment Report*

SSP scenarios are evolving for the seventh Assessment Report (AR7), whose cycle is expected to run until 2029.

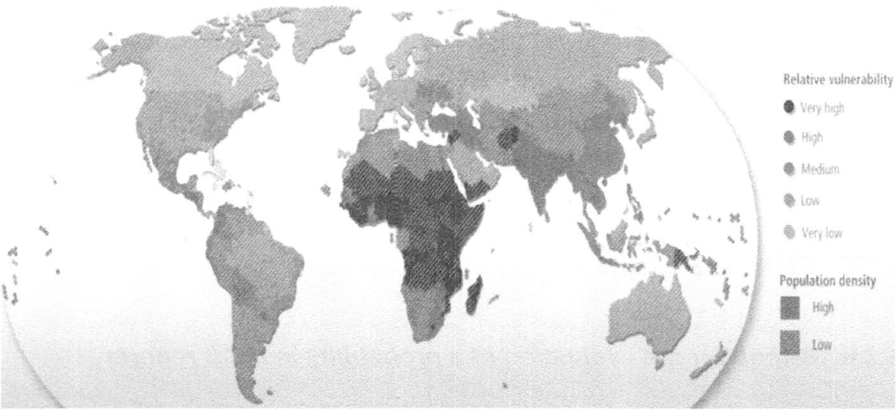

Figure 4.42. *Observed human vulnerability to climate change is a key risk factor and differs globally. Vulnerability also greatly differs within countries. Countries with moderate or low average vulnerability have sub-populations with high vulnerability and vice versa. Source: IPCC, WGII, AR6 (2021). Reproduced with permission from Birkmann et al. (2022)*

For the AR7, there will be a revision of those scenario families, and especially of certain hypotheses because we see that in almost all of the envisioned AR6 scenarios, we let a global system of societal organization persist which maintains inequities between the global North and the global South through economical and societal developments, including in the modeling, while it is now known that the fight against global warming can only be done through a joint decrease of inequities inside countries, as well as at the global scale (see Figure 4.42). Therefore, AR6 SSP scenarios will be revised in AR7, especially with the addition of scenarios including the GDP indicator stagnation/non-growth, despite the fact that all scenarios are based on a GDP growth (more or less important). The aim of the scenarios' revisions in AR7 will therefore be to better integrate all of the possible socio-economical trajectories.

In the IPCC seventh assessment cycle, the main innovations will be:

– Complementary reports to the assessment report including a special report about cities and climate change (2027), a methodological report about the inventory of climate forcings with a short lifetime (2027) and a methodological report on the elimination of CO_2 and its capture.

– A deeper work about inclusiveness, both in terms of authors' representation regarding gender, diversity of nationalities, especially for countries of the global South; as well as, in terms of content, including more "gray" literature – meaning more non classical peer-reviewed papers, regional reports, indigenous knowledge, etc.

– The handling of numerous remaining scientific questions, for instance, the IPCC Working Group 1 addressing the carbon response cycle to climate change, extreme events and tipping points (decline of the thermohaline circulation, etc.).

– An updated version of the IPCC Atlas with more indicators, more regionalized information, solution proposals (including data from IPCC Working Group I).

4.3.6. *Method for the evaluation of a probability in IPCC reports*

IPCC experts evaluate a probability starting from a high level to a very high level of confidence corresponding to the fact that there are numerous pieces of evidence in terms of quantity and quality, and that they are independent, pointing to the same conclusion for the assessed climatic phenomenon.

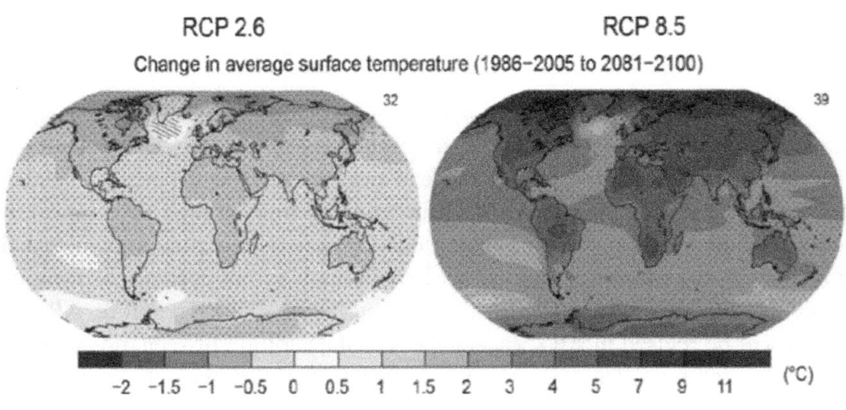

Figure 4.43. *Change in the average surface temperature obtained from respectively 32 averaged models investigating the RCP 2.6 (map on the right) and 39 averaged models for RCP 8.5 scenarios (map on the left). The small numbers on the upper-right side of each map (32 and 39 refer to the number of models averaged across). Source: IPCC, WGI, AR5 (2013). Reproduced with permission from IPCC (2013)*

Results in light of processes previously discussed are shown in Figure 4.43, as modeled for RCP2.6 and RCP 8.5, and there are two big issues: one concerning the direct consequence of warming, with many phenomena that we can explain; the other one concerns indirect consequences, which are therefore sometimes more difficult to anticipate.

The warming affecting the Earth temperature for respectively 32 models and 39 models of the fifth IPCC Assessment Report are represented in Figure 4.44, from left to right, which represent a results' comparison among many models. The color indication represents the annual average warming for two separate RCP scenarios by comparing two time periods for their averaged temperatures: a historical period, 1986–2000, which is relatively recent and posterior to a large increase in greenhouse gas emissions, on the one hand; and a future period, 2081–2100 with a projection of what may happen at the end of the century.

We note that the warming is larger over the continents because, as seen in previous chapters, the heat capacity of the continents is very small; therefore, the warming is immediate over those places and, conversely, the storage of energy is also very small. The oceans are storing the energy; and as they store the energy, the sea surface temperature is warming more slowly. We also note a very differentiated phenomenon between low and high latitudes and we must be careful not to interpret this as a fact that climate change at low latitudes is less important. This is a wrong conclusion that people who not really understand those processes may unfortunately easily jump to: "climate change will be more important in mid and high latitudes than in the intertropical regions". The scientific community knows that climate change will actually have more effect in the inter-tropical regions because what is impeding the heat rising there is the necessity of a vertical stratification in temperature.

This has been observed in the lower latitude regions; the warming is displaced upward by the convection, and this represents an important process: high-intensity dangerous storm processes bring the temperature at higher latitudes (including cyclones around the inter-tropical Convergence Zone) and near the equator. Those processes are there to impede the temperatures becoming higher near the ground but at a large cost in terms of risks. The melting of big glaciers, the fact that the temperature increase is very large is superimposed on initial temperatures which are very low (up to –40°C); therefore, the temperature is increasing in those zones from very low temperatures: the effect on climate is important but may be very different from the ones at mid-latitudes. A phenomenon occurring in all models is the fact that the Gulf stream may retreat, and there is less convection in the ocean associated with the Gulf stream that appears in almost all models.

The small dots in Figure 4.43 refer to places where the models agree in general on what is occurring in terms of sign and general amplitude. We observe that the warming is consensual among the various models. However, the amplitude of the warming is not completely consensual; it is a parameter that is analyzed everywhere, in all models.

We do not see many cooling zones, except in one zone and the associated line of dots which shows that there is no complete consensus on the cooling. There are models with a cooling, others without a cooling, and this is the difficulty with the Gulf stream in the related area; we do not know for sure what will happen, either through observations nor through models. The results in the case of RCP 8.5 data for 39 models show that the average temperature over continents of those models is over 3°C.

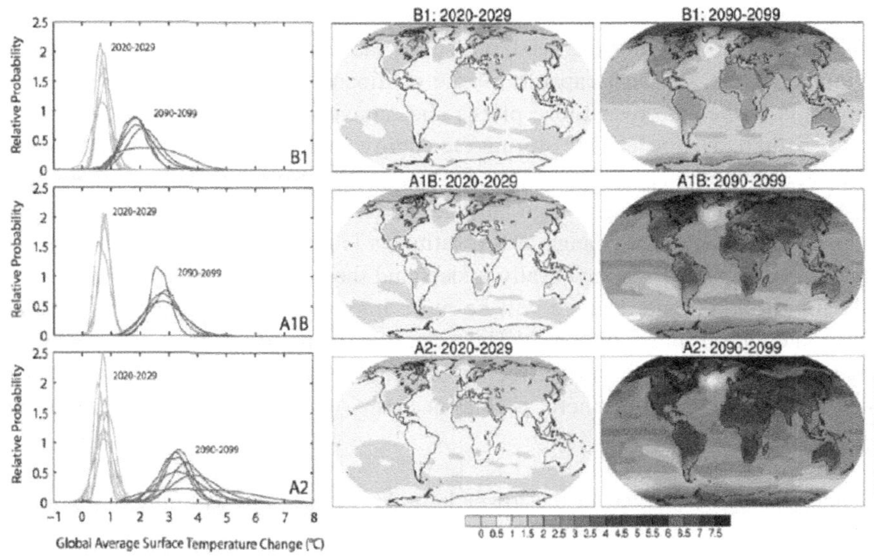

Figure 4.44. *Atmosphere–Ocean General Circulation Model (AOGCM) projections of surface temperatures. Projected surface temperature changes for the early and late 21st century relative to the period 1980–1999. The central and right panels show the AOGCM multi-model average projections for the B1 (top), A1B (middle) and A2 (bottom) SRES scenarios averaged over the decades 2020–2029 (center) and 2090–2099 (right). The left panels show corresponding uncertainties as the relative probabilities of estimated global average warming from several different models. Source: IPCC, WGI, AR4, SPM (2007). Reproduced with permission from IPCC (2007b)*

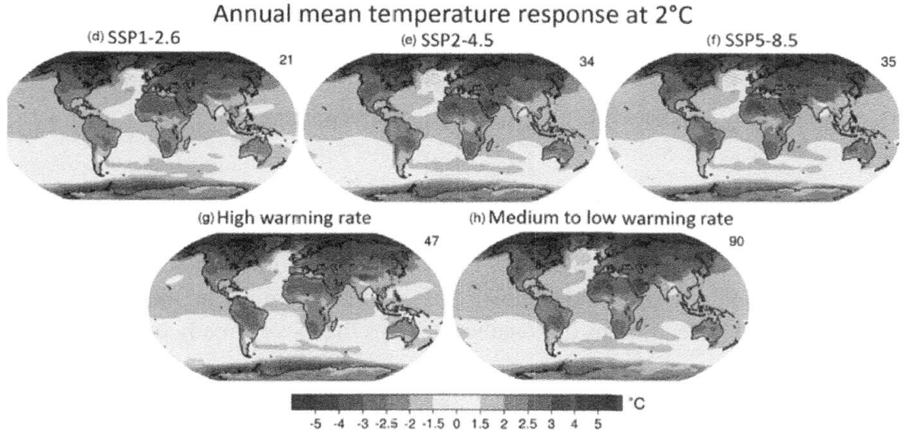

Figure 4.45. *Coupled Model Intercomparison Project Phase 6 (CMIP6) multi-model annual mean temperature change at a 2°C global warming level (GWL) (20-year mean) in three different shared socio-economic pathway (SSP) scenarios relative to 1850–1900. All models reaching the corresponding GWL in the corresponding scenario are averaged. The number of models averaged across is shown at the top-right of the panel. The maps for the other two SSP scenarios SSP1-1.9 (five models only) and SSP3-7.0 (not shown) are consistent. (g) Annual mean temperature change at 2°C in CMIP6 models with high warming rate reaching the GWL in the corresponding scenario before the earliest year of the assessed very likely range. (h) Climate response at 2°C GWL across all SSP1-1.9, SSP2-2.6, SSP2-4.5. SSP3-7.0 and SSP5-8.5 in all other models not shown in (g). The close agreement of (g) and (h) demonstrates that the mean temperature response at 2°C is not sensitive to the rate of warming, and thereby the global mean surface air temperature (GSAT) warming of the respective models in 2081–2100. Uncertainty is represented using the advanced approach: no overlay indicates regions with robust signal, where ≥ 66% of models show change greater than the variability threshold and ≥ 80% of all models agree on the sign of change; diagonal lines indicate regions with no change or no robust signal, where < 66% of models show a change greater than the variability threshold; crossed lines indicate regions with conflicting signal, where ≥ 66% of models show change greater than the variability threshold and < 80% of all models agree on the sign of change. Source: IPCC, WGI, AR6 (2023). Reproduced with permission from Chapter 11 of IPCC (2021a)*

The probability of decadal average warming over 2020–2029 and 2090–2099 is represented on the left side of Figure 4.44. The diagrams on the second column embody results for different scenarios: A2 is a scenario with an average modeled warming of between +3°C and +4°C. It is not the worst of the potential scenarios but is already a severe scenario, with a number of models which go much higher from this average. This is a mixture of different models, from those that do not react very strongly to climate change, to ones reacting more strongly, to others reacting even more strongly. In Figure 4.44, we also observe at the end of the century a picture which is slightly redder with a purple color corresponding to a +6°C or a +7°C warming in these areas.

Over 2020–2029, we obtain more or less the same results for the three scenarios.

This consensual result means that what will happen in the next decade is already written, no matter what is done now. Indeed, the warming in the next decade depends on the concentrations of greenhouse gases that have already been emitted.

Even if the world stays at +2°C as the global annual mean warming in 2100, for three different SSP matching this threshold, local temperatures over land can reach up to +4°C to +6°C, which is preoccupying (see Figure 4.45).

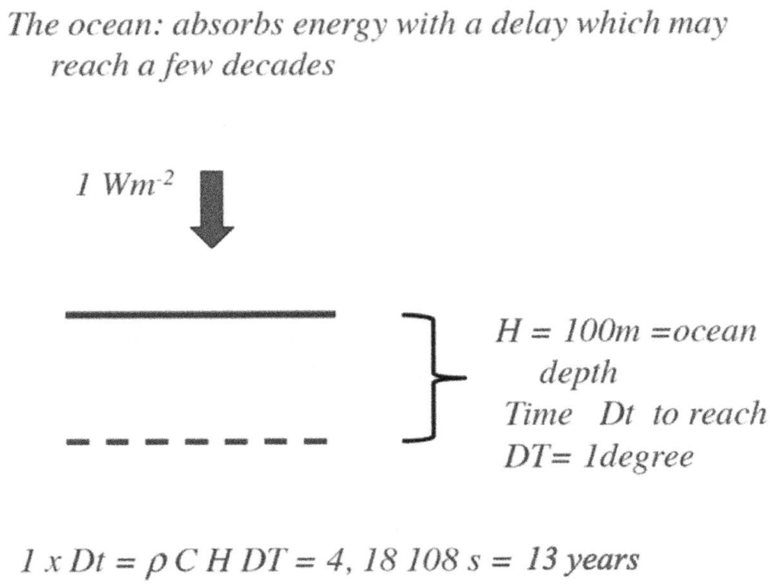

Figure 4.46. *Scheme of the ocean behavior regarding energy absorption*

There is a common political discourse stating that in order to protect the populations, a strong effort in each French city concerning the reduction of GHG emissions must be undertaken: for them, there is "an urgent need to reduce the emissions". Indeed, there is actually "an urgent need to reduce emissions", but the nature of this emergency need is usually not completely understood. The purpose of this urgent need is to protect the generation of people who are born now for the time at which they will be 30, and therefore protect the next generation. However, for the present and future of the current generation, who also deserve to be protected, a different approach to mitigation, named adaptation to climate change, must be applied. Politicians usually do not speak as much of adaptation as it is sometimes misunderstood as being contradictory with mitigation efforts.

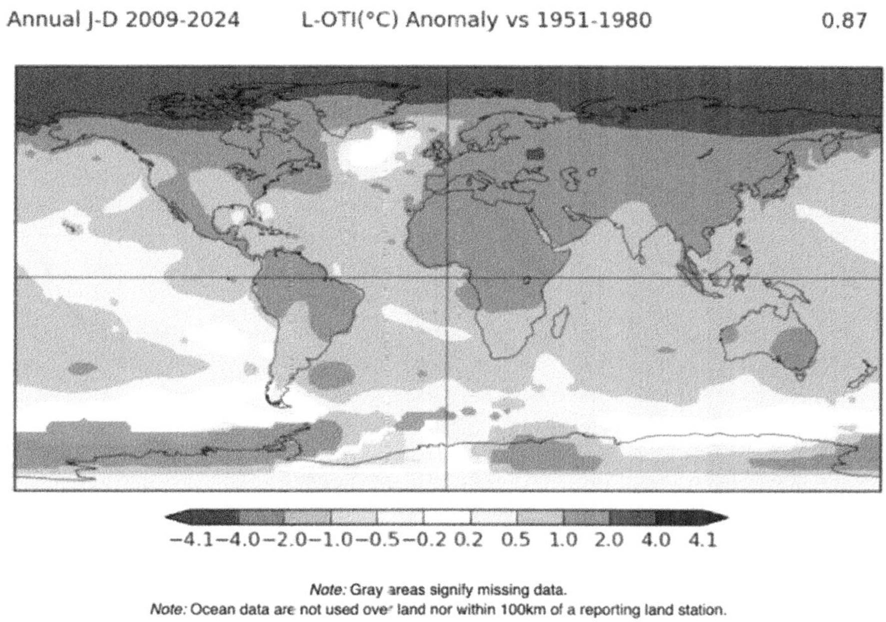

Figure 4.47. *Map of land and sea observed surface temperature changes annually in degrees Celsius over 2009–2024 as compared to 1951–1980 expressed in anomaly in degrees Celsius. Source: NASA (n.d.). See: https://data.giss.nasa.gov/gistemp/maps/*

We add a small digression to comment on the delay between present emissions and the real-world oceanic absorption timeline for emitted carbon (see Figure 4.46). Many aspects are affected by this process. This small computation shows the difficulties about the flux's estimations at the ocean–atmosphere interface.

The amount of energy needed to change the temperature of a slab of ocean of 100 m by +1°C is very small. If there is 1 m^2 of difference at the top of the ocean, and we want to increase the temperature of the slab of ocean of 100 m schemed in Figure 4.46, 13 years are needed: this time, to get out of equilibrium is rather quick, but this refers to energy stored in the ocean.

Observations of surface temperature changes (see Figure 4.47) serve as a precious comparison tool for results coming from models such as the IPSL, and to assess if the models realistically simulate temperature behavior. This comparison is easier to do by staying with scenarios which are not extreme. What we observed in the real world in 2009–2025, when we compare predictions and observations, is that the match between the two is excellent. Therefore, we know the warming is happening the way models have predicted.

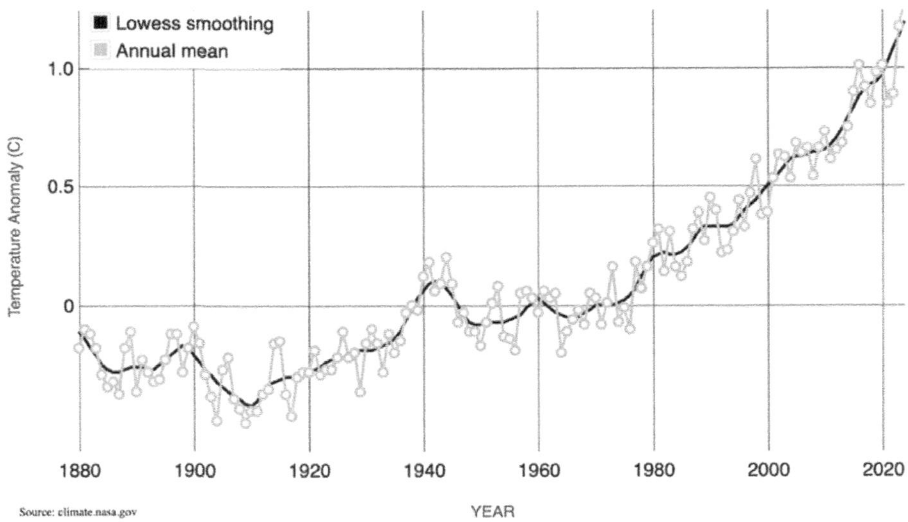

Figure 4.48. *Global land–ocean temperature anomaly in degrees Celsius as a function of time. Source: NASA's Goddard Institute for Space Studies (GISS). Credit: NASA (n.dc)*

As mentioned previously, the anthropogenic GHG emissions really started around 1950, before there were also GHG emissions from human activities, but the situation is complex to understand in terms of data availability and reliability as measures were not completely reliable. Temperature records have been available since 1880 (see Figure 4.48). The 1950s showed the beginning of surface ocean measures with the last manifestations of the little ice age of the 19th century, and the beginning of some warming which is difficult to diagnose from the rest. Therefore, this is extremely difficult to precisely describe. There is a pause between 1960 and 1970, and this pause could be interpreted by the following potential explanation: at the beginning, there are both GHG and aerosol emissions, and they tend to compensate a little between a warming and a cooling effect. But at a point, GHG becomes stronger in their effect because they accumulate in the atmosphere contrary to the aerosols which do not accumulate in the atmosphere. Therefore, this phenomenon takes some time to go over the aerosol effect. This is a potential explanation; there is some time lag before an equilibrium is reached with the aerosol increase, and then the effect of the combination between GHG and aerosols really starts. This is superimposed as shown by the black line in Figure 4.48 which is a lowess smoothing, more or less showing some continuous increase. This is superimposed on yearly changes which show a larger variability, called natural variability.

Although other types of natural variability exist at time scales longer than the yearly one, when studying the way this natural variability modulates the signal, some results may appear really disturbing. The hiatus that we have already mentioned is also visible in Figure 4.48: before 2000, there was a strong El Niño that began at a moment where there was apparently no real warming. However, the global picture from the time we started to emit and measure anthropogenic GHG to now shows that, after the El Niño, the temperatures began to grow again.

We are already witnessing climate change; we reached for the first time in 2024 a +1.5°C warming in terms of the global mean surface temperature (Bevacqua et al. 2025). What is really feared is what will happen between now and 2050. In terms of impact when talking about climate change, decades ago, there was a first stage during which scientists were able to say from models that there should be some warming in the future; this had some effect at that time, but generally only in educated communities. Now that we observe a warming, the effect is much larger on society. What matters is the future, the impacts in ten or 20 years from now, more than the warming that is already happening now. That is the reason why figures should be looked at more in detail by politicians. Generally, it is stated that there is a range of +1.5°C to +4.5°C uncertainties in our laboratories; therefore, it is difficult to give a right "definitive number", and tails risks are often completely disregarded by politicians.

There are direct consequences to warming which are probably the best way to assert what is occurring. For example, one of the most spectacular is probably the ice melting in the Arctic, as shown in Figure 4.49. All of the information concerning the sea ice and glaciers is provided on the NSIDC website (see: https://nsidc.org/).

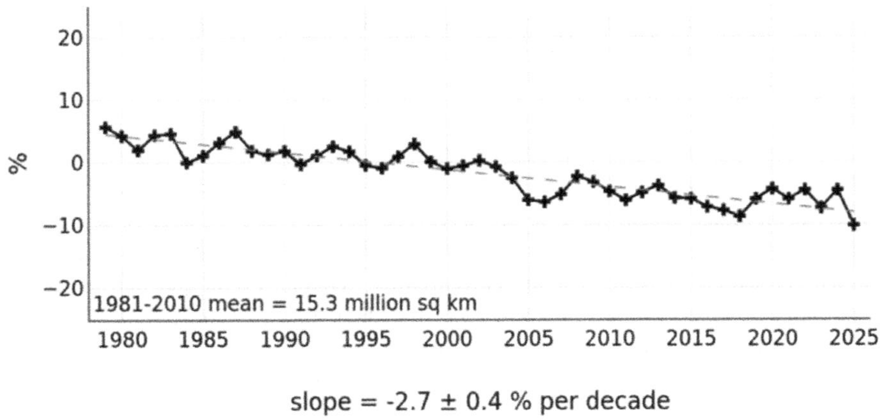

Figure 4.49. *Evolution of the Arctic ice–Northern Hemisphere extent anomalies February 1979–2025. Source: National Snow and Ice Data Center (NSIDC) (2025)*

The evolution of the Arctic ice at the end of the summer for the Northern hemisphere extent anomalies as a function of time (see Figure 4.49) shows an average value over 1981–2010 of 15.3 million sq km for February, and we are now at a slope rate of about 2.7% per decade of loss above this reference level. Further, the perennial ice has been diminishing by about 50%. These data are alarming. The sea level ice diminution happens very quickly. It is expected to no longer have sea ice in summer over the Arctic before the end of the century. This is a phenomenon consensually occurring in all models. This is an extremely impressive result because humankind has not faced this type of situation in hundreds of thousands of years.

Similarly, the global sea level has been increasing, as shown in Figures 4.50 and 4.51. This is actually a metric that is very difficult to measure, and it has been one of the huge successes in altimetry (see Figure 4.50(b)).

This figure shows that global sea level measures from different sources have converging increasing trends. What is more difficult to discern in this diagram is the chosen point for computing the tangent to the curves. In Figure 4.50(c), we see the sea level represented in yellow, and the thermosteric component in green referring to the way the heat stored in the ocean is changing the volume taken by the ocean (this represents the increasing volume of the ocean due to temperature).

The difference between the two is the glaciers' contribution to sea level. There are many ways to check that with satellite altimeters (see Figure 4.50(d)).

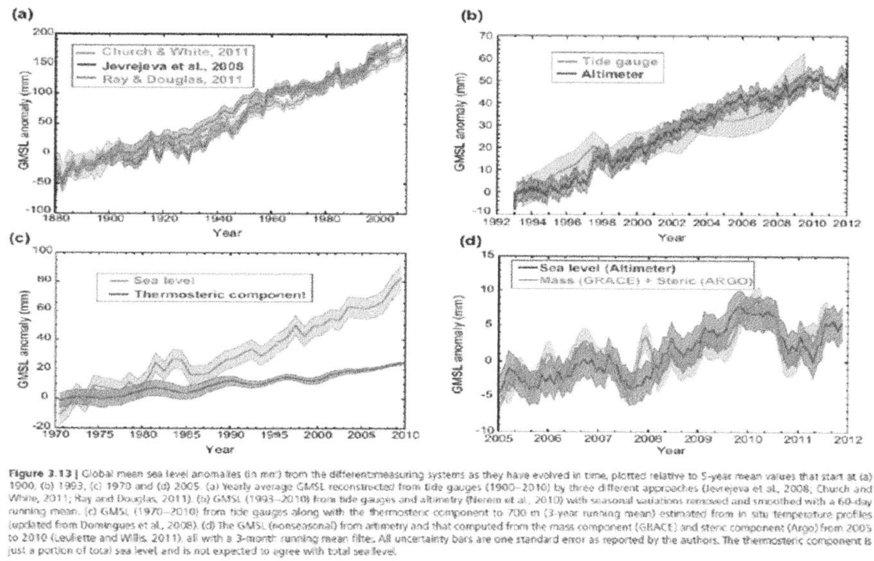

Figure 4.50. *Global mean sea level anomalies (in mm) from the different measuring systems as they have evolved in time, plotted relative to five-year mean values that start at (a) 1900; (b) 1993; (c) 1970; and (d) 2005. Source: IPCC WG1, AR5 (2013). Reproduced with permission from Rhein et al. (2013)*

More or less, we know that this is really a phenomenon due to glacier melting. We will not get into the details regarding this consideration, but this is also a direct consequence of global warming.

Another way to study this topic is to note that the sea level rise is very unequal depending on the Earth location (see Figure 4.52).

Therefore, the real effect may be much larger than the average effect.

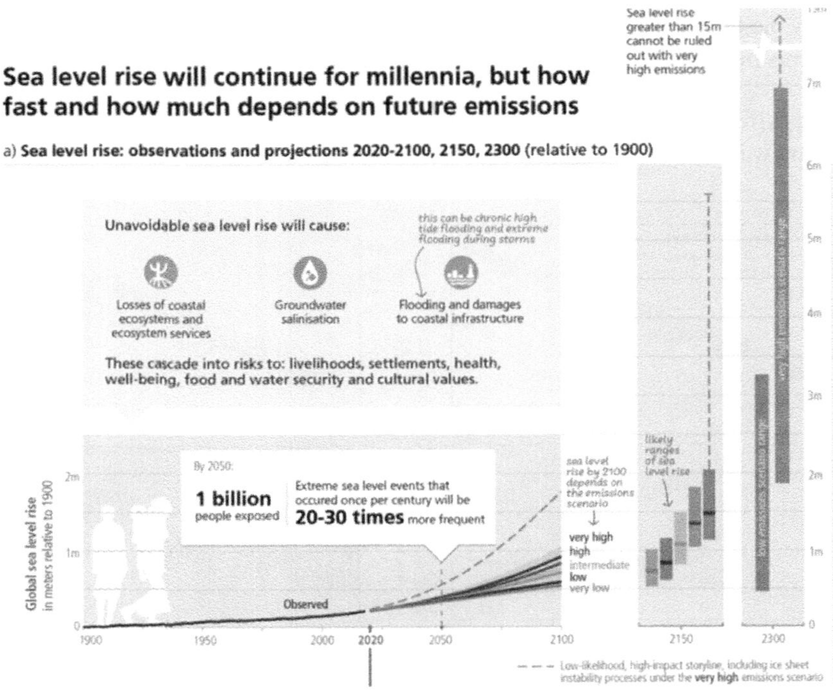

Figure 4.51. *Sea level rise: observations and projections 2020–2100, 2150, 2300, in meters relative to 1900. The historical changes (black) are observed by tide gauges before 1992 and altimeters afterward. The future changes to 2100 and for 2150 (colored lines and shading) are assessed consistently with observational constraints based on the emulation of CMIP, ice sheet and glacier models, and median values and likely ranges are shown for the considered scenarios. Relative to 1995–2014, the likely global mean sea level rise by 2050 is between 0.15 and 0.23 m in the very low GHG emission scenario and 0.20 and 0.29 m in the very high GHG emission scenario; by 2100, between 0.28 and 0.55 m under a very low GHG emission scenario and 0.63 and 1.01 m under the very high GHG emission scenario; and by 2150, between 0.37 and 0.86 m under the very low GHG emission scenario and 0.98 and 1.88 m under the very high GHG emission scenario (medium confidence). Changes relative to 1900 are calculated by adding 0.158 m (observed global mean sea level rise from 1900 to 1995–2014) to simulated changes relative to 1995–2014. The future changes to 2300 (bars) are based on the literature assessment, representing the 17th–83rd percentile range for the very low and low GHG emission scenarios (0.3–3.1 m) and the highest GHG emission scenarios (1.7–6.8 m). Red dashed lines: low-likelihood, high-impact, including ice sheet instability processes. Source: Synthesis Report, IPCC, AR6 (2023). Global and regional sea level projections are also hosted at https://sealevel.nasa.gov/ipcc-ar6-sealevel-projection-tool. Reproduced with permission from IPCC (2023b)*

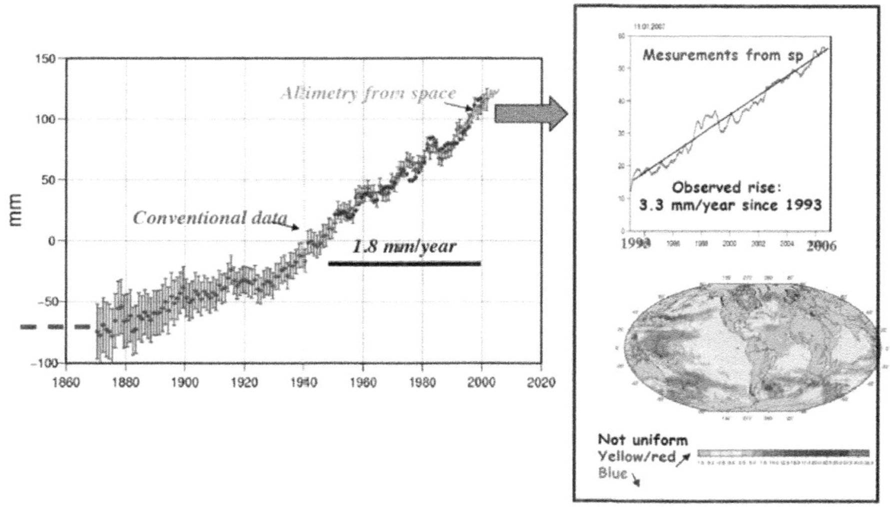

Figure 4.52. *Sea level variations throughout the 20th century at the end of the summer. Source: adapted from IPCC, WGI, AR4 (2007). Reproduced with permission from Bindoff et al. (2007)*

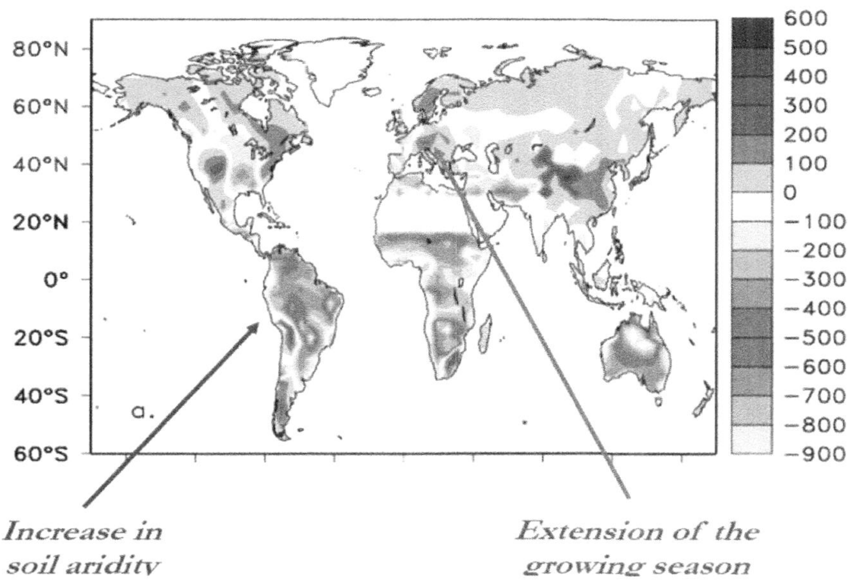

Figure 4.53. *Net climate production: climate dependency. Source: Berthelot et al. (2002)*

The direct warming impact on the vegetation has been studied since a simplified first model with a coupling of carbon cycle and vegetation model (which the scientific experts in that field generally find very relevant) done by a group of IPSL scientists in 2002 (see Figure 4.53).

The impact of the warming on the net primary production of vegetation is negative in the region colored in pink/orange on the map, between –30 South and +30 North (direct effect of temperature). Superposed on that, the effects of precipitation that will also be discussed are represented.

It is also important to make a difference between what we call climate warming and climate change in general. In climate change, there are indeed two parts: one part which is really directly linked with the warming; and another part, which is much more indirectly linked to the warming that we can call "climate complexification", which is not the usual word. The word usually used is actually deregulation but it means that climate change manifestations are not completely predictable in the details. That is linked with the role of the water cycle effects among others.

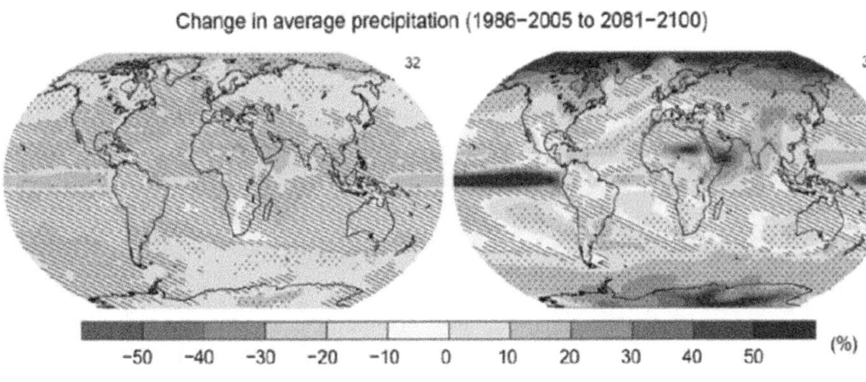

Figure 4.54. *Changes in global average precipitations between 1986–2005 to 2081–2100. The number of models averaged across is shown at the top-right of each panel. Source: IPCC, WGI, AR5 (2013). Reproduced with permission from IPCC (2013)*

Changes in average precipitations for a warming that is not too large (associated with a +2°C) and for 32 models averaged across are represented on the left side of Figure 4.54; and on the right side, the diagram shows the changes in average precipitations for a warming associated with between +3°C and +5°C warming in the future, for 39 models. We discuss here those AR5 results as they serve as a basis for the Paris Agreement in 2015 and are still valid, especially for pedagogical purpose, with updates available in AR6.

Two types of information are displayed in Figure 4.54: first, the colors reflect an average amplitude. Second, the bars of points represent the level of agreement between models.

Let us first consider the amplitude: one issue we would like to point the attention toward is that here, averages are considered. For each model, we basically take an average of 20 years between 2081 and 2100, to which we subtract other 20 years which are roughly one century before, and we average that over 39 models. It is a huge averaging, which means that all of the small changes that are not nicely located will disappear from what is considered. It is not a problem of having small details which represent some ruptures of the disequilibrium: if we have some small kinetic processes, we will not be able to see them with this type of computation. What we get here is, however, a nice consistent image, with an amplitude that we may get in the future (such as +50% more water in some regions, which is a huge result). There are also regions of droughts with –20%, –30% precipitation, which is a huge change when considering that this represents an average over long periods of time and over several models. The key take-away message is that the change in precipitations due to climate change will be very large.

Let us now consider the small dashes representing the level of agreement between models. When many dashes are present, it means that the agreement between the models is not very large, whereas points represent a large agreement among models. We note that a large agreement among models occurs more easily when there is a larger change, but it is not the main configuration: it means that for precipitations, we consider a process that could be large in terms of change, but that may differ from one place to another.

We intend to deliver some simple explanation of this phenomenon in Figure 4.55.

The explanation of the phenomenon described in Figure 4.55 is that in a given area of the world, manifestations of precipitations may have different patterns. When considering an average value over an area, the result is a bit more difficult to interpret. Figure 4.55 shows the result of a work done by Hervé Le Treut with Jean-Christophe Cassou (Toulouse) at CNRS, one of the best experts in France about climate variability. This diagram corresponds to the winter climate. We observe what are called "climate regimes" which are differentiated statistically: climatologists are trying to find structures that repeat themselves/patterns and to describe them by a few key structures. By proceeding this way, we get statistically to four key structures. These patterns correspond to the dumping of a system which is on the contrary increased otherwise.

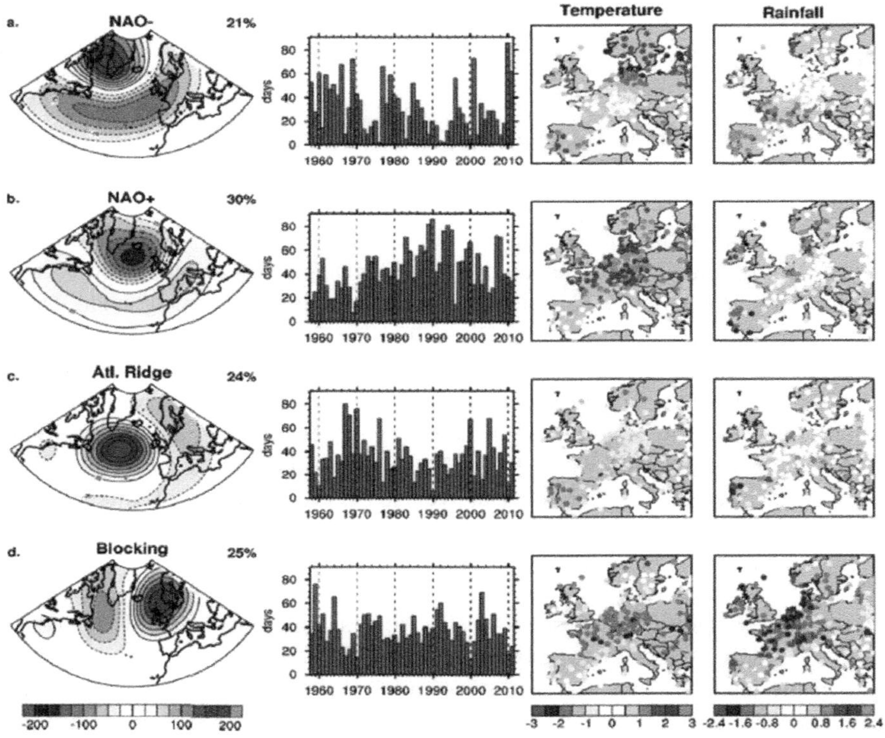

Figure 4.55. *Climate natural variability affects future change (NAO stands for North Atlantic Ocean). Source: Cassou (2010)*

Let us begin with the second point, where we increase the original system. The original system is where there is an anticyclone named the Acores. The depression above is the Icelandic depression. What may happen is that the anticyclone is reinforced: the diagram shows the pressure, with the high-pressure area which will become an even higher-pressure area. A low-pressure area is also visible, which will be transformed with even lower pressures.

There is also a reverse modification: the anticyclone is a bit less marked, and the depression (cyclone) would also be dampened.

In some places, the anticyclone is more over the oceans and there are other situations where the anticyclone is more over the continents. Figure 4.55 shows a representation of very cold areas in winter. On the contrary, during the summer, the same kind of blocking will be associated with dense air coming from the South and moments with very strong heat waves. If we consider the percentage of time the climate system spends, we see respectively from up to down approximately 21%, 30%, 24% and 25%, meaning mostly about a quarter everywhere. What this means is that in fact climate change may mean a modification of one of those four kinds of processes. Therefore, in each region, we may have climate change occurring in one of these situations, with different results.

In a situation with moist conditions, climate change will mean even moister conditions. But in a situation like represented in case (d) in Figure 4.55, with a blocking over the continents, climate change may mean a blocking which is moist. Generally, climate change tends to always accentuate the situations. This would be more severe, which means less precipitations as well as a cooler climate.

Thus, there are different situations and different possible answers. Some patterns may also increase, for example, frequency may increase with time in the future. During the summers, it is clear that there will be situations with heatwaves, and important droughts may also increase in the future.

Therefore, the kind of change happening locally due to precipitations is much more complex than the change in temperatures. It is a multi-tenancy of potential answers, and we are not able to say directly which effect will prevail in the future. We may only describe what will happen if we have the same kind of patterns, keeping the same percentage of what is happening in the future (we know that some patterns will be increasingly intense). However, the way the system exactly changes is more complex and less understood.

As for associated risks, of course it is necessary to adapt not only to one risk but also to a pattern of different risks. This complexity is at the source of the problem named "maladaptation" (for instance, colleagues in North Africa find that on average precipitations will decrease around Southern Europe/Northern Africa). North African countries legitimately want to protect themselves against a lack of water, but the one thing that is often omitted in related adaptation plans is the risk of very intense flooding events: even if it may seem counter-intuitive, as an answer to climate change, this area may have important floods additionally to important droughts.

When it comes to water, the changes indeed may be dual: there may be several phenomena occurring in the same regions, just because the patterns of climate change manifestations are not the same in that case. Matters related to water are more under deregulation that can be "bracketed" a little. Work has to be done to communicate more about lists of different risks addressed in terms of protection, including "tail" risks. The scientific community has information to deliver, but it does not necessarily come with one nice simple unique value/figure, in a political world seeking certainty statements.

The assessment of the changes in Coupled Model Intercomparison Project Phase 6 (CMIP6) multi-model mean precipitation change at a +2°C global warming level in three different SSP scenarios relative to 1850–1900 all show that the trend is an increase over land, with disparities and locations affected by up to 40–50% increase (see Figure 4.56).

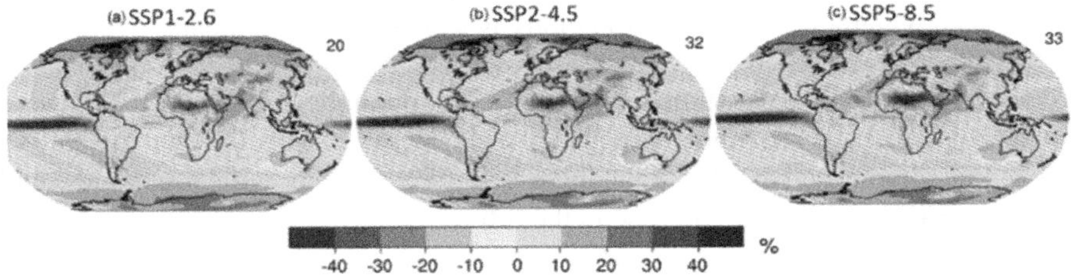

Figure 4.56. *Changes (a–c) Coupled Model Intercomparison Project Phase 6 (CMIP6) multi-model mean precipitation change at a 2°C global warming level (GWL) (20-year mean) in three different shared socio-economic pathway (SSP) scenarios relative to 1850–1900. All models reaching the corresponding GWL in the corresponding scenario are averaged. The number of models averaged across is shown at the top-right of each panel. The maps for the other two SSP scenarios SSP1-1.9 (five models only) and SSP3-7.0 (not shown) are consistent. Uncertainty is represented using the advanced approach: no overlay indicates regions with robust signal, where ≥ 66% of models show change greater than the variability threshold and ≥ 80% of all models agree on the sign of change; diagonal lines indicate regions with no change or no robust signal, where < 66% of models show a change greater than the variability threshold; crossed lines indicate regions with conflicting signal, where ≥ 66% of models show change greater than the variability threshold and < 80% of all models agree on the sign of change. Source: IPCC, WCI, AR6 (2021). Reproduced with permission from Chapter 11 of IPCC (2021a)*

NOTES ON FIGURE 4.58.– Values correspond to the average of significant explained variance fractions based on HadCRUT, GISTEMP, BerkeleyEarth and CRU-TS (for SAT) and GPCC and CRU-TS (for precipitation). Significance is tested based on

F-statistics at the 95% level confidence, and a slash indicates that the value is not significant in more than half of the available datasets. The color scale corresponds to the sign and values of the explained variance, as shown at the bottom. DJF: December–January–February. MAM: March–April–May. JJA: June–July–August. SON: September–October–November. In (b), Northern Annular Mode (NAM) and El Niño–Southern Oscillation (ENSO) teleconnections are evaluated for 1959–2019. Southern Annular Mode (SAM) for 1979–2019, Indian Ocean Basin (IOB), Indian Ocean Dipole (IOD), Atlantic Zonal Mode (AZM) and Atlantic Meridional Mode (AMM) for 1958–2019, and Pacific Decadal Variability (PDV) and Atlantic Multi-decadal Variability (AMV) for 1900–2019. All data are linearly detrended prior to computation. Source: IPCC, TS, WGI, AR6 (2021).

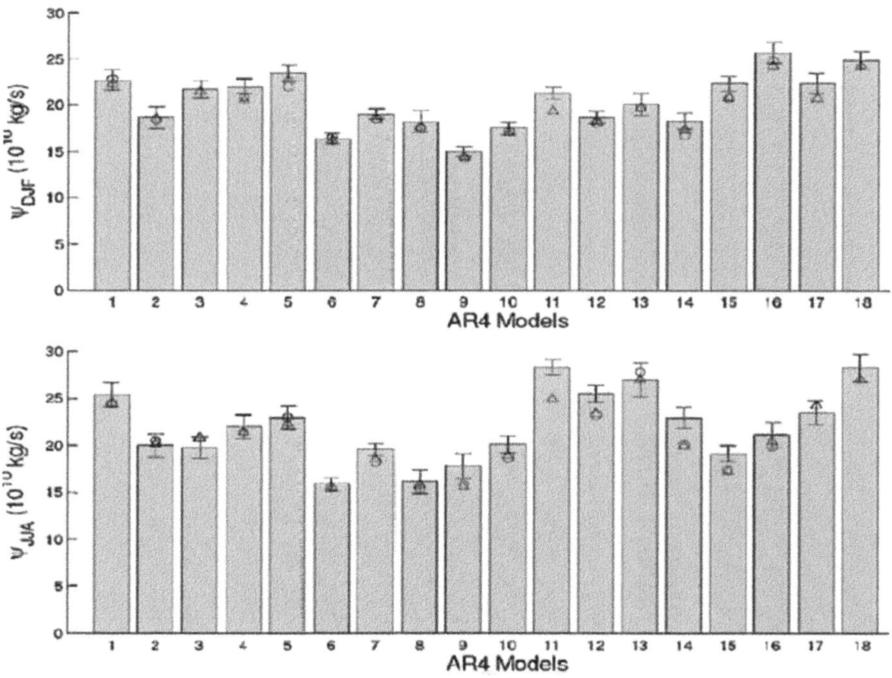

Figure 4.57. *Changes in the Hadley cell circulation.*
Source: Gastineau et al. (2008)

Changes on the Hadley cell circulation, and the amplitude of the Hadley cells in the Poles, upon which all options depend, are represented in Figure 4.57.

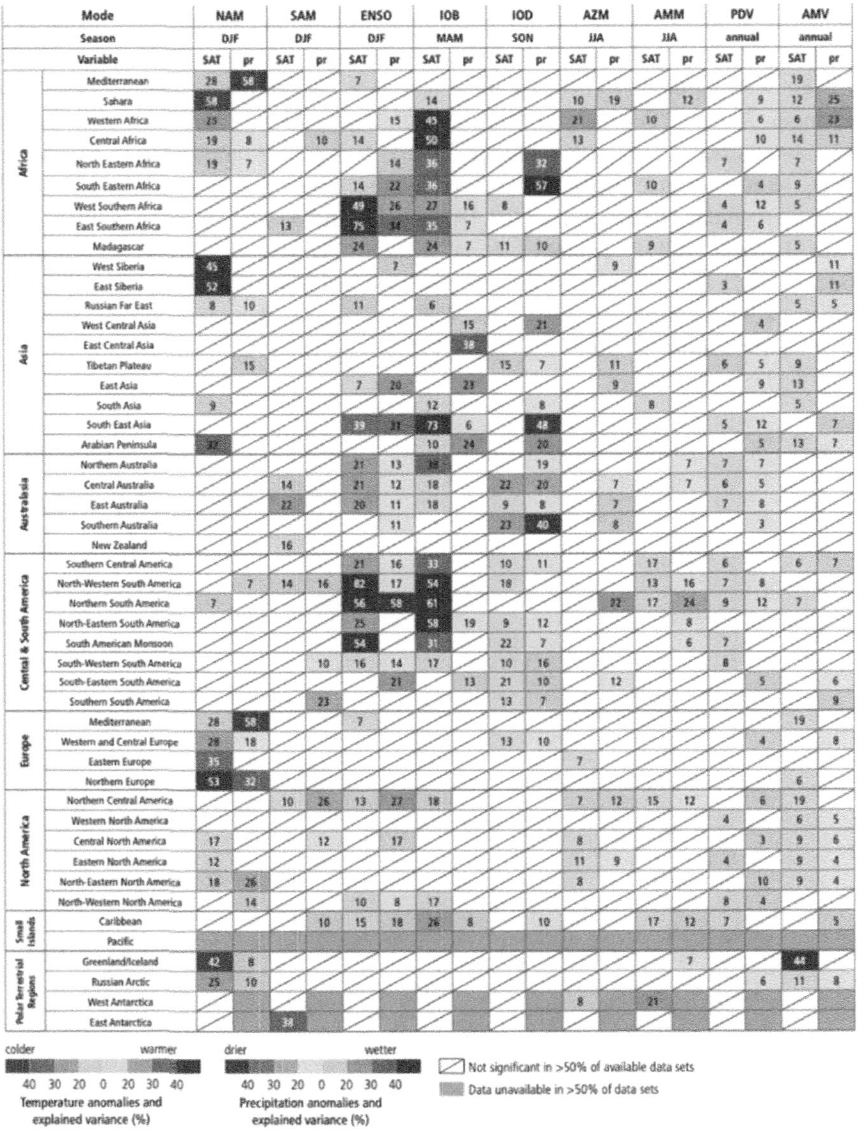

Figure 4.58. *Fraction of surface air temperature (SAT) and precipitation (pr) variance explained at the interannual time scale by each mode of variability MoV for each AR6 climate region (numbers in each cell; in percent)*

The technical summary of the Sixth IPCC Assessment Report delivers a detailed assessment of climate variability as represented in Figure 4.58 and delivers the following synthesis about it:

> Since the late 19th century, major modes of variability (MoVs) exhibited fluctuations in frequency and magnitude at multi-decadal time scales, but no sustained trends outside the range of internal variability [...]. An exception is the Southern Annular Mode (SAM), which has become systematically more positive (high confidence) and is projected to be more positive in all seasons, except for December–January–February (DJF), in high CO_2 emissions scenarios (high confidence). The influence of stratospheric ozone forcing on the SAM trend has been reduced since the early 2000s compared to earlier decades, contributing to the weakening of its positive trend as observed over 2000–2019 (medium confidence).
>
> In the near term, projected changes in most of the MoVs and related teleconnections will likely be dominated by internal variability. In the long term, it is very likely that the precipitation variance related to El Niño–Southern Oscillation will increase.

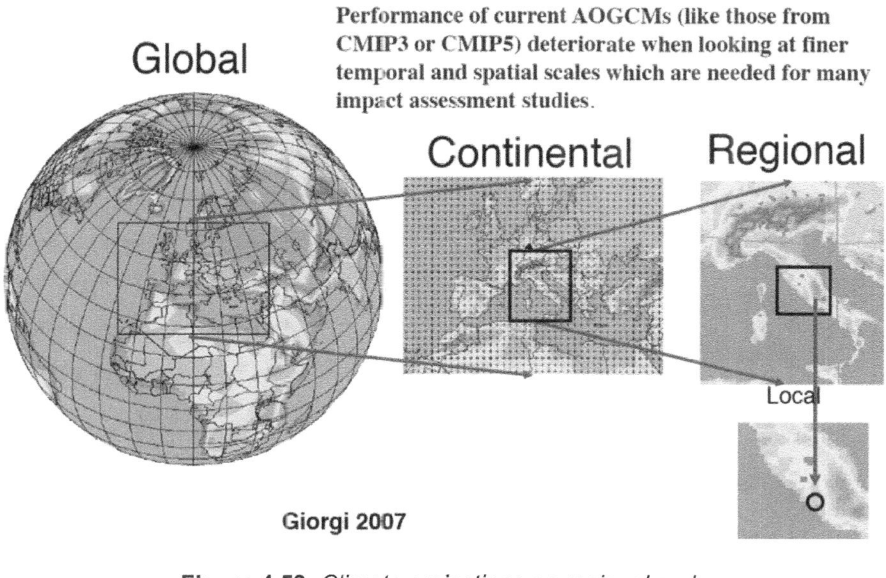

Figure 4.59. *Climate projections on regional and local scales. Source: Giorgi and Lionello (2008)*

There is also a local view of the climate getting from top to bottom (top-down approach). From a global model, some limits are taken, and another model is being entrained by what happens at the scale of the atmosphere with a better orography; a small model is created this way, from top to bottom. Giorgi and Lionello (2008) showed this in the results in Figure 4.59. In order to study local consequences of what has been discussed, a local approach may bring information about small-scale phenomena such as Mistral, requiring further steps than those of global models. However, there are things that can be derived from global models, including local features concerning orography. But if there is no consistency between the wind's dynamics of a small model and the dynamics of the model surroundings, it leads to problematic not completely consistent situations.

The problem of deregulation is opening a wide area for issues connecting the global scale to the local scale, a very important interface. Again, in this chapter and also in the rest of the book, the ocean response is very briefly discussed, but is, however, very important. The authors recommend reading the work done by Laurent Bopp on this topic.

In this chapter, we discussed an image of the global climate. In Chapter 5, we will emphasize that a local view is also important and needed as schematized on the right part in Figure 4.59. Chapter 4 was indeed mainly an introduction for the next chapter related to the interface between science and decision-making.

5

From Scientific Diagnostic to Decision-Making

This chapter is focused on what is scientifically known from climate research and might be useful for decision-makers. We will first discuss time scales and the way they may affect the decision-making process related to climate policy. We will then discuss the relationship between local and global geographical scales.

To the important and complex question: "how much should we reduce greenhouse gases to keep the Earth sustainably livable and with what agenda?", the simplest scientific answer is that humans should stop all emissions and reach zero emissions (no fuel, gas or coal) as soon as possible. But socio-economic externalities have to be considered in the agenda for action.

The question of keeping a carbon budget under the goals from the Paris Agreement is also extremely important. It is important to note that the related estimates defining the constraints to design future trajectories for GHG reductions derive from computations and assumptions which embody uncertainties. In this chapter, we will discuss general related points, taken from IPCC reports.

Decision-making processes related to climate policy do not necessarily use much science. The main issue corresponding to this is related to how we deal with uncertainties. Climate science needs to be simplified both for computing models and above all to be used by non-experts, including political decision-makers who are usually not climate scientists. These simplifications generally do not involve the presence of climate scientists, even though it would be valuable. Therefore, the rationale behind simplifications are usually forgotten. At COP21 for instance, a

---

For a color version of all of the figures in this chapter, see www.iste.co.uk/letreut/energetics.zip.

Energetics of the Climate System,
by Hervé LE TREUT and Mounia MOSTEFAOUI. © ISTE Ltd 2025.

high-level ministerial representative forbade his team from using the word uncertainty with the media, a word that is, however, extremely important in physics to specify a range of possibilities associated with a result. In the decision-making and communication communities, it is, however, not an appreciated concept. Dealing with this tension between science precision and political and communication aims and habits is a delicate balance.

This chapter comprises two parts: first, the role of uncertainties in assessing global warming, accounting for natural variability and estimating climate impacts. The purpose of this first section is to name and to distinguish the different types of uncertainties, and how it is possible to deal with them. The work of the IPCC has notably consisted of delivering common metrics, assessing and grouping them. The aim of the second section of this last chapter is to deliver a downscaled view, from global to local issues, a scale shift that is at the heart of the link between science and decision-making.

5.1. The role of uncertainties on assessing global warming, accounting for natural variability and estimating climate impacts

A main IPCC result taken into account in the Paris Agreement is that we have a limited time window to stay below dangerous levels of global warming, and to define an agenda for GHG reduction. This takes into account that at some point in the future, we should reach zero emissions, with a related roadmap, determined by the idea that there is a limited allowance based on the work shown in Figures 4.35 and 4.40, with uncertainties attached to them.

In the representation of global surface temperature changes in degrees Celsius since the 1850–1900 period, as a function of carbon dioxide (CO_2) cumulated emissions expressed in $GtCO_2$ between 1850 and 2019 (see Figure 4.35), the colored areas refer to model results corresponding to some range of anthropogenic CO_2 emissions and an associated temperature. The straight line indicates that there is a linear correlation between the estimate of the two. But this should not be taken as the one and only truth, as some spread is visible in Figure 4.35. Furthermore, the diagram in Figure 4.35 is based on CO_2 emissions only, while there are other different greenhouse gases and mixes; therefore, this figure represents only a simplified view. Figure 4.35 also shows the way things have been translated to show different possible pathways including one consistent with +2°C warming as discussed during the Paris Agreement: those pathways are constrained according to carbon allowances. The basic principle is that by summing overall years, we should respect the total allowance that is left to stay below a +2°C warming target in 2100, without resorting to the capture of CO_2 in the atmosphere. To stay below +1.5°C, all emissions must have stopped before the middle of this century.

There are many joint issues and a resulting calendar deriving from the idea that we have some allowance; we are not actually exactly sure of the quantitative allowance: there is a bit of flexibility shown by the spread and related to uncertainties, also depending on what is done now.

The net global greenhouse gas trajectories until 2100 for past emissions and for three scenarios are summarized in Figure 4.40: in orange-red, implemented policies (resulting in a warming of +3.2°C in 2100 with a range between +2.2°C and +3.5°C); in green, a trajectory for limiting global warming to +2°C; in blue, a trajectory for limiting global warming to +1.5°C in 2100, and in black, past emissions (over 2000–2015).

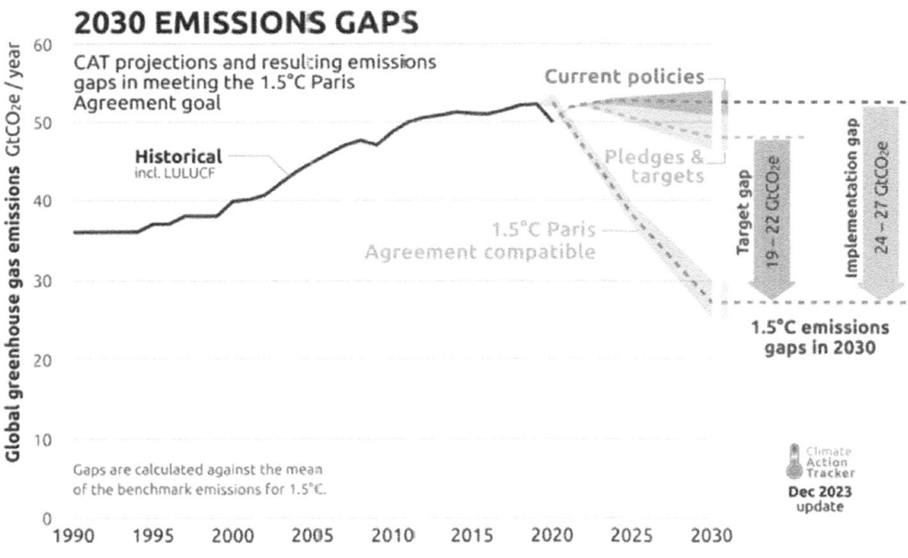

Figure 5.1. *2030 emission gap: CAT projections and resulting emission gap in meeting the 1.5°C Paris Agreement goal. Source: Climate Action Tracker (2024a). Available at: https://climateactiontracker.org/global/cat-emissions-gaps/*

Online tools such as Climate Watch or Climate Action Tracker deliver quantitative and qualitative indicators to assess countries' progress in terms of their nationally determined contribution (NDC) commitments as shown in Figure 5.1.

Current national commitments generally go up to 2030 only as required, and not after. From a strictly political view, it is objectively challenging for democracies with five to seven years of political mandates to commit to policies at a time scale much larger than their electoral mandate durations. However, current commitments are not sufficient to reach the goal of the Paris Agreement: current policies are not in line with a +1.5°C nor even in line with a +2°C pathway at the 2100 horizon. From a business as usual baseline, an effort two to four times larger than the current trajectory should be made.

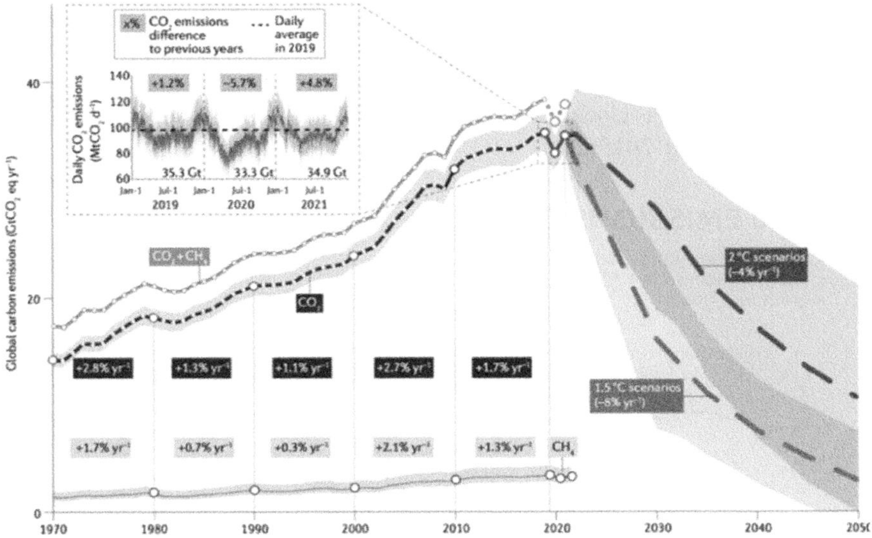

Figure 5.2. *Global CO_2 and CH4 emission trends. Temporal evolution of historical CO2 emissions (navy; including emissions from fossil fuel combustion and the process of cement production), near-real-time CO_2 emission mitigation pathways (dark blue and aqua) and historical fossil CH_4 emissions (light blue; 1970–2018 data from (red), projected CO_2 EDGARv6.0, scaled to 2021 with IEA data). Solid/dashed lines and shading represent the median and range respectively. The inset depicts daily near-real-time CO_2 data over 2019–2021, and the corresponding year-on-year changes in annual CO_2 emissions. Current emission trends will use up the allowed future emissions for limiting anthropogenic warming to 1.5°C (the remaining carbon budgets) within 10 years. Source: Liu et al. (2022). See: https://www.nature.com/articles/s43017-022-00285-w*

If all countries stick to their current NCD (which is not even ensured yet), this would actually lead to about a +3.2°C warming (see Figure 4.41). There is no real capacity to precisely determine the current and future progress (or not) of countries regarding the trajectories in terms of change of temperature, and the range of uncertainties is quite large. Developing countries and especially the African continent are facing significant challenges in terms of monitoring, reporting and verifying the respect of the Paris agreement (Mostefaoui et al. 2024).

The rate of reduction in percentage per year required for different pathways as compared to current trends is presented in Figure 5.2: we see that a +2°C trajectory would require a global annual reduction rate of –4% yr^{-1}, and a +1.5°C trajectory would require a –8% yr^{-1} reduction, while the trend over the last 50 years has been a constant emission increase globally. Therefore, these goals seem very challenging. As recalled by Liu et al. (2022), starting from 2020, the IPCC has estimated that the global carbon budget to stay below a +1.5°C global warming is 400 $GtCO_2$, with a 67% likelihood and 900 $GtCO_2$ with an 83% likelihood. The global carbon budget to stay below +2°C is 1150 $GtCO_2$ with a 67% likelihood, or 300 $GtCO_2$ with an 83% likelihood (IPCC Climate Change 2021: The Physical Science Basis 2021).

These projections change if we consider the capacity to remove CO_2 from the atmosphere. Different methods exist for this, as shown in Figure 5.3.

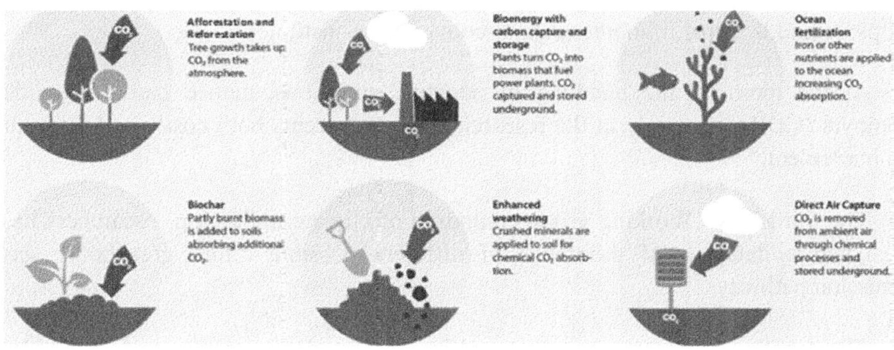

Figure 5.3. *Negative emissions and international climate goals – learning from and about mitigation scenarios. Source: Hilaire et al. (2019)*

Different techniques exist for carbon capture/negative emissions depending on where the storage is done, and on how the capture is done, as detailed in Figure 5.3.

Direct capture generally refers to methods in which the carbon is captured in plants: for example, by letting algae/plants grow, using carbon as a source of fuel and therefore taking out this energy as a fuel. This fuel comes from photosynthesis, not from the ground, and the CO_2 is taken out by the energy produced by these organic molecules and stored in the ground: this is geological storage.

It is also possible to use agriculture and afforestation/reforestation and direct biomass to store the CO_2 directly deep in the ground.

At the current time, the geological storage of CO_2 is not really well known and devices are still at the prototype stage with projects such as CO_2 storage in the Norwegian sea.

The main challenges come from the acceptability by the local populations that could be affected by the implementation of these technical solutions and the reluctance for full-technology solutions by environmental NGOs. Several studies have been done so far but not at a large scale. As for using agriculture or afforestation/reforestation to store carbon near the Earth's surface, a few studies and projects exist, among which is the effort from the cross-disciplinary French research Convergence Institute created in 2017, directed by Philippe Ciais and funded by the French government in order to "better understand key feedbacks and assess risks and sustainable options for integrated management of land ecosystems" (see: https://cland.lsce.ipsl.fr/about-us/cland-convergence-institute).

At the moment, the handling of negative emissions, named carbon dioxide removal (CDR), is mainly at the research stage and seems both costly and difficult to implement.

The third IPCC Working group composed of climate mitigation researchers has designed a database of thousands of different possible future greenhouse gas emission pathways.

These pathways are mainly produced by integrated assessment models (IAMs) that include assumptions regarding choices for energy production, use, technologies, land-use and land-use changes and forestry evolution, as well as socio-economical hypotheses. The IPCC climate scenarios are categorized into eight different labels from C1 to C8, depending on the level of warming that will be reached in 2100 compared to the 21st century. The level of temperature change is calculated by climate emulators such as MAGICC (Model for the Assessment of Greenhouse gas-Induced Climate Change). Under these eight different categories, another classification exists to represent the results of different policies thanks to seven different illustrative pathways (IPs). Two of these scenarios illustrate a "pessimistic" projection that would result from no change in the current business-as-usual policies: the current policy scenario, referred to as (Cur-Pol) and representing the effects of sticking to the policies that are currently in place; and the moderately increased action scenario, referred to as (Mod-Act), where 2030 national commitments include only a moderately increased ambition.

Among those seven IPs, five deep mitigation pathways aim at limiting global warming to 2°C or 1.5°C by 2100 as synthetized by the Carbon Brief from the IPCC in 2022 (Box 5.1).

Presentation of five deep mitigation pathways aiming at limiting global warming to 2°C or 1.5°C by 2100 as synthetized by the Carbon Brief from the IPCC (2022).

The shifting development (SD) scenario illustrates mitigation in the context of a broader shift towards sustainable development, including by reducing inequality, and with a phase-out of fossil fuels.

The low-demand (LD) scenario puts a strong emphasis on energy-demand reductions and a phase-out of fossil fuels.

The Renewables (Ren) scenario is based on a heavy reliance on renewable energy.

The extensive use of net negative emission (Neg) scenario relies on the deployment of carbon dioxide removal that results in net negative global GHG emissions.

The gradual strengthening (GS) of current policies is based on a scenario with a gradual strengthening of climate policies implemented in 2020, with rapid reductions mid-century and a massive reliance on net negative emissions by the end of this century.

Box 5.1. *Presentation of five deep mitigation pathways. Source. IPCC, AR6 – Carbon Brief (2022)*

The trajectories needed for all GHGs derived from inverse models in order to keep under either +2°C or +1.5°C in 2100, including scenarios with negative emissions (by carbon capture and sequestration (CCS) or its equivalent with biological material, bioenergy with carbon capture and storage (BECCS) methods) and for each GHG are displayed in Figure 5.4: efforts have to be made to avoid any delay. Efforts should also be made for many years: these model results are not necessarily well considered/understood in the real world. Not only is reducing GHG emissions and pollutants an effort that needs to be made as soon as possible, there also need to be shifts in behaviors such as agricultural practices. However, all this will not reasonably be enough to reach more than +1.5°C at the end of the century. Other ingredients are needed to reach the Paris Agreement goals, requiring technologies and structural changes, for instance in terms of transportation networks and urbanism adjustments: this is a problem for which economists are not able to build robust models in terms of structural change impacts at the time scales of climate modeling and for periods over five years. They hardly follow the required timelines for climate actions delivered as the results of climate models in order to avoid the dangerous consequences of global warming.

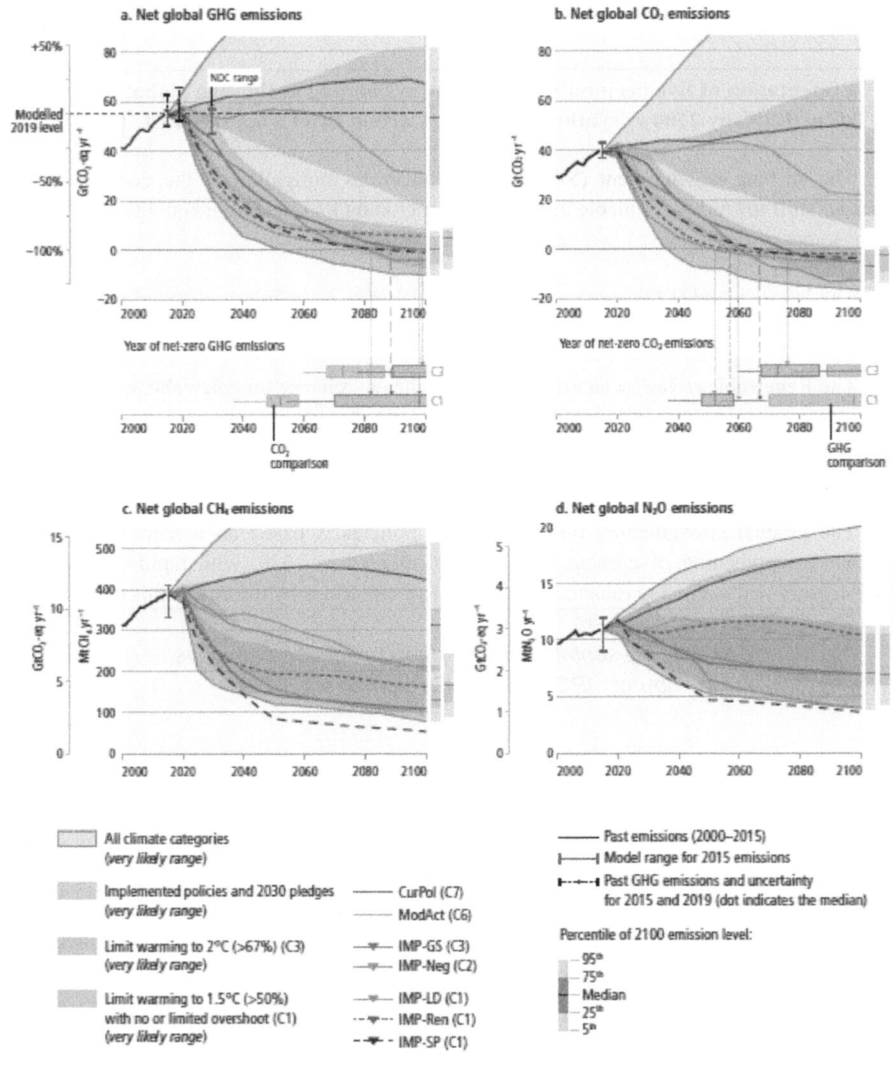

Figure 5.4. *Net CO_2 emissions (positive emissions minus CDR) (left) and net GHG emissions (right) under the seven illustrative pathways. Reproduced with permission from IPCC (2023b)*

Another aspect of climate complex reversibility is the ocean. As seen previously, the ocean is a component of the climate system where energy is absorbed. At the same time, this absorbed energy means that we are not at equilibrium in the climate system: this is an additional complexity when designing future trajectories and goals. We always consider the global equilibrium, and for regions such as France for instance and beyond, we work towards a carbon equilibrium. But a carbon equilibrium is not an easy concept to define.

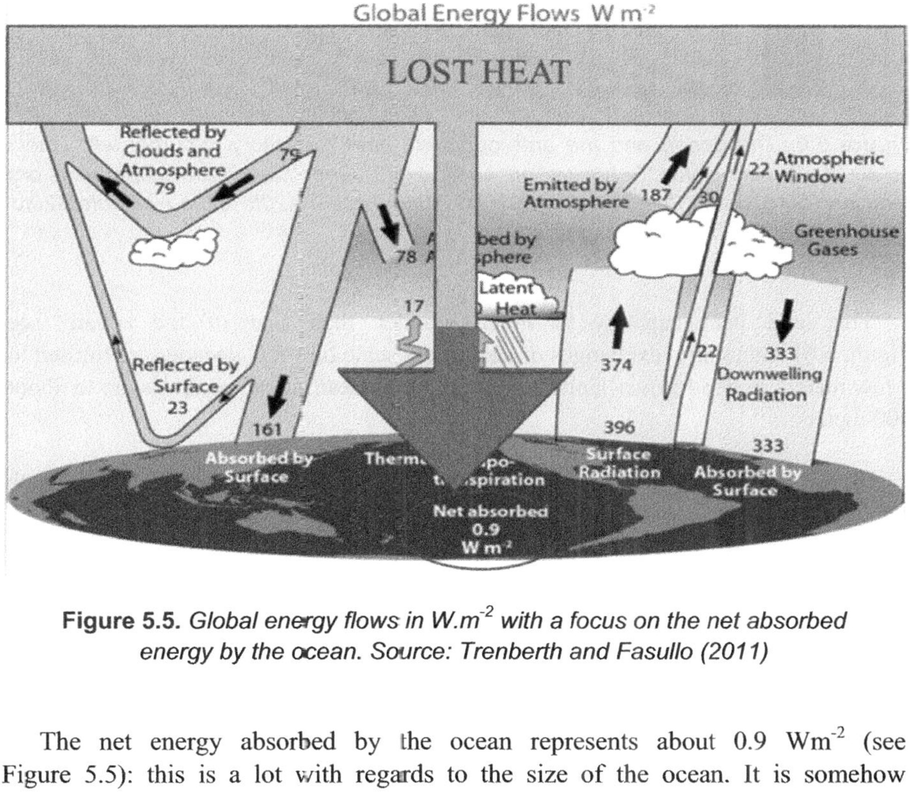

Figure 5.5. *Global energy flows in W.m^{-2} with a focus on the net absorbed energy by the ocean. Source: Trenberth and Fasullo (2011)*

The net energy absorbed by the ocean represents about 0.9 Wm^{-2} (see Figure 5.5): this is a lot with regards to the size of the ocean. It is somehow irreversible as it is stored in the ocean; we are not sure how and when it will be released, and this storage is a result of the energy unbalance at the top of the atmosphere.

The lost heat is visible in the diagram in Figure 5.6. The heat is stored in different reservoirs, but the ocean is by far the one that can accumulate the largest amount of heat.

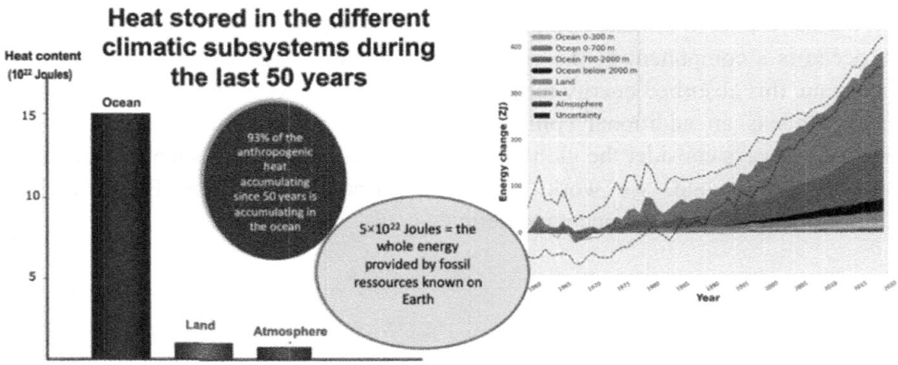

Figure 5.6. *The ocean and the anthropogenic heat repository. Source: left: Rhein et al. 2013. Right: von Schuckmann et al. 2023. See: https://essd.copernicus.org/articles/15/1675/2023/#:~:text=This%20is%20equivalent%20to%20a,available%20for%20melting%20the%20cryosphere*

The land heat capacity is much smaller than that of the ocean (see Figure 5.6). Generally, exchanges of heat at the scale of a few decades are limited to a few meters in depths over lands, whereas in the ocean, it can penetrate up to about 100 meters.

Figure 5.6 also shows that the absorption capacity of the atmosphere is much smaller in comparison with the ocean. The 5×10^{22} Joules in Figure 5.6 represent all of the energy provided by fossil resources known on Earth. What is being stored little by little is a lot because of course, through the greenhouse effect, there is a change which is much larger than the direct thermic effect. And the energy which is accumulating in the ocean is much larger (about three times more) than the one being caused by the direct burning of fossil resources (coal, oil, etc.).

The map of the absolute geostrophic current illustrates that the ocean, with its dynamics, is a very turbulent fluid (see Figure 5.7).

The representation of the oceanic circulation as it was seen in the early 2000s (see Figure 5.7) has strongly changed, with more turbulence and small gyres: it is a very complex fluid, with some kind of filament mixed with a lot of other phenomena, the singularity of which is increasingly being investigated by the climate research community.

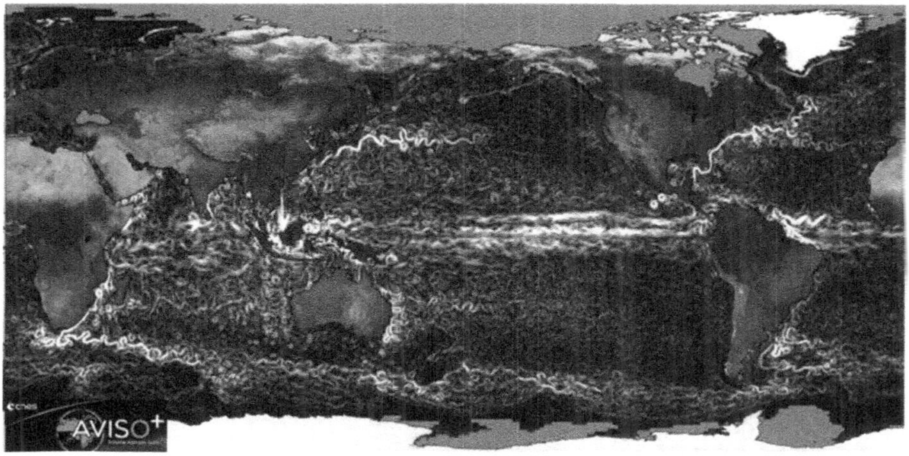

Figure 5.7. *Map of absolute geostrophic current, showing that the ocean dynamics is a very turbulent fluid. Source: CNES, AVISO+ Satellite Altimetry Data (2013)*

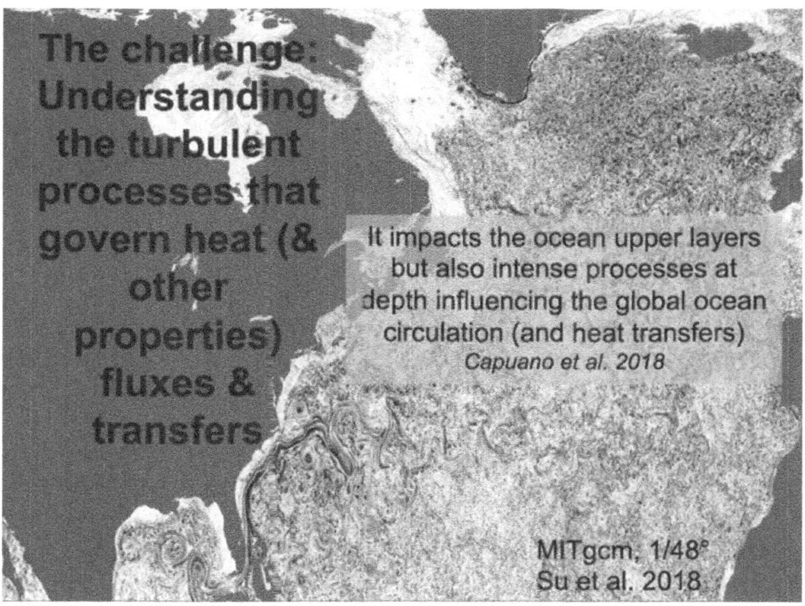

Figure 5.8. *The research challenge of understanding the turbulent processes governing heat fluxes and transfers. Source: Capuano et al. (2018) and Su et al. (2018)*

The understanding of turbulent processes governing heat fluxes and transfers is a main research challenge (see Figure 5.8). Oceanic transport is much more difficult to ascertain nowadays than before. It is indeed an important challenge for current oceanographers to precisely ascertain the phenomena currently occurring, how the ocean heat content is being absorbed and transferred in depth in the ocean.

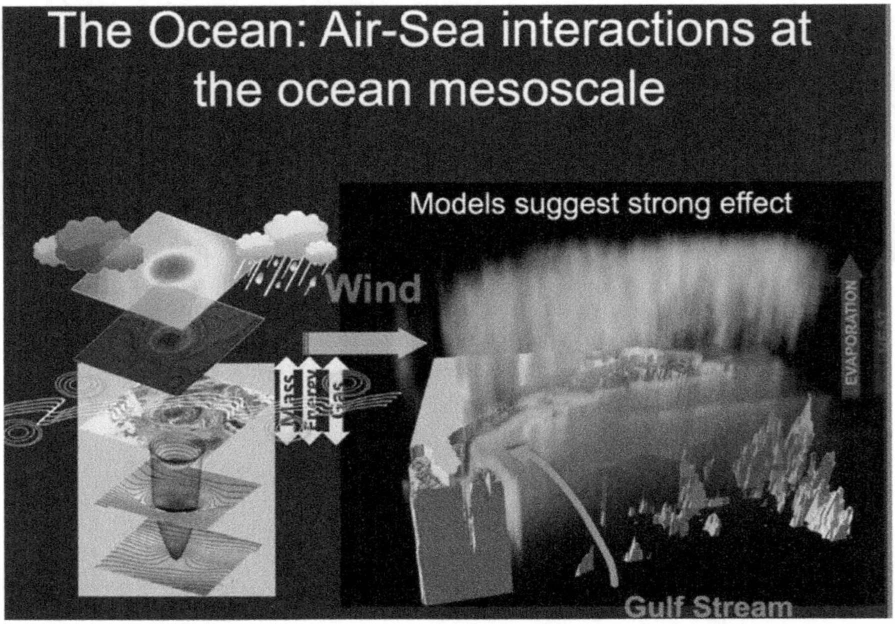

Figure 5.9. *The importance of mesoscale effects. Source: Speich (2017)*

Small-scale and mesoscale oceanic circulation phenomena are certainly the vectors of energy exchanges. At those scales, the impact of turbulence on oceanic models is important in terms of heat taken by the ocean (see Figure 5.9).

Some studies also have shown the significant impact of hurricanes (with the observation case of hurricane Harvey for instance (see Figure 5.10)) in drawing heat and water from the ocean and in transferring heat and energy from the ocean to the atmosphere. Pr. Kerry Emmanuel from MIT has been one of the worldwide pioneers to bring attention to this since the 1980s (see: https://texmex.mit.edu/pub/emanuel/PAPERS/sanders77.pdf).

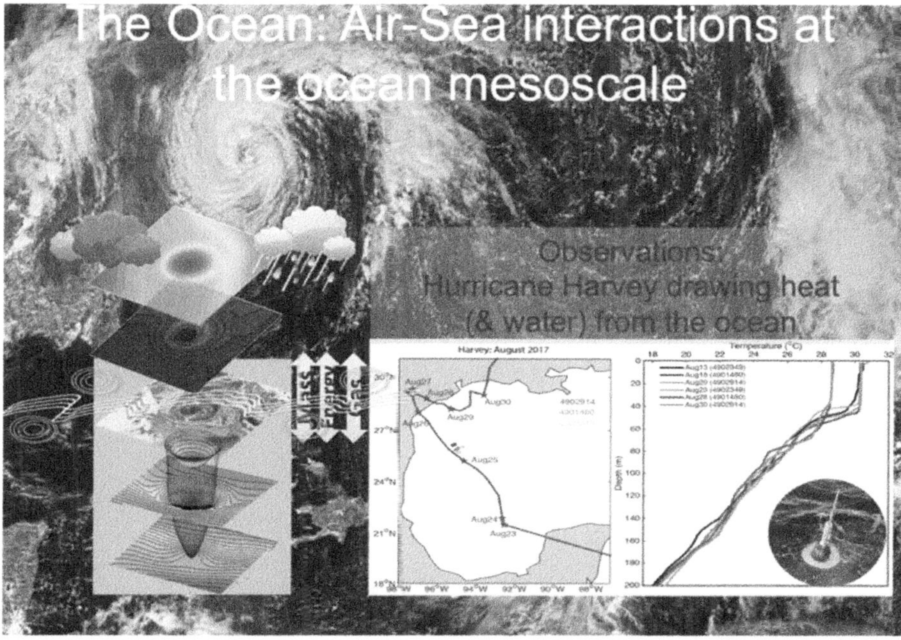

Figure 5.10. *Air–sea interactions at the ocean mesoscale and the example of the hurricane Harvey effect. Source: Speich (2017)*

This phenomenon has not been considered in models since the heat exchange is enhanced by hurricanes which take their energy from warm (about +26°C) sea surface heat. This heat has a role in the elevation of the moist air near the oceanic surface, and that might be a missing piece in the kind of budgets we build for the ocean. Recent studies (Warner et al. 2023) have also shown the effect of hurricanes in pushing heat even deeper into the ocean than what was thought, enhancing long-term ocean warming.

As a summary, there is no scientific doubt on the point that energy is being stored in the ocean, but it is still a complex topic of study, and the energy stored in the ocean is a testimony of the fact that we are not at equilibrium. It also warns us that we are not in a reversible world in many respects regarding climate change.

Another important point for assessing how flexible the situation regarding global warming is or is not, is certainly to study how much we are managing to change the different types of energy production.

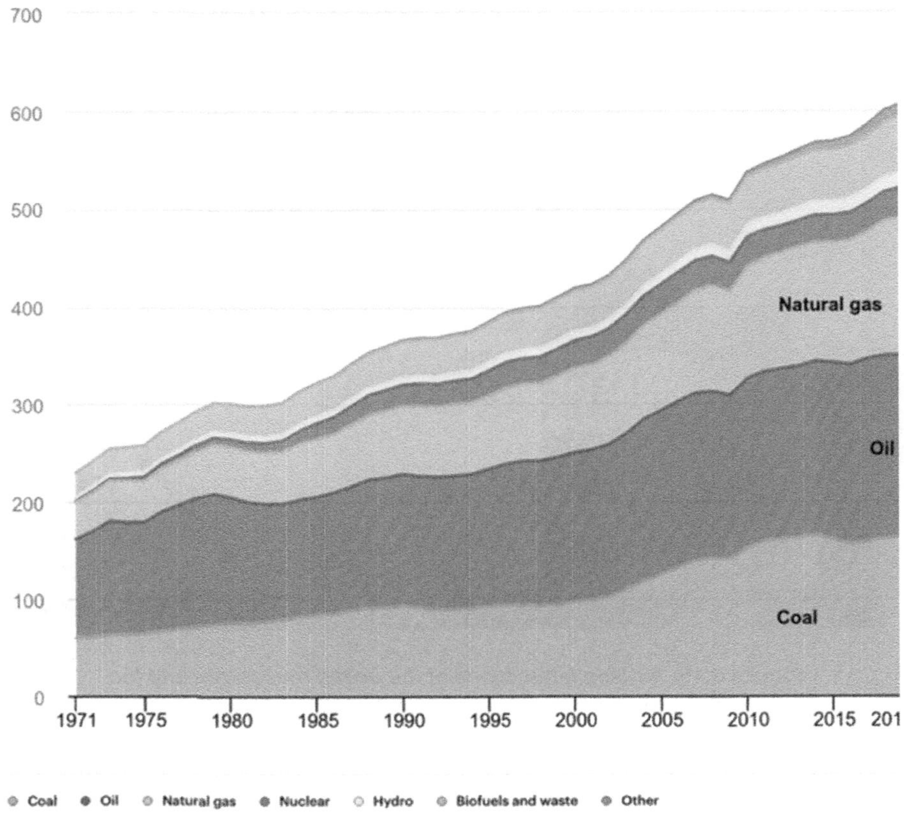

Figure 5.11. *World energy mix of total primary energy supply (TPES) from 1971 to 2019 by source in exajoules (EJ) with 1 exajoule = 278 Terawatt hours. Source: IEA, World Energy Balances (2021). World total energy supply by source, IEA, Paris (see: https://www.iea.org/dataand-statistics/charts/world-total-energy-supply-by-source-1971-2019*

The International Energy Agency (IEA) regularly assesses the World energy mix (see Figure 5.11). The large majority (about 80%) of the recent world's energy mix for the total primary energy supply (TPES) from anthropogenic production systems is based on fossil fuels, especially on oil, coal and natural gas. The diagram shows that in 1971, this distribution was quite similar and oil, coal and natural gas represented already a bit more than 80%: there has been no major change over those decades in the global use of fossil fuels so far. Nuclear energy (in dark green) part has not increased very much. In yellow, "hydro" mostly represents hydroelectricity (hydro can also be used to store energy in the future), in orange are represented biofuels and waste covering everything that comes from the ground and also deforestation (not only positive energy). The small dark orange line for "other" is the addition of solar and wind energies. Solar and wind are limited to a little more than 1%. That was the situation in 2021; we are a bit further along in 2025, but there are only a few years left to make this change by 2030 at the time of the writing of this book, through both a combination of a shift in the energy mix and sobriety/energy economies. This seems extremely challenging. This situation has consequences, and various aspects of the problem must be considered, including radical adaptation without delay, as we cannot reasonably rely only on the idea that internationally stated climate goals will be achieved.

The CO_2 produced for the energy mix of the world and coming from different countries is mixed in the atmosphere as it is a long-lived gas, meaning that for instance about 1.5% of the world's CO_2 emissions is produced in France. CO_2 increases everywhere else in the world, which means that in France, we depend for 98.5% on the efforts of other countries. Hence, we definitely do not have our destiny at our national hands for the topic of climate change, even if every effort counts, including by leading by example to inspire the required global actions. But it is not enough to solve this global problem, involving all countries, and mainly caused by a few major emitters.

In previous chapters, we discussed the feedbacks constraining the changes in temperature and considered physical feedbacks (such as cloud, surface and water vapor feedbacks), which we also call Charney's sensitivity. This Charney's sensitivity induced by physical feedbacks only delivers a range from +1.5°C to +4.5°C, with a middle value of +3°C, but as we already underlined, the range of temperature is still open. Climatologists have not managed so far to determine what will occur, among the lower possible sensitivity and the larger possible climate sensitivity for a given forcing. The forcing that we usually consider is $2 \times CO_2$ since the pre-industrial era, which represents approximately, without any feedback, a +1°C warming.

Among other scientists, the climatologist Philippe Ciais has been working on additional feedback. The diagram shown in Figure 5.12 delivers a very simple view of the multiple interactions among tropospheric chemical processes, biogeochemical cycles and the climate system. This schematic representation illustrates climate sensitivity with other feedbacks, linked with biodiversity, agriculture and forestry, and includes parameters such as changes in nitrogen cycle, phosphorus cycle, etc.

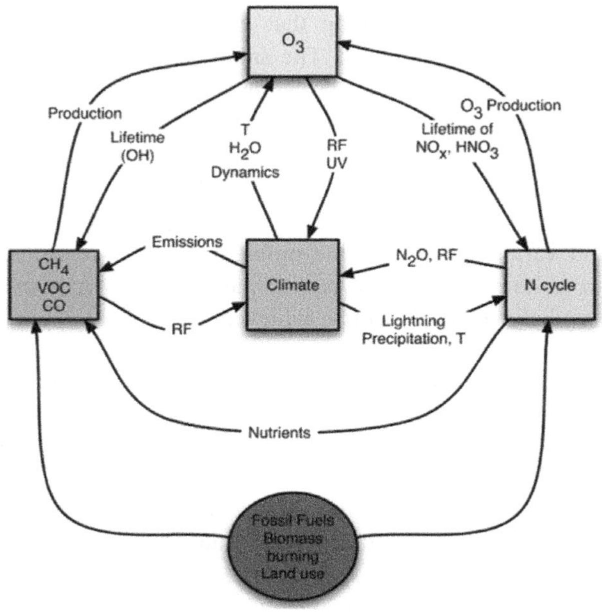

Figure 5.12. *Schematic representation of the multiple interactions between tropospheric chemical processes, biogeochemical cycles and the climate system. RF represents radiative forcing, UV ultraviolet radiation, T temperature and HNO_3 nitric acid. Reproduced with permission from Denman et al. (2007)*

The diagnostic of this work is about the long-term climate sensitivity: in the course of the next century, we might obtain a temperature change quite larger than expected because of those additional feedbacks which are not or not well considered in most models. This represents an important limit in terms of diagnosing the future as we stick mostly to a physical world representation. But when considering multiple interactions of biogeochemical cycles and tropospheric chemical processes with the climate system, important additional retroactions and additional phenomena may appear, which we have not precisely quantified/considered yet in models. We will not discuss this important point any longer, but research efforts to more precisely quantify this topic are very important and growing. This "known" unknown in terms of possible huge additional effect should also enhance the urgency for acting quickly and boldly to tackle global warming.

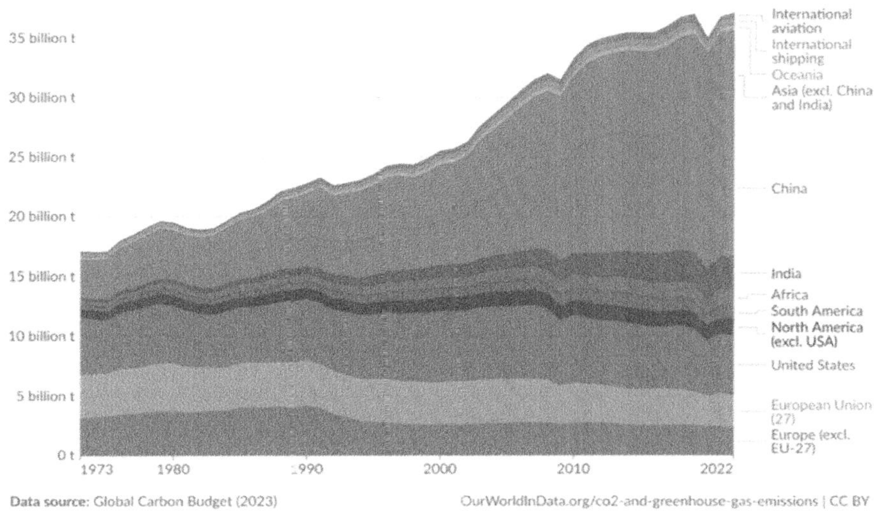

Figure 5.13. *Annual CO_2 emissions by world region. Source: Global Carbon Budget 2023 and Our World in Data (2023)*

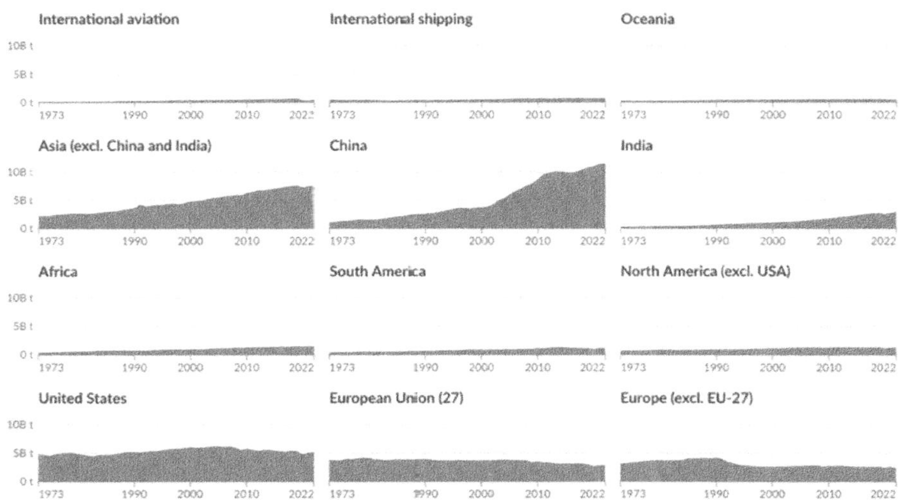

Figure 5.14. *Evolution of regional shares of CO_2 emissions from fuel combustion over the same period. Source: Global Carbon Budget (2023) and Our World in Data*

China, the rest of Asia and the United States dominate the annual CO_2 emissions by world region (see Figure 5.13). The announced American withdrawal from the Paris Agreement in 2025 is therefore particularly alarming.

China in particular has been largely increasing its participation in the global CO_2 emissions from fuel combustion over 1973–2022, and is currently the main global emitter, followed by India and the rest of Asia (see Figure 5.14). The global distribution of these emissions displays large regional differences: from 1973 to 2022, the total global CO_2 emissions from fuel combustion largely increased from 17.08 $GtCO_2$ to 37.15 $GtCO_2$.

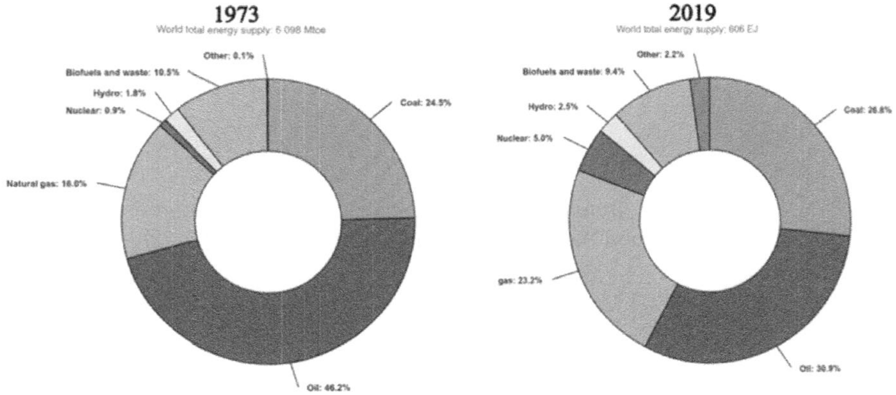

Figure 5.15. *Evolution of the share between fuels. Source: IEA, World Energy Balances (2021)*

The share comparison between different sources of fuels between 1973 and 2019 (see Figure 5.15) shows that in both years, oil represents by far the largest share: 46.2% in 1973 and 30.9% in 2019, but it has decreased over those 47 years.

On the contrary, coal, the second most important energy supply, has a slightly increasing share in 2019 (26.8%) as compared to that in 1973 (24.5%). Regarding natural gas, the share also increased from 1973 (16.0%) to 2019 (23.2%). However globally, the total share of those three fossil fuels does not reduce much as it represented 86.7% in 1973 and 80.9% in 2019.

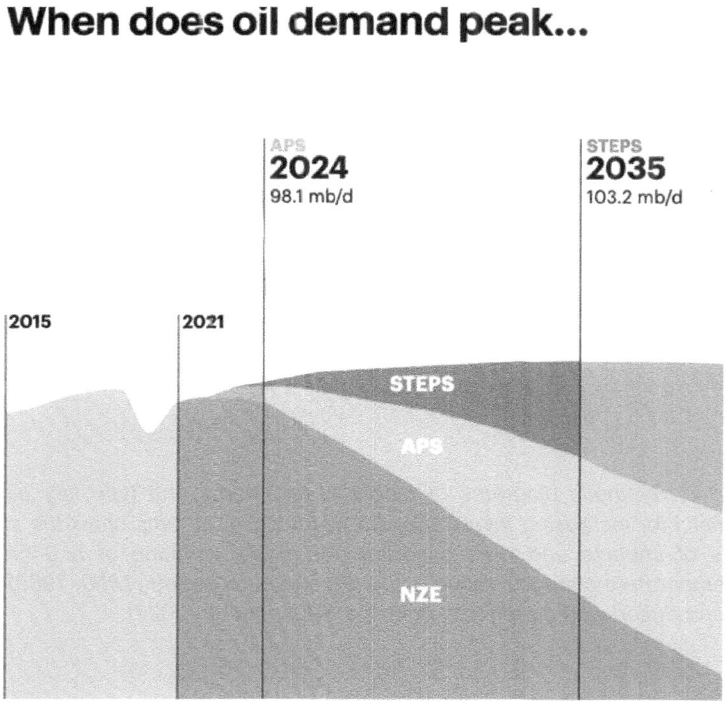

Figure 5.16. *A scheme of oil demand peak dependent on scenarios: announced pledges scenario (APS), stated policies scenarios (STEPS) or net zero emissions by 2050 (NZE). Source: World Energy Outlook 2022, International Energy Agency (IEA) (see: https://iea.blob.core.windows.net/assets/830fe099-5530-48f2-a7c1-11f35d510983/WorldEnergyOutlook2022.pdf)*

Another key point that has been discussed across the decades is whether or not the problem would exist because of peak oil (see Figure 5.16).

Some people thought that peak oil would solve the problem of global warming.

In fact, the problem is that the definition of oil is not the same for everybody: we had a definition of oil that used to correspond to what we thought we would be able to exploit, but the capacity to extract petrol has extended, especially with the use of unconventional oil and gas (UOG), referring to fossil schist. The production of oil was first technologically limited, the part due to high technology came later (by 2020s): in any case, it cannot help for the climate change issues that are occurring at a much shorter time scale. Therefore, the idea that peak oil can help has failed as it is shown that it is not possible to rely on that.

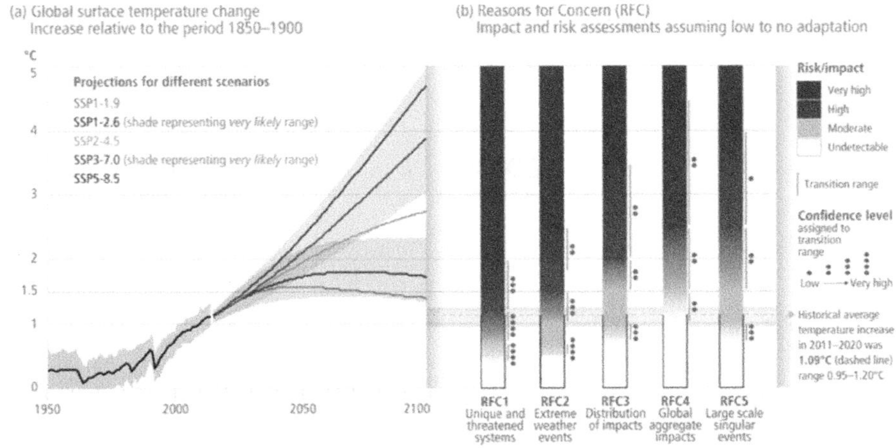

Figure 5.17. *Synthetic diagrams of global assessments and type key global and regional risks for increasing levels of global warming. Diagrams show the change in the levels of impacts and risks assessed for global warming of a 0–5°C global surface temperature change relative to the pre-industrial period (1850–1900) over the range. Reproduced with permission from Ara Begum et al. (2022)*

For decision-making and political choices, a powerful tool is to consider various scenarios of future trajectories and their associated outcomes, past trajectories and current state (see Figure 5.17). But this is very difficult to diagnose only through observations. The diagram on the left shows five different scenarios of the Sixth Assessment Report of the IPCC (2022), and their associated range of temperature evolution. The SSP 8.5 scenario for instance is the worst-case scenario; the SSP 2.6 is the emission stabilization scenario to stay between +1.5°C and +2°C, etc.

The take-away message from these extremely different scenarios is the overlapping period during quite a long phase (until about 2030s), and therefore quite a long period of time for which observations are not enough to suppress the unknown for predicting the future trajectory. At the same time, the kind of nonlinear issues discussed previously, such as the fact that there is also stored energy in the ocean, etc., makes these trajectories different already; by looking just at the surface temperatures, the scenarios do not distinguish the very short-term trajectories, and the future pathway is actually complex to assess from the present and only from observations. Indeed, current observations are actually consistent with various IPCC pathways, and this situation will continue for a few years until around 2030: this is very important for the analysis, and it underlines that the situation cannot merely be assessed by considering surface temperatures only.

Five different nuances of reasons for concerns (RFC) are also discussed by the IPCC and the assessment of their associated impacts and risks according to four types (undetectable, moderate, high or very high), assuming that there is low to no adaptation (see Figure 5.16(b)). The confidence level from low to very high is also represented by dots. This is a systemic view used for the diagnosis of possible future global and regional risks for increasing levels of global warming. The simple underlining idea is that basically the risks increase with the rise of the surface temperature, but this diagram is a simplification, lacking direct information.

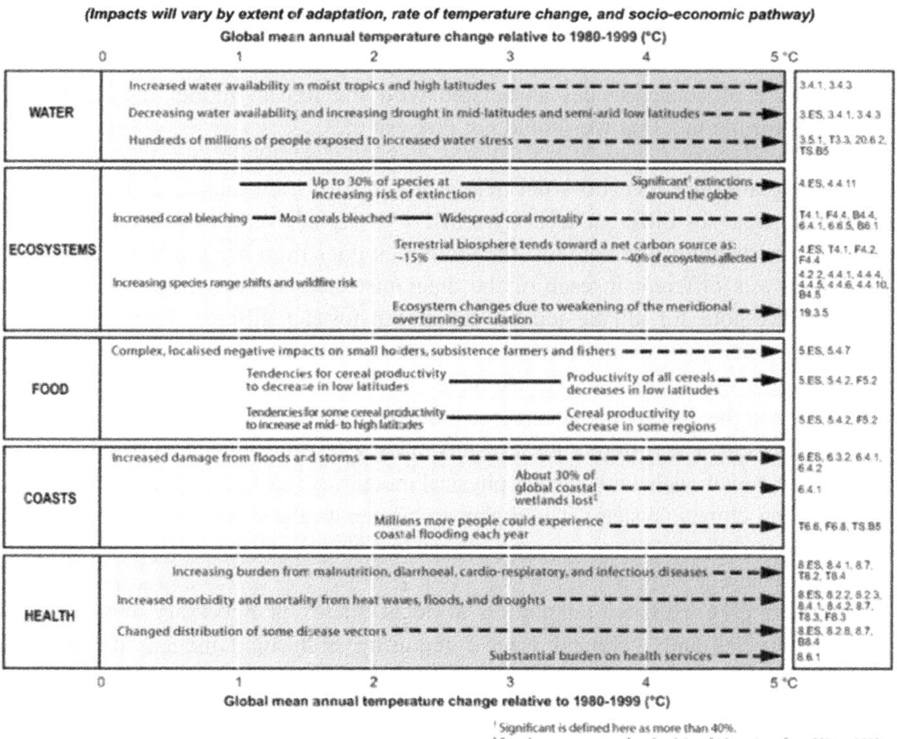

Figure 5.18. *Why the 2°C target? Illustrative examples of global impacts projected for climate changes (and sea level and atmospheric carbon dioxide where relevant) associated with different amounts of increase in the global average surface temperature in the 21st century. Reproduced with permission from IPCC (2007a)*

A more complete historical diagram dating back to AR4 is shown in Figure 5.18: it is probably one of the most important results that led to the decision of setting the global target of the Paris Agreement to stay below +2°C at the end of 2100. The +2°C target was designed mostly by people wanting to take actions against global warming, given that we were already approximately +1°C at the time when this target was chosen, and the idea was to refuse to go beyond a doubling of this figure. This goes together with the first experiment trying to assess local impacts. As previously mentioned, going from high-scale models to local diagnosis is still challenging and highly associated with uncertainties, but it is possible to try to do this exercise through a number of studies using models named "impact models", related either to water, ecosystems, food, coasts or health as presented in Figure 5.18, where a scale shows the results of the impact of those models expressed in terms of temperature (in degrees Celsius).

The range of consequences associated with the impact model results deliver indications for the fact that we should not go over +2°C.

This diagram, however, does not deliver the idea of a threshold: it merely shows different impacts for different local systems, with critical values modeled by dotted arrows on the diagram. We note that the moment it goes from black full-line arrows to dotted arrows is different in each of the diagrams: there is no one unique common identified threshold for all the local systems, but instead different thresholds when considering consequences delivered by impact models for each system, and those thresholds may vary quite a lot. For example, the thresholds are different for the water availability and the different effects on the biosphere, and differ also for the ecosystems used for food, etc. The different local results of those various impact models are being mixed in the new models with current physical models to constantly assess the different aspects of the climate system: it is urgent to accelerate the work on this coupling to have a complete view of climate consequences, including local ones.

In a changing climate, adaptation at local scales is necessary but it raises socio-economical and technical issues, requiring both fundamental and applied research. A main problem when going from large scale to small scale for adaptation is the limit of what is predictable or not, which is actually not well understood. A large number of uncertainties is actually due to small-scale phenomena. There are many things we do not know for sure, but there are a few well-known predictable results, such as the fact that the climate will change and will be warmer and many other results delivered with high confidence levels in the successive IPCC reports.

We need to explore climate change as the changes of a complex system, which requires multidisciplinary approaches. A more comprehensive approach of the link between global and local perspectives is also a source of innovations in many domains. ("Big Data" in particular fundamental and applied research.) The models currently developed are not the "ultimate" models, and different designs may enable a better modeling of local-scale phenomena, not only with Navier–Stokes equations, but with additional tools, including artificial intelligence, etc.

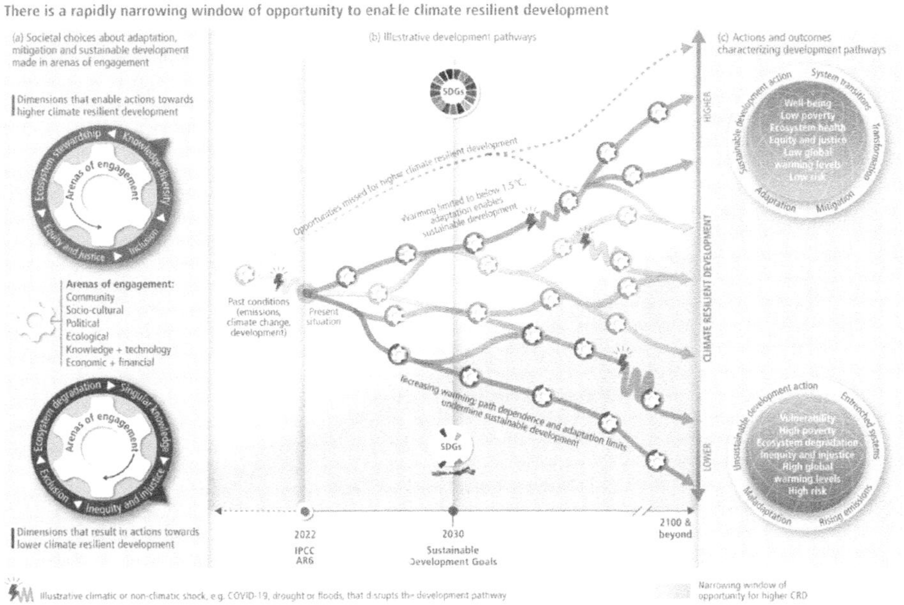

Figure 5.19. *Climate Resilient Develooment Pathways are development trajectories that successfully integrate GHG mitigation and adaptation efforts to support sustainable development for all. (a) Climate resilient development is a process that takes place through continuous societal choices towards higher CRD (illustrative green pathways) or lower CRD (illustrative red pathways). (b) CRD is described by five development dimensions – people, prosperity, partnership, peace, planet – on which the SDGs build (18.2). Reproduced with permission from Schipper et al. (2022)*

Climate Resilient Development (CRD) Pathways development are trajectories that successfully integrate GHG mitigation and adaptation efforts to support sustainable development for all (see Figure 5.19). This diagram taken from the Working Group II to the Sixth Assessment Report of the IPCC shows five development dimensions consistent with the United Nations Sustainable Development Goals (SDG): people, prosperity, partnership, peace and planet, and the level of climate actions and outcomes characterizing development pathways with their levels of risk and resilience (from lower in red to higher in green), resulting from societal choices about adaptation, mitigation and sustainable development made in arenas of engagement.

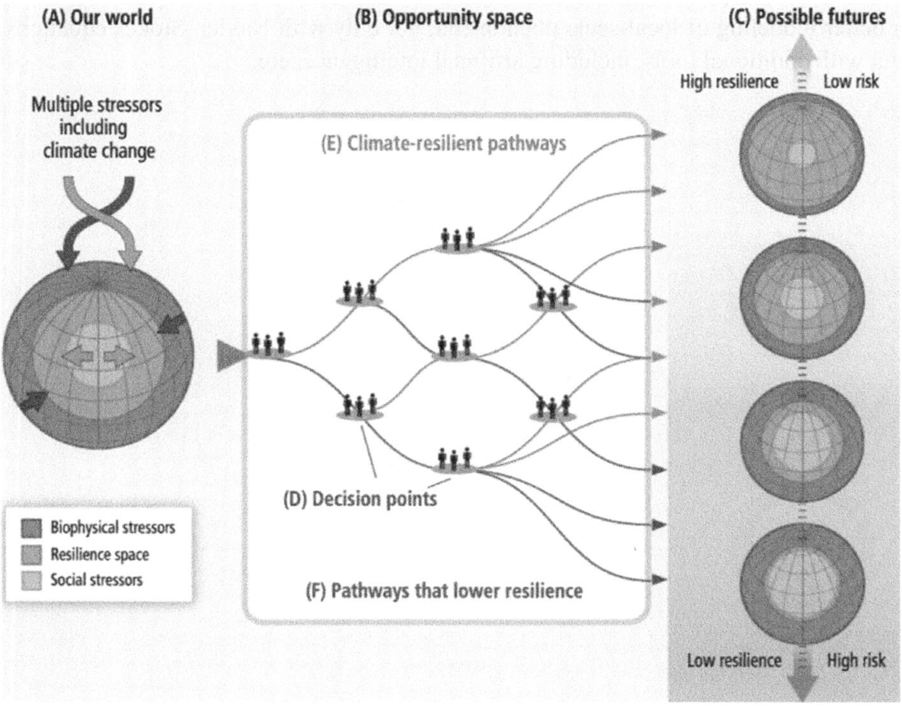

Figure 5.20. *Opportunity space and climate-resilient pathways. (A) Our world [sections A-1 and B-1] is threatened by multiple stressors that impinge on resilience from many directions, represented here simply as biophysical and social stressors. Stressors include climate change, climate variability, land-use change, degradation of ecosystems, poverty and inequality, and cultural factors. (B) Opportunity space [sections A-2, A-3, B-2, C-1 and C-2] refers to decision points and pathways that lead to a range of (C) possible futures [sections C and B-3] with different levels of resilience and risk. (D) Decision points result in actions or failures-to-act throughout the opportunity space, and together they constitute the process of managing or failing to manage risks related to climate change. (E) Climate-resilient pathways (in green) within the opportunity space lead to a more resilient world through adaptive learning, increasing scientific knowledge, effective adaptation and mitigation measures, and other choices that reduce risks. (F) Pathways that lower resilience (in red) can involve insufficient mitigation, maladaptation, failure to learn and use knowledge, and other actions that lower resilience; and they can be irreversible in terms of possible futures. Reproduced with permission from IPCC (2014a)*

The opportunity space and climate-resilient pathways realized in AR5 by the Working Group II of IPCC that focused on impacts, adaptation and vulnerability (see Figure 5.20) is also interesting to look at because it served for decisions made at COP21 even if it was updated in AR6 (see Figure 5.19). It was mostly built by sociologists and managed to summarize floating ideas and influence the negotiations. Figure 5.20 shows on the left a representation of the Earth with different circles, establishing various types of overlapping priorities, and shows a way to analyze the climate problem from the perspective of the Conferences of the Parties (COP). In green are represented the physical barriers to action for resilience, i.e. the physical obstacles towards climate action. This has been changed to a more optimistic view, with the idea of a "space" for action on climate change instead of a barrier. Social stressors are represented in yellow, and non-physical biophysical stressors, largely representing biodiversity, are shown in brown. The three main items representing a view of the climate system are therefore represented with those three colors for the physical world, the biosphere and the social stressors.

The message of this diagram is that it is necessary to align the different problems to make them convergent. For example, when a policy for an issue affecting social stressors is designed, it should be done in a positive phase with physical and biophysical climate action as well.

The opportunity space modeled at the center of Figure 5.20 is a step-by-step process, an iterative method that was used during COP21 especially for the Nationally Determined Commitments (NDC), during which proposals and decisions were made one by one to check what would happen consequently and whether adaptation would be required.

The process is also iterative in the sense that major milestones of global reviews (named the global stock take) will take place every five years under the United Nations Convention on Climate Change (UNFCCC) to review all national commitments in order to always increase the efforts and ambitions of the ratifying countries.

This view is of course idealized, and to date in the real world, most countries do not comply with their climate commitments.

However, this representation is interesting and it also interestingly includes the weight of the pressure of civil society. The idea of finding aligned solutions, common to the different aspects of the climate problem, is also an original perspective, adding an important value to this figure.

5.1.1. *Process studies and uncertainties*

There are many approaches to uncertainties: one of them is an approach based on process understanding. Another method for assessing uncertainties is based on multi-models and multi-actor statistics. Furthermore, the IPCC approach aims at building consistent metrics using a specific dual vision to qualify the nature of information.

An underlying question behind process studies and uncertainties is: How do we verify models? This is complex as the aim of climate science is mainly to predict future trends, for which, of course there are no available observation data. For several years, there was the idea of a relative easiness with the expectations of having the ability to check processes little by little, but in fact the complex nature of the system means that verifying the models is not an easy task. As an example, we will discuss the case of the verification for global climate sensitivity.

There are three important questions to assess in terms of uncertainties: First, what is the amplitude of global warming? Second, how natural variability is considered? And finally, how much do we trust the notion of climate impacts?

In each case, these questions mobilize notions seen in the previous sections of this book.

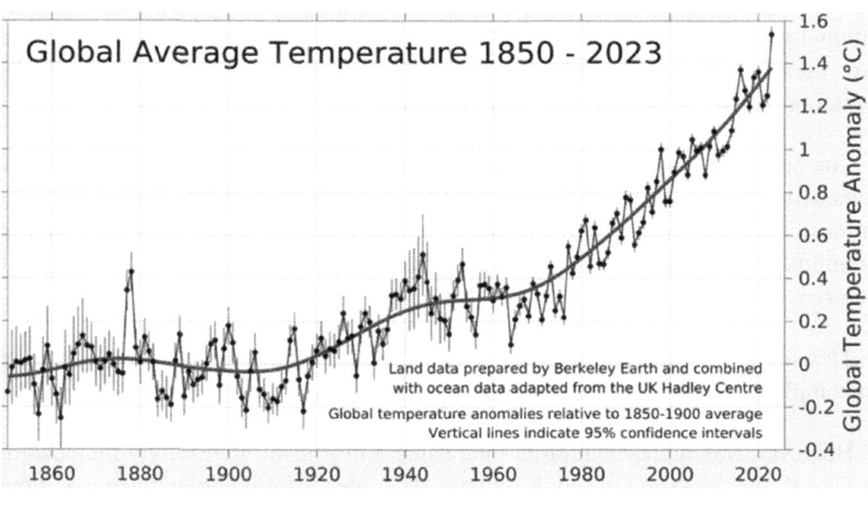

Figure 5.21. *Global average temperature 1850–2023. Global temperature anomalies relative to the 1850–1900 average. Vertical lines indicate 95% confidence intervals. Source: Land data prepared by Berkeley Earth and combined with ocean data adapted from the UK Hadley Centre (2024)*

When considering global surface temperatures since the 1960s for instance, and when comparing them to current ones (see Figure 5.21), we see well the fluctuations in black corresponding to natural variability. The question is: how can we use the knowledge about past trends to deduce future trends? We have the capacity to access direct estimates of temperature trends over time. We know the quantity of emitted greenhouse gases between the 1960s and nowadays, to a fair certainty level.

Aerosols represent another important parameter which we do not know well. Clouds and cloud feedbacks are also a key element, representing a main source of uncertainty in climate models. The mixing of aerosols and clouds is also an important phenomenon strongly affecting the radiative budget of the Earth, the understanding of which needs to be largely improved. When gathering the different pieces of information at our disposal, it might seem that we have a rather good estimate of the radiative forcing by greenhouse gasses.

However, all the parameters previously mentioned (aerosols, clouds/cloud feedback, the mixing of aerosols and clouds, among others) represent the main sources of uncertainties. Due to these uncertainties, making a simple connection between the quantity of anthropogenic greenhouse gases in the atmosphere and the associated change in terms of temperature, i.e. precisely defining the temperature sensitivity of the Earth, is complex, not direct, and includes uncertainties that are not side details, as those parameters are not only not well understood, but they also may change with time. Therefore, the relationship between what we know quite well (greenhouse gases, temperatures) and parameters that are less understood (aerosols, clouds/cloud feedbacks, the mixing of aerosols and clouds, etc.) may vary with time, and possibly strongly.

In the next paragraph, these uncertainties will be discussed in more detail.

The levels of confidence associated with radiative forcing (RF) estimates for different emitted compounds or processes that result in a combination of drivers are represented in Figure 4.16 from the IPCC Fifth Assessment Report. The RF for the gases are presented, as well as the RF for the gases associated with their chemistry or their effect. An updated version of the numerical values of this figure is available in the IPCC Sixth Assessment Report, reproduced in Figure 4.17. However, for pedagogical purposes, we choose to present the AR5 diagram which is more complete.

For CO_2, which is a long-lived GHG, the level of confidence is very high. Methane's radiative forcing is associated with a high level of confidence, as it is a bit more of a complicated assessment involving different chemistry mechanisms, therefore a more complex behavior than CO_2. For the RF for halocarbons, the confidence level is high, etc. Ozone for instance appears at different places in this diagram as some gases tend to have a diminishing effect on ozone, while others tend to increase its quantities. Ozone's chemistry is complex as it is mixed with the rest of the chemistries both in the stratosphere and in the troposphere: it may be increased or decreased by different gases.

For NOx, for instance, the negative radiative forcing shows the associated distribution by gases and aerosols.

For aerosols and precursors, the processes are mainly well-known (with a high level of confidence), but for quantification, there is a huge associated range of values. Therefore, the complete effect of aerosols including the greenhouse effect of black carbon may be negative or positive, we do not know that so far.

The complete effect of cloud adjustment due to aerosols is also huge with a range from about -1.33 $W.m^{-2}$ to almost 0. We are therefore considering extremely uncertain estimates. It is very complex to get a global estimate to derive from global features, which means that despite all temptations to simplify, it is not reasonable to avoid digging into the details of process studies.

To one of the driving questions of this section, "How do we verify models?", an answer is that they cannot be diagnosed at the global scale; the verification instead requires process studies.

Two of the main sources of uncertainties are aerosols and clouds, as well as the interactions between them.

5.1.1.1. *Aerosols*

Regarding aerosols, the important and complex parameters at stake to quantify their radiative forcing is their size, nature, distribution and altitude (very complex). Very large uncertainties are associated with these features.

5.1.1.2. *Clouds*

As for clouds, they have a large feedback effect that tends to largely affect the amplitude of climate change by several aspects. The radiative forcing assessment of clouds notably depends on various parameters including their height, optical thickness (related to water or ice content), temperature, structure and distribution. There is of course a difference between what is associated with the greenhouse effect and the reflection of solar energy (as clouds are affected by both effects), and there is also a compensation between those two effects.

Therefore, to properly assess the order of magnitude of the total anthropogenic forcing, at least one of the two main sources of uncertainties (aerosols and clouds) must be estimated correctly. Otherwise, the following problem arises: when considering the evolution diagram of temperature over time, which is something known, the aerosol component is superimposed, which is unknown with precision. When assessing what are the causes of the global temperature increasing over time, there are basically two forcings related to human activities: climate sensitivity (reaction to greenhouse gases and feedbacks/clouds), and aerosols. A main unknown is the share between climate sensitivity and the effect of aerosols.

In the hypothesis that there is currently a significant effect of aerosols in the climate system, as an important uncertainty in the present diagnostic, it would mean that we observe temperature increases instead of a strong cooling by aerosols. This means that if the effect of aerosols is significant in the present, climate sensitivity is also significant, which is a bit counter-intuitive.

If the continuous increase of temperatures has been dampened by aerosols, it means that without aerosols, this increase would be even larger. And as we know that aerosols are bound to increase more slowly than greenhouse gases because aerosols do not accumulate in the atmosphere (they are short-lived gases that stay in the atmosphere for around a few weeks), aerosols increase at a smaller rate than greenhouse gases, which means that if the aerosol effect is significant in the present, the greenhouse effect will be more significant in the future because the effects of greenhouse gases and aerosols are not compensated at the same rate, and of course reversely.

Due to aerosols, we cannot easily extrapolate the climate sensitivity only from greenhouse gases.

The radiative forcing for aerosols has been assessed by different models (see Figure 5.22), and individually for organic matter, generally due to forest fires and all kinds of decomposition of organic matter, carbon soot which is also due to fires and has some greenhouse effect, it is not only reflecting solar radiation, as well as for sulfate radiative forcing. Sulfate emissions are mainly located in the northern hemisphere linked with the use of oil, motor engines, etc. The radiative forcing for all those aerosols has also been assessed, as simulated by different models which contributed to IPCC reports (bottom of Figure 5.22). We observe huge differences from one model to another; for the recent period especially, the results are not very conclusive. Therefore, there is no “safe way” to characterize model results regarding aerosol radiative forcing.

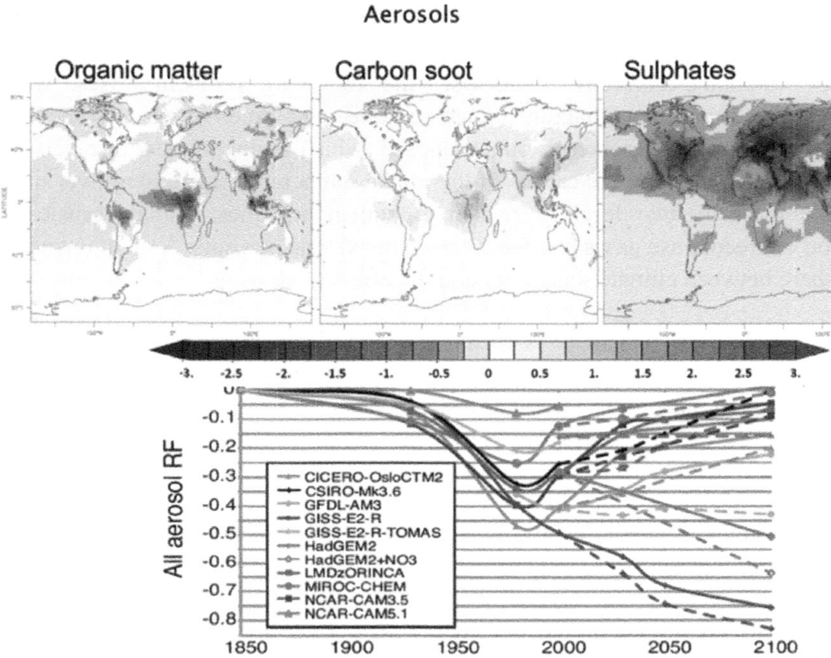

Figure 5.22. *Assessment of aerosol radiative forcing for organic matter (top left), carbon soot (middle panel), sulfate radiative forcing (top right) and all aerosols (bottom diagram), expressed in W.m^{-2}. Source: Shindell et al. (2012)*

The results also suffer from another bias: of course, for the recent period of time, measures of temperatures are available and researchers know what trajectories models should follow in terms of temperatures until now. All models currently have a scheme concerning aerosols that is, however, difficult to calibrate. A very common tendency in the scientific community is to tune the models (a very common practice – many parameters can be tuned, such as feedbacks for ocean–atmosphere models, there is also for instance a little solar absorption associated with clouds that is possible to tune in terms of quantity). It is possible to adjust models with the aim of correctly representing past and present phenomena, but they may contain errors due to incorrect aerosol modeling.

This means that papers should be taken with care, especially concerning reaching the +2°C target in 2100 for instance. The political and decision-making spheres need simplified figures as points of reference and for easy communication to the global public, but at the same time, the underlying complexities must be well understood and represented.

The simplified problem can be described as follows: if we consider for instance the year 2025, and want to assess how much emissions are left in order to avoid going beyond a +2°C in terms of global temperature elevation in 2100, the method is to start in 2100 with a stabilized temperature and to go backwards in models to estimate which pathway would allow us to stay below this limit. Therefore, the exercise performed is actually a backward trajectory assessment: from the future to now. If this exercise is done by true mathematicians, the uncertainties actually cumulate and the result really depends on the chance of more optimistic or more pessimistic probabilities occurring. That is the reason why these trajectories should be considered with distance and as an engineering design exercise, not as an ultimate and perfect tool to fully assess the problem.

NOTE.– In recent IPCC reports, the research communities organize themselves very strongly with regard to the scenarios for GHG emissions, respecting a certain value of radiative forcing by 2100. However, for aerosols, the research communities were free to choose any aerosol scenario, which means differences, as shown in Figure 5.22. Measures are made even more complex by the presence of clouds constantly moving.

The understanding of the impact of aerosols is improving but has still not been fully mastered. A set of various satellites composing the so-called "A-train", each looking at a sol part of the Earth, are used for the assessment of aerosols and cloud parameters as shown in Figure 4.26 in Chapter 4. The currently active satellites, excluding failed and deactivated/decommissioned ones, are:

– OCO-2 launched by NASA in 2014, which replaced OCO after its failure. Its mission is mainly to study atmospheric CO_2 at regional scales;

– Aura also launched by NASA in 2004, with the aim of measuring ozone quality, and other atmospheric parameters including water vapor and aerosols;

– GCOM-W1 with the nickname "shizuku", launched by JAXA in 2012, with the successor of the instrument initially carried by Aqua. Its mission is to observe the water cycle.

The idea behind the launch of these satellites was of course to gain information about climate sensitivity through a good understanding of climate processes and to settle the issue.

But the climate system is more complex than the current state of available measurement tools, and a way to deal with this high complexity is to conduct analyses about different elements of the climate problem and therefore to divide it into smaller, more understandable parts.

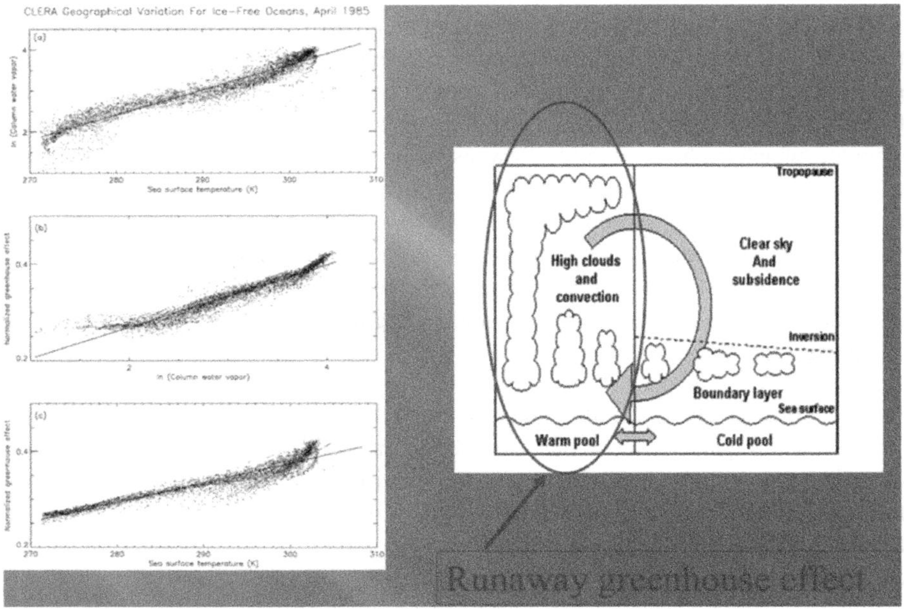

Figure 5.23. *The runaway greenhouse effect. Scatterplots and regression lines from the geographical distributions of CLERA data over the ice-free oceans for April 1985. (a) Natural logarithm of column water vapor (kg m^{-2}) against sea surface temperature (K). (b) Normalized greenhouse effect against natural logarithm of column water vapor (kg m^{-2}). (c) Normalized greenhouse effect against sea surface temperature (K). Source: Slingo et al. (1999)*

For the potential runaway greenhouse effect that could happen in the Tropics (see Figure 5.23), already discussed in a previous section, it would be stabilized but disappearing in the intertropical region: in fact, cloud feedback has a very different effect in the tropical area versus elsewhere in the planet.

If we consider a very simplified view of the climate system, limited to the intertropical area, with the warm pool of water near the equator, and the cold pool of water in the subtropical region, at about between 30°S and 30°N (see Figure 5.24), in the warm pool, there are large cloud formations, due to the convection/hot water and to different types of clouds.

Two different kinds of feedbacks are possible: they can be due to deserts over lands, or they can be due to the boundary layer clouds over oceans. Indeed, if we heat the system, there are two responses: one feedback for the region on the left part of the equator as represented in Figure 5.24, and one part of the region on the right part of the equator line in Figure 5.24.

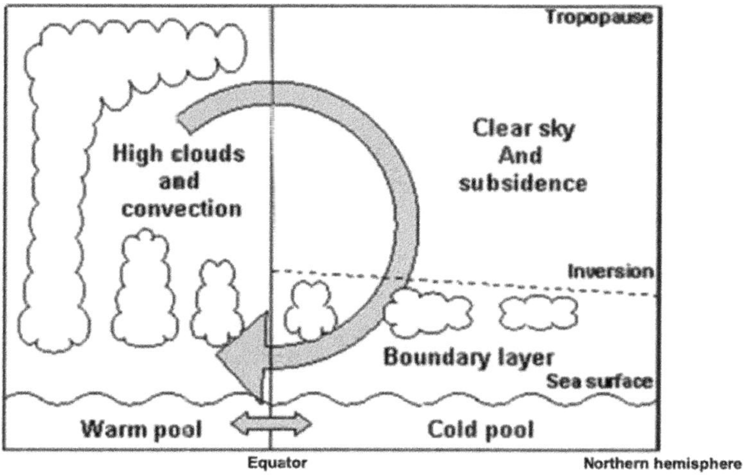

Figure 5.24. *A schematic simplified view of the climate system. Source: Le Treut (2018)*

Part of the studies on this topic, largely led by Jean-Louis Dufresne and Sandrine Bony at the Pierre Simon Laplace Institute, show that it is not possible to look at the problem of global warming and to assess whether the model is right or wrong due to the global feature of climate change.

However, if we can isolate specific feedbacks and assess them individually and for different regions of the world, it is possible to restore a more realistic big picture. This view validates different types of models: for instance, binary models similar to the scheme in Figure 5.23, with two boxes.

By using such binary models and analyzing how feedbacks occur, it is also possible to study their responses to events such as the simulation of volcano eruptions or to seasonal cycles, etc. There are several situations for which it is possible to question the capacity of a simple model or of more complex models to simulate real-world events.

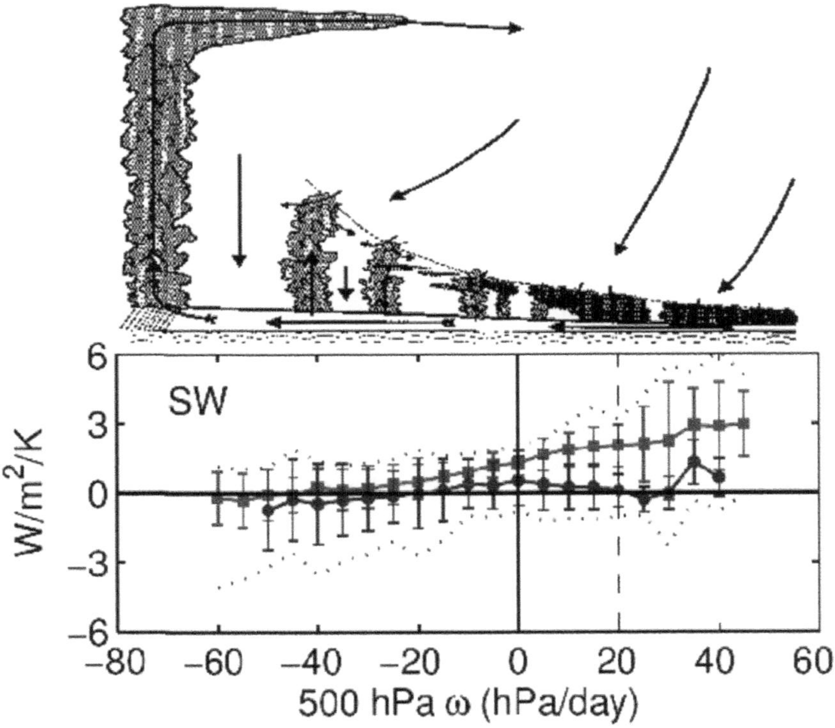

Figure 5.25. *A scheme of a method to split the cover of the Earth by studying regions corresponding to ascending or descending components of the atmosphere: relationship between radiative forcing and temperature (y-axis) as a function of the vertical velocity measured in hPa/day (x-axis). Source: Bony and Dufresne (2005)*

The researchers Sandrine Bony and Jean-Louis Dufresne decided to study not a two-column model, but rather models with a large number of columns. The bottom part of Figure 5.25 shows a way to split the cover of the Earth by studying regions corresponding to ascending or descending components of the atmosphere. The velocity is measured with a daily pressure expressed in hPa/day. Pressure reduces with height, which means that a negative velocity shows an upward direction. For each of the models presented in the bottom part of Figure 5.25, there is a double derivation: a bin is chosen for collecting the results of all models in this bin, and the feedbacks in this bin are analyzed. Then, there are a number of occurrences of vertical velocity at –40 hPa/day for instance, and the idea is to assess the dependence of this value with temperatures (in $W/m^2/K$). For the velocities with one given value, we can study how the radiative forcing depends on temperature. The principle is to analyze the relationship between forcing and temperature for a given atmospheric dynamical feature, which is the vertical velocity measured in hPa/day. This method is an important result, and it has been used as a way to intercompare models. Indeed, the models delivering a strong value in terms of climate sensitivity, which means in terms of global amplitude of climate change, were the models where there is also a larger feedback for the short-wave radiations/solar reflection.

As for clouds, they occur both at very high vertical velocity and at very low vertical velocity: the differentiation is probably at the origins of the discrepancies between the models. These differences between the models are actually large: we already saw that it is about ±1°C. Again, if we want to draw a backward trajectory starting from the end of the century with a +2°C elevation of temperature target, it means that in the more optimistic case, we may be able to reach an objective of a reduction that is higher than the Paris Agreement target; but in the worst-case scenario, it is necessary to also be prepared for this configuration too.

The tendency is to look at scenarios only, instead of having as a comparison two sets of models: on the one hand, the models, and, on the other hand, the feedbacks that can be observed in the real world.

Observed feedbacks can be very different and are related to natural changes including:

– seasonal cycles (a model that would not give the right feedback at the seasonal cycles would be excluded from any inter comparison exercise to assess the level of future warming of the Earth);

– effects such as El Niño, which is a strong modulator of the climate system and there are feedbacks to El Niño;

– volcanic eruptions: this is a phenomenon not much studied at the Institute Pierre Simon Laplace.

The satellites of the A-Train are now about 15 years old, and it took a long time to launch those missions. During this satellite mission time period, the research teams have been quite unlucky as there have not been many El Niño events. It would have been very helpful to use some observed feedbacks to this as a strong El Niño event would leave a large mark on the Earth, in order to assess how models react to that.

There are also many efforts to study models as if there were really considering the information given by satellites, therefore having satellites measuring the models, as they would measure Earth's parameters.

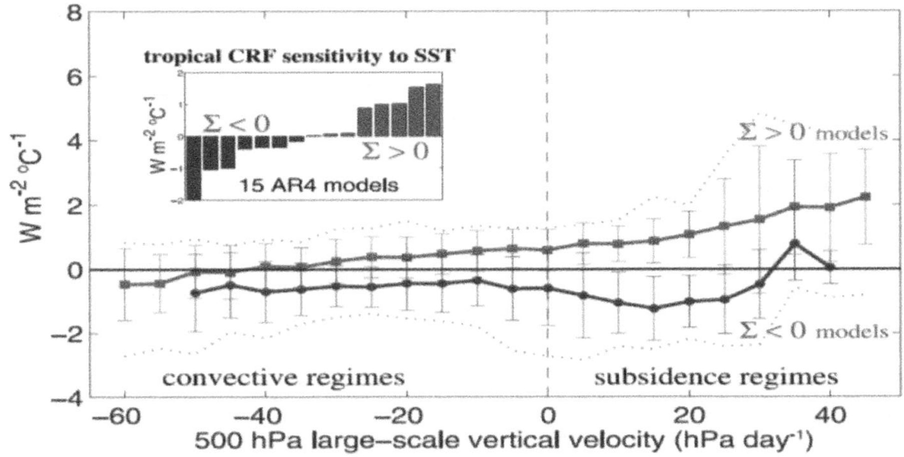

Figure 5.26. *Sensitivity of the tropical net cloud radiative forcing to sea surface temperature (SST) changes associated with global warming. Source: adapted from Bony and Dufresne (2005)*

The same type of results from models shown in Figure 5.24, but with different results of sensitivity for, respectively, convective and subsidence regimes are displayed in Figure 5.26. Many other points could be added on this part, but we will stop the discussion of this topic at this point.

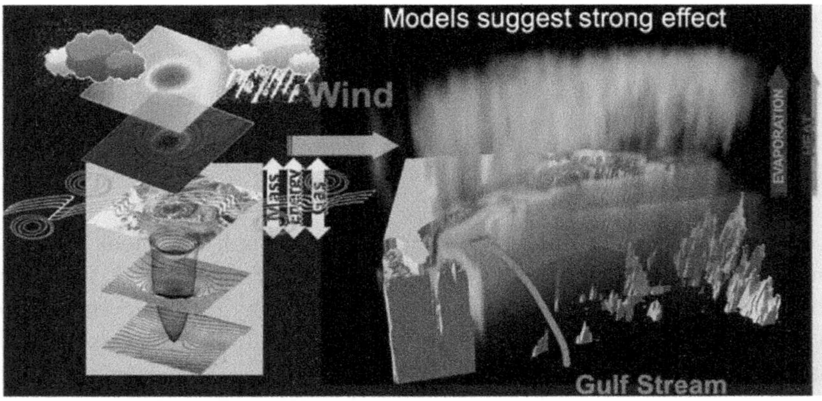

Figure 5.27. *Multi-model assessment of changes in marine ecosystems – a simplified scheme of the ocean: air–sea interactions at the ocean mesoscale. Source: Speich (2018)*

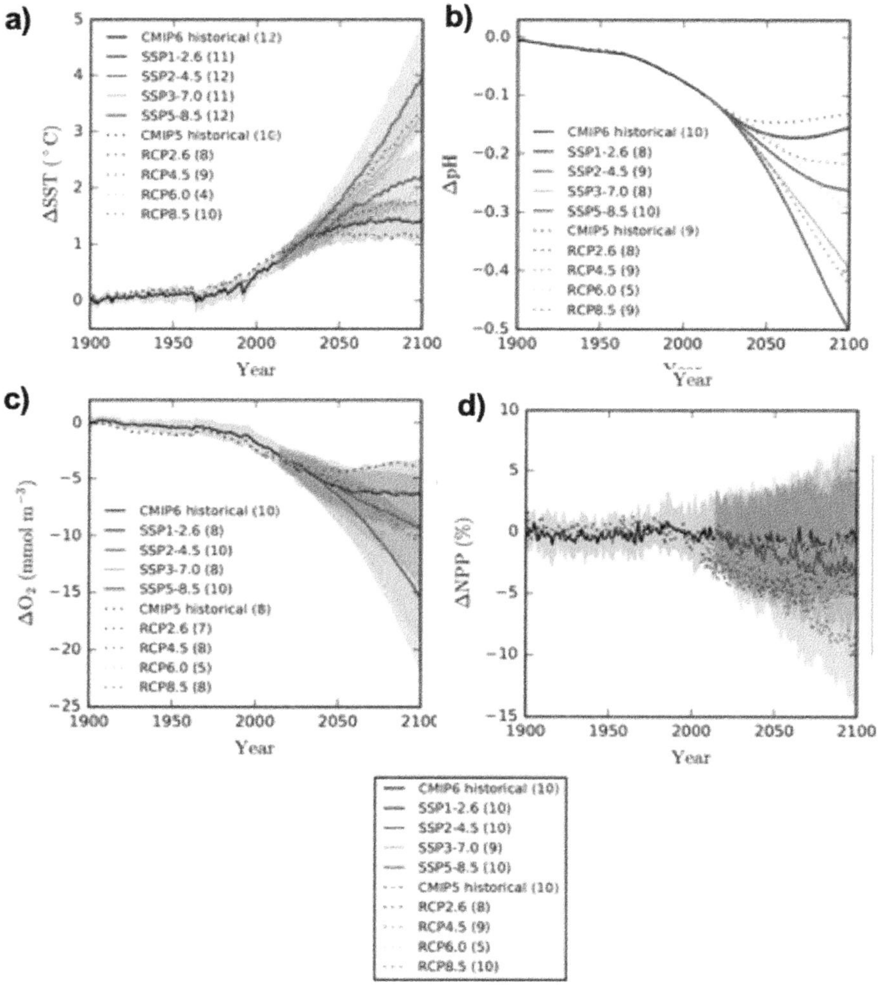

Figure 5.28. *Global mean projections of upper-ocean impact drivers. Global mean projections of (a) sea surface temperature (°C), (b) surface ocean pH, (c) subsurface dissolved O_2 concentration (averaged between 100 and 600 m, mmol.m^{-3} and (d) depth-integrated net primary production (%). Values are anomalies relative to the 1870–1899 reference period. CMIP6 mean anomalies for the historical and SSP simulations are shown as solid lines, with shading representing the inter-model SD. CMIP5 projections only show the multi-model mean. The model ensemble size for each scenario is given in parentheses. Source: Kwiatkowski et al. (2020)*

We have not much emphasized the Earth–ocean interaction represented in Figure 5.27, but it is now well known that oceanic air–sea interactions are intimately linked with turbulences over the ocean. This new view of the oceanic air–sea interactions is also very important, especially knowing that about 90% of the supplementary heat due to greenhouse gas is stored in the ocean.

In the end, considerations on feedbacks have consequences on many results coming from models. For example, Figure 5.28 shows results from IPSL models involving marine biochemistry and results from the team of Laurent Bopp (Kwiatkowski al. 2020) with the comparison of CMIP6 and CMIP 5 results. These results show the surface temperature and the surface pH from different scenarios, as well as the oxygen and the net primary production (NPP). Changes in NPP due to climate change are still affected by several uncertainties.

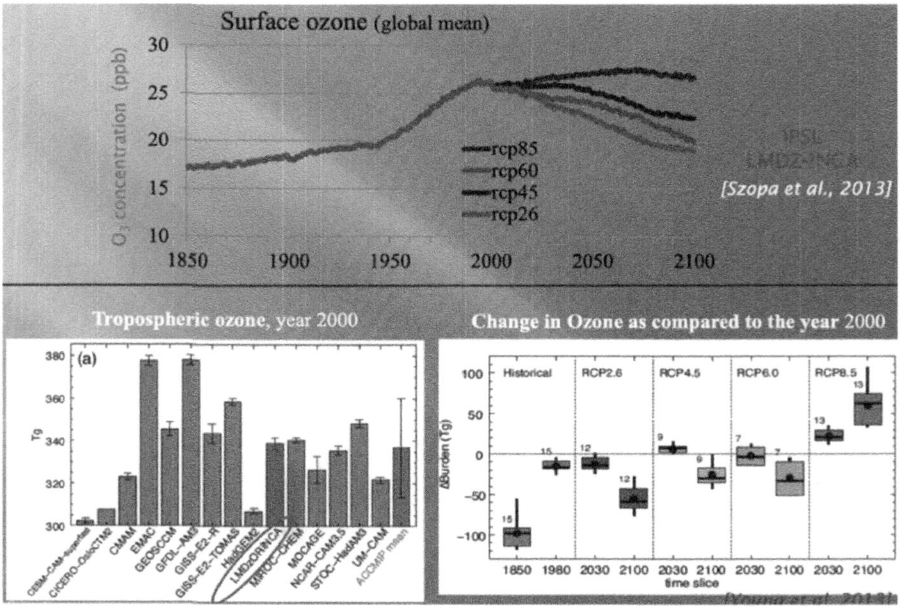

Figure 5.29. *Changes in the tropospheric ozone. Sources: IPSL, LMDZ-INCA, Szopa et al. (2013) and Young et al. (2013)*

When considering results from tropospheric models regarding the tropospheric ozone, i.e. ozone associated with the decomposition of surface gases, which are pollutants (see Figure 5.29), we note that paradoxically, uncertainties tend to increase when results are combined for a deeper understanding using multidisciplinary sciences: the more we know, the more we know what we do not know.

Two key takeaway points from the study by Cassou shown in Figure 4.56 (see Chapter 4) about natural variability are very important: the major amplitude is at the scale of big anticyclones and cyclones. This is important as we may think that if we go to small scales, it is increasingly complex, with more and more uncertainty. This is not completely true in the climate system, as there is a natural nonlinear dynamic between complex systems, which is causing natural variability and this dynamic, which is unpredictable. This is one of the reasons that causes difficulties in making meteorological forecasts over a few days, and a main driver of uncertainties. At smaller scales, the forecast is sometimes easier. For example, by following the position of an anticyclone over the Atlantic, it is predictable to forecast rain in Europe. At an amplitude of large synoptic-scale processes lies a large part of the uncertainties.

There are numerous climate fluctuations, such as the North Atlantic Oscillation, and a few more related to the North Atlantic region (during the summer, it is difficult to find real oscillation signature, and it is better organized in winter): this signal is partly chaotic and uncertain in terms of predictability.

As a synthesis of Figure 4.56, the regional climate oscillates towards different frequently occurring patterns. And the issue is that the climate goes from one pattern to another. The amount of time spent in days per year is displayed for each pattern. Figure 4.56 shows results over about 50 years from now, and we note that there are no "stick oscillations". This kind of approach has been developed over the last 30 years, but still the results are not completely understood. What we know is that there are different kinds of potential changes and it is an important result as it is crucial to avoid what is called "maladaptation". For instance, there was a first rush when the first results from Cassou were released, stating that southern Europe would be drier: then, politicians focused on new ways to anticipate the future on the angle of dryness, but it is not the way it will necessary be happening in central and western Europe for instance; floods are even possible some years, with intermittency. What can happen is very differentiated between episodes. The purpose of Figure 4.56 is to show that what we call "uncertainties" in mid-latitude regions are not completely stochastic: there are four well-quantified, regimes, and it is possible to "bracket" the variability of climate depending on location. It is also necessary to imagine future infrastructure in relation to that.

El Niño and the North Atlantic Oscillation are generally two well-known phenomena associated with natural variability, but there are many others: for instance, the oscillation over the South Atlantic Convergence Zone (SACZ) represented in blue in Figure 5.30, which changes the climate, and especially affects Argentina and its agricultural sector.

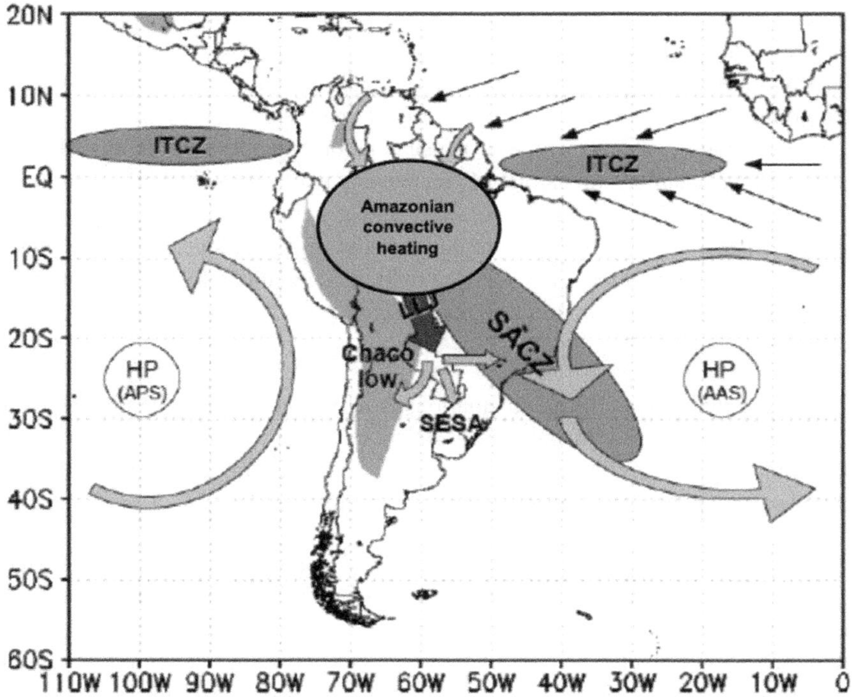

Figure 5.30. *A scheme of the South Atlantic Convergence Zone. Source: Le Treut (2018)*

If we want to use models for specific regions, the 100 models participating in CMIP6 for instance, it is necessary to select the capacity to obtain realistic results for natural variability in the current world. This means that when studying intercomparison models, for anything, we put a range of values which are most of the time the results of a direct use of the whole range of models: this is less and less what is being done because if we try to separate models integrating some oscillations such as the North Atlantic Convergence Zone, and models that do not have such oscillations, we have access to the results of models that integrate those oscillations, which is a much more coherent way to proceed.

This method tends to indicate that climate change has produced some new features over Argentina, and that the climate over this part of America may perhaps have really changed because of human activities.

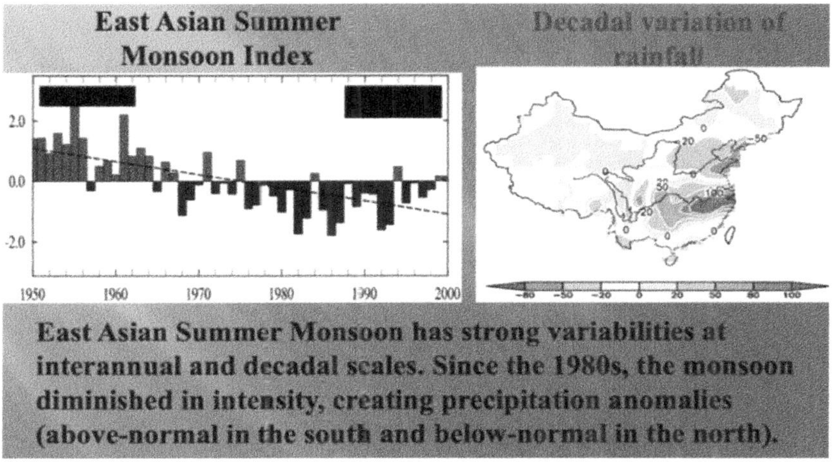

Figure 5.31. *Variation of the East Asian Summer Monsoon during the second half of the 20th century. Source: Hsu et al (2013), after Li et al. (2010)*

Another example of a less well known natural oscillation is shown in Figure 5.31. The right side of the figure represents an oscillating system, and the metric represented with colors is the empirical numerical orthogonal function (ENOF) over China. There is a significant change between the monsoon in southern China and the monsoon in northern China. A positive sign of monsoon index means that the monsoon in northern China is large. The trend of this index (left side of Figure 5.31) seems to show that the monsoons have been decreasing since the 1950s, therefore already for decades. China has built two canals for water transportation from the south to the north. The first canal was built under the era of the Chinese emperors and modernized later on. A second canal was finished just a few years ago, and investments have also been gathered for a third one, representing a huge financial investment, much larger than for the Three Gorges river dam for instance. During recent years, monsoons in northern China have come back, but they may disappear again.

When dealing with natural variability, the concept of risks is key, and the question is how they are considered or not. China decided to make the necessary investments in order to get water in the north, in a resilient way towards natural conditions.

Figure 5.32. *Risk in the IPCC assessment through time. (a) An explicit risk framing emerged in the IPCC Special Report on Managing the Risks of Extreme Events and Disasters to Advance Climate Change Adaptation (SREX) and WGII AR5 (IPCC 2014a, b). (b) In the current assessment, the role of responses in modulating the determinants of risk is a new emphasis (the "wings" of the hazard, vulnerability and exposure "propellers" represents the ways in which responses modulate each of these risk determinants). The difference between risk and vulnerability has been emphasized in IPCC Managing the Risks of Extreme Events and Disasters to Advance Climate Change Adaptation report on extreme events. Reproduced with permission from Ara Begum et al. (2022)*

They are different ways to look at vulnerability, which link with the risk detailed in Figure 5.32: a weather/climate risk can be defined at the convergence of an unaddressed vulnerability and exposure during weather and climate events.

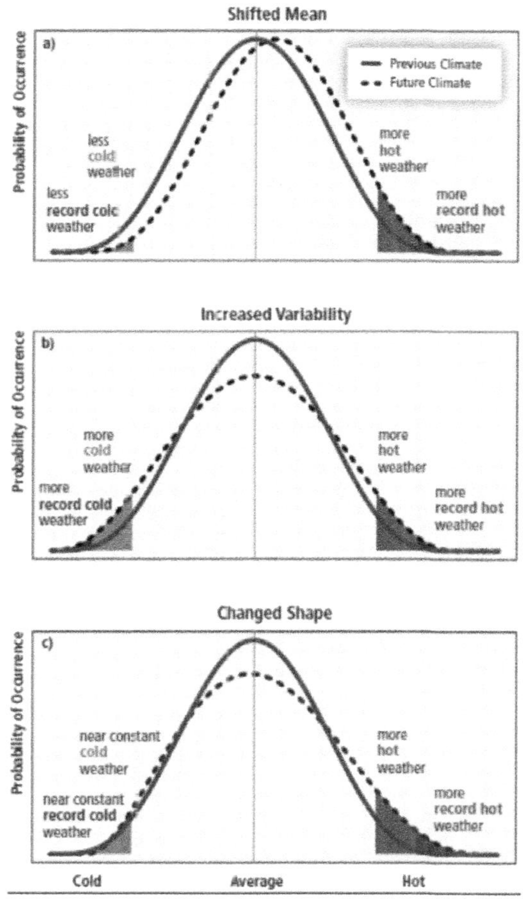

Figure 5.33. *The largest uncertainties affect rare extreme events. Reproduced with permission from IPCC (2012)*

The curve of probability of having a temperature at a given point of the globe is a Gaussian (see Figure 5.33), and we want to study how the probability of occurrence of temperature may change in the future as compared to the past climate. The result from a global increase is materialized by a shift of the temperature curve.

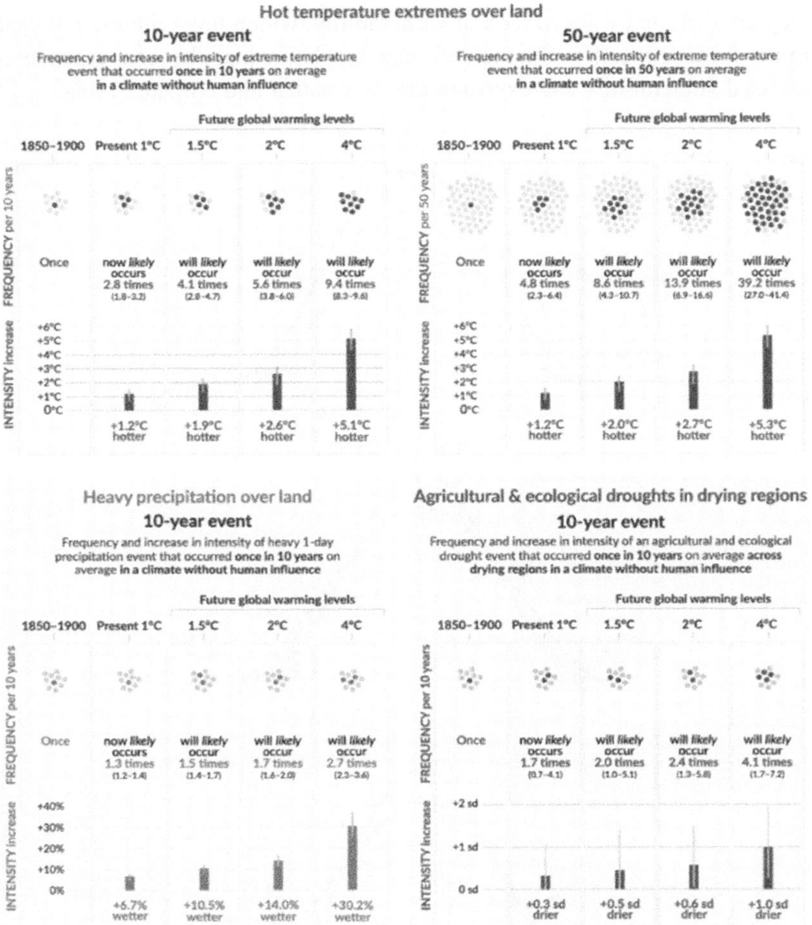

Figure 5.34. *Projected changes in the intensity and frequency of hot temperature extremes over land, extreme precipitation over land, and agricultural and ecological droughts in the drying region. Reproduced with permission from IPCC (2021b)*

NOTES ON FIGURE 5.34.– Projected changes of annual soil moisture are shown at global warming levels of 1°C, 1.5°C, 2°C and 4°C and are relative to 1850–1900, nine representing a climate without human influence. The figure depicts frequencies and increases in the intensity of 10- or 50-year extreme events from the base period (1850–1900) under different global warming levels. Hot temperature extremes are defined as the daily maximum temperatures over land that were exceeded on average once in a decade (10-year event) or once in 50 years (50-year event) during the 1850–1900 reference period. Agricultural and ecological drought events are defined as the annual average of the total column soil moisture below the 10th percentile of the 1850–1900 base period. These extremes are defined on a model grid box scale. For hot temperature extremes and extreme precipitation, results are shown for the global land. For agricultural and ecological drought, results are shown for drying regions only, which correspond to the AR6 regions in which there is at least medium confidence in a projected increase in agricultural and ecological droughts at the 2°C warming level compared to the 1850–1900 base period in the Coupled Model Intercomparison Project Phase 6 (CMIP6). In the "frequency" section, each year is represented by a dot. The dark dots indicate years in which the extreme threshold is exceeded, while light dots are years when the threshold is not exceeded. Values correspond to the medians (in bold) and their respective 5–95% range based on the multi-model ensemble from simulations of CMIP6 under different shared socio-economic pathway scenarios. For consistency, the number of dark dots is based on the rounded-up median. In the "intensity" section, medians and their 5–95% range, also based on the multi-model ensemble from simulations of CMIP6, are displayed as dark and light bars respectively. Changes in the intensity of hot temperature extremes and extreme precipitation are expressed as degree Celsius and percentage. As for agricultural and ecological drought, intensity changes are expressed as fractions of the standard deviation of annual soil moisture. Source: IPCC, AR6, WGI, SPM (2021).

There is an almost systematic effect in these diagrams showing the probability of occurrences, which is the large increases of tail risks/extreme events. This means that extreme events will have a higher probability of occurrence/will be more frequent, with for instance a decadal rather than centennial return period. Figure 5.33 represents a Gaussian. However, we saw previously that the probability of occurrence of temperature over a given region is not Gaussian; generally, it has at least two rules: one for blocking events, and the other for all the other events. Therefore, these are important concepts, but it is not exactly as simple in reality. The main message from Figure 5.33 is that the largest uncertainties affect rare extreme events.

In the Sixth IPCC Assessment Report, there is, however, a trend pointing to the fact that the projected changes in extremes are larger in frequency and intensity with every additional increment of global warming (see Figure 5.34).

5.1.2. *From natural variability to climate impacts*

The aim of this section is to discuss how to go from natural variability to climate impacts.

It has been extremely challenging so far to deliver precise climate projections for a given region. A way to work around this problem is to look at "vulnerabilities" towards temperature changes for different ecosystems, and to settle things in terms of processes considering these vulnerabilities in terms of water, ecosystems, food, coastal areas and health as shown in Figure 5.21. Experiments have been carried out on this, with studies on systemic results and derivations of temperatures reflecting the concepts of vulnerability, exposure and risk. This depends a lot on models.

In fact, two approaches tend to be balanced as represented in Figure 4.59 (in Chapter 4). One approach consists of "cutting" a location from a global view, leading to a different model with related specific associated physics, because clouds and processes are represented differently with the possibility to downscale a global model into smaller areas. Therefore, these models are not only driven by their boundaries by the global models, they are also specific in terms of physical processes: they can have more explicit cloud physics inside because when using a different model, it is possible to rewrite this differently, and it is just driven by its own boundary conditions, which gives the capacity to go into the detailed physics.

Another approach that we have in France is to have models with an adaptive scale such as the LMDZ model as represented in Figure 5.35. This version is regular in latitude and longitude, but it is possible to increase the steps between two lines of latitude and longitude by modulating the mesh of the model grid differently in latitude and longitude, thus creating a zone of regional focus with a very high resolution. The difficulty is to use the same physics when proceeding this way. But there is also a necessity to save computer time, and if we used high resolution for all models, it would not be possible to do the simulations with the existing computers; much more powerful computers would be needed. These are tricks climatologists use while waiting for better computing capacity. In the course of the last decades, computing times have been increasing dramatically, and further innovations can happen fast, but it is not enough right now.

The Arpege model of Meteo France also uses decomposition into spherical modes with Fourier transforms, and these spherical modes are used to reconstruct the Earth's behavior: the problem is that with spherical modes, adding resolution on one side tends to give less resolution to the other side; therefore, there are problems everywhere and with all methods.

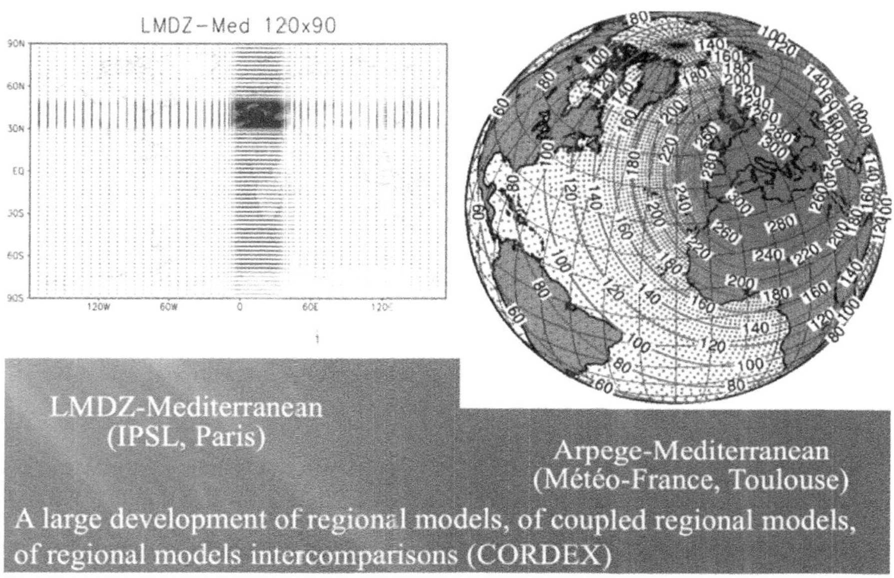

Figure 5.35. *The French "zoomed" climate models, a large development of regional models, of coupled regional models, of regional model intercomparisons (CORDEX). Source: IPSL/Meteo France (2007)*

There is a regional model intercomparison program named CORDEX. CORDEX has made huge numbers of experiments in all regions of the world. Over South America, CORDEX treats a huge domain, with mountains, the interface between oceans and continents, changes in some models going through the equator. This means having this being driven by the boundary conditions, as there is complex physics inside. Research is still ongoing in the area of local issues that still need improvements.

In the next section, we will say a few words about how the IPCC makes use of all the tools previously described to deliver confidence estimates and uncertainty assessments fitting the purposes of their analyses.

5.1.3. *Zoom about uncertainties and the IPCC*

In the treatment of scientific work, there are many different kinds of uncertainties to handle, from unknown processes to progress to be done.

An excerpt of the IPCC Working Group II Fifth Assessment Report (2013) delivers the following definitions of two ways to handle uncertainties, by relying on:

> Confidence in the validity of a finding, based on the type, amount, quality and consistency of evidence (as, for example, mechanistic understanding, theory, data, models, expert judgment) and the degree of agreement. Confidence is expressed qualitatively.

And when possible, in addition, on:

> Quantified measures of uncertainty in a finding expressed probabilistically (based on statistical analysis of observations, model results, or expert judgement).

Another important work of the IPCC is to try to precisely match words and the likelihood of a possible outcome as shown in Table 5.1.

Term	Likelihood of the outcome
Virtually certain	99–100% probability
Extremely likely	95–100% probability
Very likely	90–100% probability
Likely	66–100% probability
More likely than not	> 50–100% probability
About as likely as not	33–66% probability
Unlikely	0–33% probability
Very unlikely	0–10% probability
Extremely unlikely	0–5% probability
Exceptionally unlikely	0–1% probability

Table 5.1. *Communication of confidence level in the Working Group II Fifth Assessment Report. Reproduced with permission from Chapter 1 of IPCC (2021a)*

A confidence scale defined by the terms which likelihood has been specified in Table 5.1 is itself characterized by both a level of agreement (low, medium, high) and by evidence (type, amount, quality, consistency, which can be limited, medium or robust), as detailed in Figure 5.36.

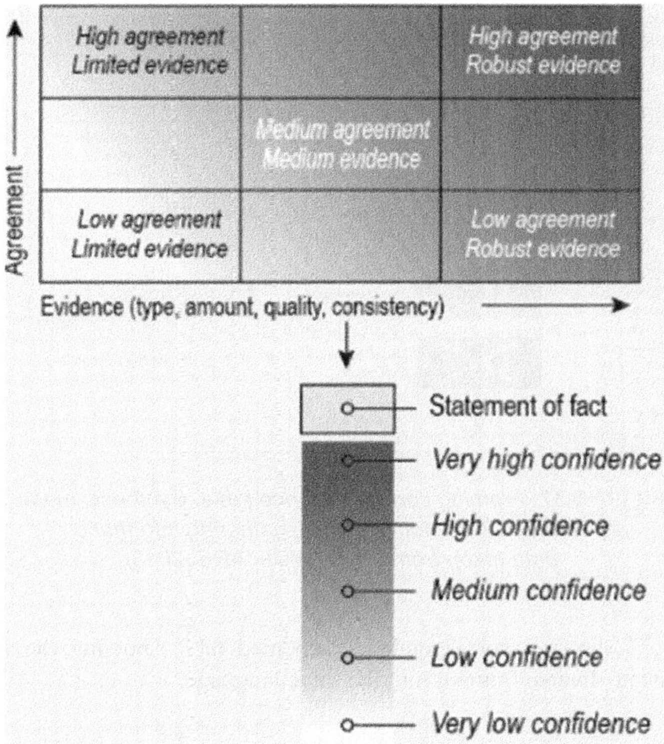

Figure 5.36. *Component of confidence scale as defined in the Working Group I Sixth Assessment Report. Reproduced with permission from Chapter 1 of IPCC (2021a)*

The confidence scale detailed in Figure 5.36 is used as a way to specify the confidence in assertions delivered by the IPCC. This IPCC vocabulary is extremely sophisticated, and unfortunately there is no use of that language outside of science, even if it could also be very helpful in the world of media, in the judicial sphere, etc.

5.1.4. *Process of uncertainty evaluations by the IPCC as detailed in the Fifth Assessment Report (AR5)*

Additionally, the idea that the future assessment of climate will be increasingly multidisciplinary is now clear, and it will be necessary to strengthen the mixing of many different sciences, including social sciences, agricultural science, epistemology, etc.

Table 1-2 Notes: Attribution in the top-half of the table is to human forcing of the climate, primarily through the increase in greenhouse gases (WGI). Attribution in the bottom-half, impacts, adaptation and vulnerability is to the observed local or regional climate change (WGII).

Trend		Confidence Assessment		Likelihood Assessment	
∧	Increasing overall	HC	High or Very High confidence	****	*Virtually Certain* 99% - 100%
∨	Decreasing overall	MC	Medium confidence	***	*Extremely Likely* 95% - 100%
∧∨	More regions increasing than decreasing	LC	Low confidence	**	*Very Likely* 90% - 100%
∨∧	More regions decreasing than increasing	X	Very low or no confidence	*	*Likely* 66% - 100%
~	Regionally varies or no clear trend	X	No assessment made		

Figure 5.37. *Framing correspondence table. Synthesis made by Hervé le Treut for this book, gathering information data taken from IPCC, WGII, AR5 (2013)*

Figure 5.37 shows a correspondence schemed table showing the IPCC team effort to translate diverse framing into the same language.

Phenomenon	Change	Observed to 2010 X-axis Fig 1-6	Attributed (humans or obs. climate change) Y-axis ■ ■	Projected 2050-2100 Y-axis ○	Source
Greenhouse gases: CO_2, CH_4, N_2O	∧	****	****	**** (RCPs: CO_2,N_2O)	AR5 I-2, I-10, I-11, I-12
Global Mean Surface Air Temperature (GMST)	∧	****	***	****	AR5 I-2, I-10, I-11, I-12
GMST over all continents except Antarctica	∧	****	*	****	AR5 I-2, I-10, I-11, I-12
Global mean sea level	∧	****	**	****	AR5 I-3, I-10, I-13
Arctic sea ice cover	∨	****	**	**	AR5 I-4, I-10, I-11, I-12
Hot days and nights over land (warmth, frequency)	∧	**	**	****	AR5 SPM-1
Cold days and nights over land (warmth, frequency)	∨	**	**	****	AR5 SPM-1
Extreme high sea level (incidence, magnitude)	∧	* (since 1970)	X	**	AR5 SPM-1
Heat waves and warm spells over land (frequency, duration)	∧∨	MC	*	**	AR5 SPM-1

Heavy precipitation events	^v	*	MC	**	AR5 I-2, I-10, I-12
Drought (intensity, duration)	^v	MC (some regions)	LC	*	AR5 SPM-1, SREX-4
Tropical cyclones (intensity, frequency, some basins)	~	LC	LC	MC (intensity increase, some basins)	AR5 SPM-1
Global mean precipitation	^	LC	LC	****	AR5 I-2, I-10, I-11, I-12
Contrast between wet and dry regions	^	X	X	HC	AR5 I-12
Snow cover (NH, extent)	v	HC	HC	HC	AR5 I-4, I-10, I-12
Permafrost regions (degrade)	v	MC	X	MC	AR5 I-4, I-12
Storm tracks (shift poleward)	^	*	X	*	AR5 I-2, I-12
Wave heights (different oceans)	^	MC (N. Atlantic)	X	** * (Arctic) (Southern)	AR5 I-3, I-13
Upper ocean (warming)	^	****	***	***	AR5 I-3, I-10, I-11, I-12
Ocean acidification	^	****	***	****	AR5 I-3, I-10, I-6
Oceanic oxygen	v	MC	MC	**	AR5 I-3, I-10, I-6
Floods (magnitude, frequency)	~	LC	LC	LC	SREX-3
Mountain phenomena (slope instabilities mass movement, glacial lake outbursts)	^	HC	HC	HC	SREX-3, AR4 SyR
Monsoons	~	LC	LC	LC	SREX-3
Plant and animal species (move poleward or up in altitude)	^	HC	HC	HC	AR4 II-SPM, AR4-SyR
Mountain phenomena (slope instabilities, mass movement, glacial lake outbursts)	^	HC	HC	HC	SREX-3, AR4 SyR
Timing of spring events (earlier leafing, greening, planting, bird migration, ...)	^	HC	HC	HC	AR4 SyR
Marine / freshwater biological systems (shifts in algal, plankton and fish range)	~	HC	HC	HC	AR4 SyR
Human health (heat-related mortality, infectious disease vectors)	^	MC	MC	X	AR4 SyR
Water resources	v	X	X	(many regions)	AR4 SyR-SPM
Mountain glaciers	v	HC	X	HC	AR4 II-SPM
Coral degradation, bleaching	^	HC	X	HC	AR4 II-SPM, SyR-SPM
Economic losses from weather- and climate-related disasters	^	HC	X	HC	SREX-4
Annual costs of climate change	^	X	X	**	Ar4 SyR-SPM

Figure 5.38. *Confidence in the observation, attribution of changes in climate system phenomena. Synthesis made by Hervé le Treut for this book, gathering information data taken from IPCC, WGII, AR5 (2013)*

Since the IPCC Fifth Assessment Report, well-known phenomena are characterized by a high confidence level (see Figure 5.38). Some processes have been observed in the past until now. For instance, the increase of greenhouse gases in the atmosphere. Other criteria include the attribution to anthropogenic causes, change and projection.

When studying how a parameter may evolve in the future in light of climate change, the method of assessment is to combine three criteria, answering the following questions: Is it changing? Can we attribute this change to human activities? Do model projections deliver consistent patterns? Figure 5.39 shows this principle, with the same three categories answering the same questions just mentioned, once again: Can we observe changes? Can we attribute the changes? Are projections convergent?

Changes in Phenomenon	Uncertainty in observed changes (since about the mid-20th century)			Uncertainty in projected changes (up to 2100)		
IPCC Assessment Report	TAR	AR4	SREX	TAR	AR4	SREX
Higher maximum temperatures and more hot days	*Likely* over nearly all land areas	*Very Likely* over most land areas	*Very Likely* at a global scale	*Very Likely* over nearly all land areas	*Virtually Certain* over most land areas	*Virtually Certain* at a global scale
Higher minimum temperatures, fewer cold days	*Very Likely* over nearly all land areas	*Very Likely* over most land areas	*Very Likely* at a global scale	*Very Likely* over nearly all land areas	*Virtually Certain* over most land areas	*Virtually Certain* at a global scale
Warm spells/heat waves. frequency, length or intensity increases	-	*Likely* over most land areas	*Medium Confidence* in many regions	-	*Very Likely* over most land areas	*Very Likely* over most land areas
Precipitation extremes	*Likely*[1], over many Northern Hemisphere mid- to high latitude land areas	*Likely*[2] over most areas	*Likely*[3]	*Very Likely*[1] over many areas	*Very Likely*[2]	*Likely*[2,4] in many land areas of the globe
Droughts or dryness	*Likely*[5], in a few areas	*Likely*[6], in many regions since 1970s	*Medium Confidence* in more intense and longer droughts in some regions, but some opposite trend exists	*Likely*[5], over most mid-latitude continental interiors (Lack of consistent projections in other areas)	*Likely*[6]	*Medium Confidence*[7] that droughts will intensify in some seasons and areas; Overall *low confidence* elsewhere
Changes in tropical cyclone activity (i.e. intensity, frequency, duration)	Not Observed[8], in the few analyses available	*Likely*[9], in some regions since 1970	*Low confidence*[10]	*Likely*[8], over some areas	*Likely*[9]	*Likely*[11]
Increase in extreme sea level (excludes tsunamis)	-	*Likely*	*Likely*[12]	-	*Likely*	*Very Likely*[13]

Figure 5.39. *Changes in the phenomenon, uncertainty in observed changes (since about the mid-20th century), and uncertainty in projected changes (up to 2100). Reproduced with permission from Cubasch et al. (2013)*

All of the items listed on the first columns of Figure 5.39 are very probably changing. But when considering the changes in the tropical cyclone activity for instance in terms of intensity, frequency and duration, the related uncertainty stated is "low confidence". Going back to Figure 5.38, we also read that the uncertainty related to global mean precipitation is stated with "low confidence" too, as well as for the anthropogenic attribution of droughts, and models disagree on future changes, similarly for floods and monsoons. This is important to understand when studying local issues, but this does not mean at all that the findings should not be considered: this only means that there will be changes for those parameters, but it is not possible to say "for sure" where, how and when the changes will happen. And it is not possible to predict either that a given category 5 cyclone is directly attributable to climate change. We should not try to "decide" whether a single event is due to climate change or not; this makes no sense: one single event could be 25% due to climate change and 75% to natural variability; this has no sense if we do not consider a series of events and use statistics. This is the reason why it is important to precisely read and cite IPCC reports to make sure that statements are grounded on robust results. It is extremely important to stick to this common reference language and to what is truly demonstrated, especially for journalists and decision-makers.

Using a precise language is actually a main source of misunderstanding between the scientific community and the decision-makers and journalists: the most recent tend not to go into the complexity of processes, and weak/incomplete information may create weak decisions and poor public information.

The following quotation from the IPCC AR5 report illustrates the importance of framing and of interpretation for the concept of uncertainty and probabilistic statements, and underlines the fact that it is impossible in the end to make a language completely objective:

> Many social sciences studies have found that the interpretation of uncertainty is contingent on the presentation of information, the context with which statements are placed and the interpreter's own lexical preferences. Readers often adjust their interpretation of probabilistic language according to the magnitude of perceived potential consequences (Patt and Schrag 2003; Patt and Dessai 2005). Furthermore, the framing of a probabilistic statement impinges on how it is interpreted (Kahneman and Tversky 1979): for example, a 10% chance of dying is interpreted more negatively than a 90% chance of surviving.

NOTE.– In France, the precaution principle exists in the Constitution: a sufficient amount of indication that there is a risk should push for a better risk estimate, and if it appears reasonable to believe that this requires a protection for different assets, the risk must be considered. This principle was a sufficient reason for France to have the obligation to protect the country against what may happen in 30 years, as it has committed to do by ratifying the Paris Agreement, declined in national, legally binding texts, notably through the national low-carbon strategy.

5.1.5. *Process of uncertainty evaluations by the IPCC as detailed in the Sixth Assessment Report (AR6)*

The IPCC evaluation framework is very codified through various steps (see Figure 5.40). The first step consists of collecting proofs for a given problem. For that purpose, the IPCC experts review the scientific literature to find these proofs. As a reminder, the IPCC does not produce any research but merely evaluates the scientific literature. If there are enough proofs in terms of quantity, as well as in terms of quality, and if they are consistent, then the IPCC evaluation goes to the second step, which consists of establishing a level of confidence regarding the evaluated problem/conclusion using the evidence collected in the literature. For instance, with regards to the problem of the global mean temperature, when gathering all proofs, including temperature observations, as well as model simulations and other indicators including sea levels, the IPCC can affirm, as quoted from the AR6 summary for policymakers (2021):

> Human activities, principally through emissions of greenhouse gases, have unequivocally caused global warming, with global surface temperature reaching 1.1°C above 1850–1900 in 2011–2020.

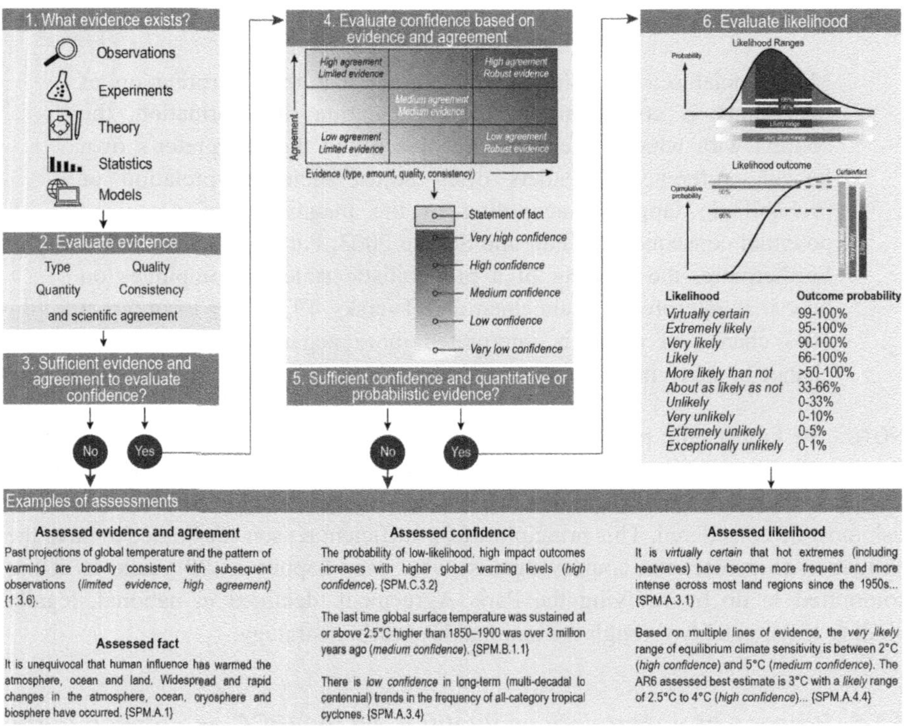

Figure 5.40. *Process of evaluation and communication of certainty in AR6 findings. Reproduced with permission from Chapter 1 of IPCC (2021a)*

The word "unequivocally" is the highest level of confidence that can be given in an IPCC evaluation; it refers to an established fact. The compilation of all evidence, their consistency, the fact that they are independent and point to the same conclusion, leads to this level of confidence.

There is not the same level of confidence for all the indicators: for precipitation for instance on an average for all the precipitations on the planet, the trend is an increase, but the quality and quantity of observations is such that it is not possible to conclude that it is an established fact: there is a probability for this increase after accounting for all evidence related to precipitation that there are two chances out of three that the global precipitations have increased for the last 50 years, as quoted from the AR6 summary for policymakers (2021):

> Globally averaged precipitation over land has likely* increased since 1950, with a faster rate of increase since the 1980s (medium confidence**)
>
> * likely = probability > 66%
>
> **medium confidence = not enough confidence to estimate a probability.

Another important question is whether precipitation increase is accelerating or not, especially over the last decades: for this indicator, the level of confidence is lower than the one for precipitations only, because there are fewer observations over the whole planet for the last decades. It is possible to deliver probabilities for the increase of precipitations in 2050, but not for the acceleration of this increase for which the level of confidence is medium, due to the quantity and quality of these evidence.

5.1.6. *Method for the evaluation of a probability in IPCC reports*

IPCC experts evaluate a probability starting from a high level to a very high level of confidence corresponding to the fact that there are numerous pieces of evidence in terms of quantity and quality, and that they are independent, pointing to the same conclusion for the assessed climatic phenomenon.

The last part of this book will be dedicated to the shift from global to local approaches.

5.2. From global to local climate features: a key shift for the interaction between science and decision-making

When going from global to local considerations, which is absolutely necessary, some results are well known while others are less understood, and this is an approach where uncertainties are key.

For many citizens, approaching the problem of climate change only through the mitigation and reduction of greenhouse gases is not acceptable. Some argue that since France represents only about 1.5% of global greenhouse gas emissions, while 98.5% of greenhouse gas emissions come from elsewhere, especially from China, the rest of Asia and from the USA, in this regard, the problem of energy for instance must be addressed globally and not only at a national scale, and also that what will happen in the next ten years has already been decided. Those statements are true. But they are not a justification for inaction, on the contrary.

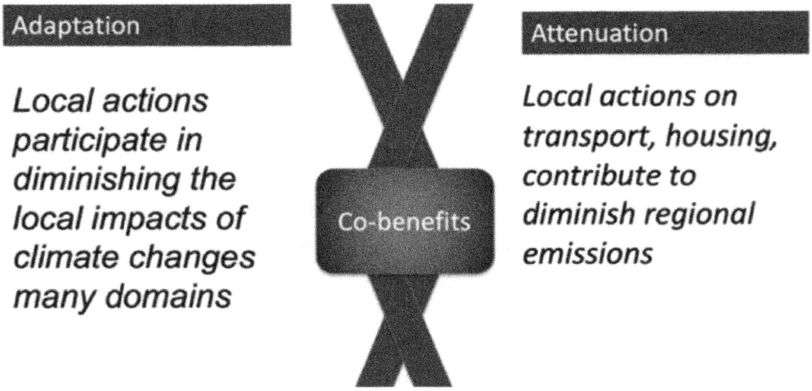

Figure 5.41. *The regions as an interface between climate adaptation and mitigation. Source: Le Treut (2018)*

The assessment of co-benefits of coupling both mitigation and adaptation efforts at the local scale is an effective incentive to local climate actions (see Figure 5.41).

The analysis of how it is possible to protect the human population from the impacts of climate change concretely is a good way to involve people's participation, especially for their region or their town, instead of only asking for mitigation efforts that may seem intangible for non-specialists. The actions at local scales are indeed closely related to lifestyles (housing, transport/infrastructure, agriculture, etc.): these activities are both essential for limiting emissions and are a main object of adaptation strategies as they will suffer from climate change. Many cities and regions have an important role to play in this matter.

In this section, we will mainly discuss the case of the Aquitaine region in France, with the wish that this demonstration serves other regions and purposes, as the Portuguese writer Miguel Torga used to say: "universal, is local minus walls" in his eponymous book published in 2012. Figure 5.41 therefore shows that regions should both work on adaptation and on attenuation to climate change in order to find local co-benefits as well, in terms not only on emission reduction, but also for biodiversity, food, health, social problems, etc. In the special IPCC report about +1.5°C, it has been recommended to make these co-benefit links, especially as the current trajectory is far from remaining below +1.5°C at the end of the century, and alternative adaptation solutions must also be created and implemented as soon as possible.

Adaptation is a good topic to gather people; it is not contradictory to attenuation. Adaptation is necessary although not sufficient: it needs to be evaluated at the local level, in light of results of both local and global model uncertainties, understanding that uncertainties at the local scale may result from different scales of the system. Adaptation requires both fundamental and applied research: the limits of what is predictable or not are not well known.

If we consider a large enough region, it is not possible to look at uncertainties separately. Results in one region cannot necessarily be extrapolated as there are large differences between different situations. A more comprehensive approach to the link between global and local perspectives is also a source of innovations in many domains, including Big Data, artificial intelligence, etc. Climate changes must indeed be understood as the changes of a complex system, which requires multidisciplinary approaches.

In terms of adaptation, it is also essential to avoid dangerously wrong "good solutions", such as the management of solar input for instance, with unpredictable and potentially irreversible effects on the Earth system balance and on biodiversity notably.

The challenges of the future are to simultaneously meet the needs of a limitation of greenhouse gases and to adapt to a future warming that reasonably does not seem to be avoided, especially in a global geopolitical context of fierce competition, not to say of war, and to develop a sufficient level of understanding in the population.

Continent	Increase factor from 2000 to 2050
Africa	**5**
Asia	2.5
Europe	1
Latin America	2
North America	1.5
Oceania	1.5

Table 5.2. *Evolution of food requirements for the different continents. Source: FAO, cited by Benjamin Sultan IPSL/IRD*

A striking result when analyzing the evolution of food requirements for the different continents (Table 5.2) is a factor of 5 between 2000 and 2050 in terms of estimates for necessary food in Africa. Africa is mostly in the intertropical region, i.e. mostly in a region particularly vulnerable in terms of climate change. In this area, adaptation strategies are absolutely necessary although not sufficient.

A local example (Bordeaux area, southwestern of France) is given with the AcclimaTerra project (see: www.acclimaterra.fr), involving about 400 senior scientists.

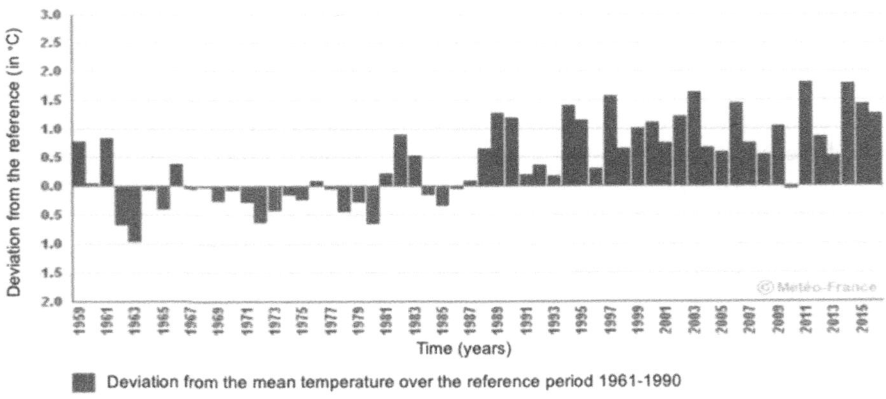

Figure 5.42. *Temperature evolution in the Nouvelle-Aquitaine region of France over the last 57 years (deviation from the mean temperature over the reference period 1961–1990). The difference in temperature is shown in red (positive) and blue (negative). Source: AcclimaTerra (2018)*

The assessment of the temperature evolution in Nouvelle-Aquitaine over the last 57 years between 1959 and 2016 (see Figure 5.42) underlines that it is +1.4°C larger in terms of spread to the mean temperature over 1959–2016 as compared to the pre-industrial era. This trend is common to all countries in Western Europe, as global change is imposed by global scales. Of course, for the temperature evolution in the Nouvelle-Aquitaine region as presented in Figure 5.42, one single person in this region cannot change a global trend, but all efforts count to make a difference.

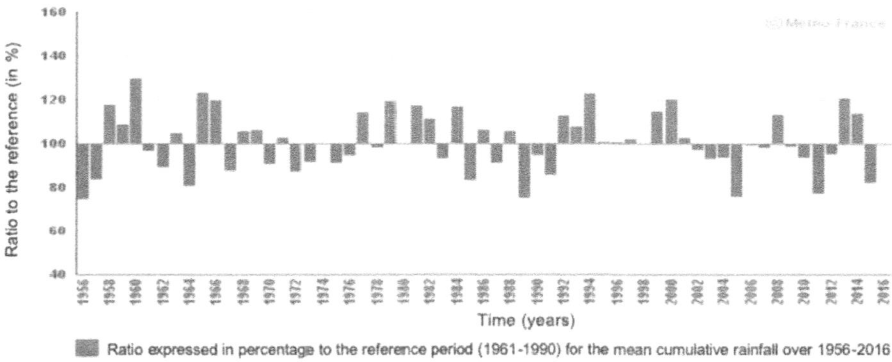

Figure 5.43. *Annual cumulative rainfall in the Nouvelle-Aquitaine region of France: ratio expressed in percentage to the reference period (1961–1990) for the mean cumulative rainfall over 1956–2016. Source: Météo-France 2018*

Conversely, some trends are very difficult to *diagnose* and understand, such as the accumulation of rainfalls over the same Nouvelle-Aquitaine region (see Figure 5.43). The diagram shows changes from one year to another over the period 1956–2016: they tend to oscillate differently; the interpretation of this evolution has not been decisive so far, no significant issue was found to explain it.

The study of agricultural droughts over the period 1959–2016 (see Figure 5.44) points to a significant number of droughts at a recurrent period of the year, most probably resulting from intense evaporation.

The trend shows a rise of the surface of agricultural droughts of about 6–7% since 1959. Droughts over the Nouvelle-Aquitaine territory are extremely critical as this region is the largest agricultural region in France, and its economy also relies a lot on tourism.

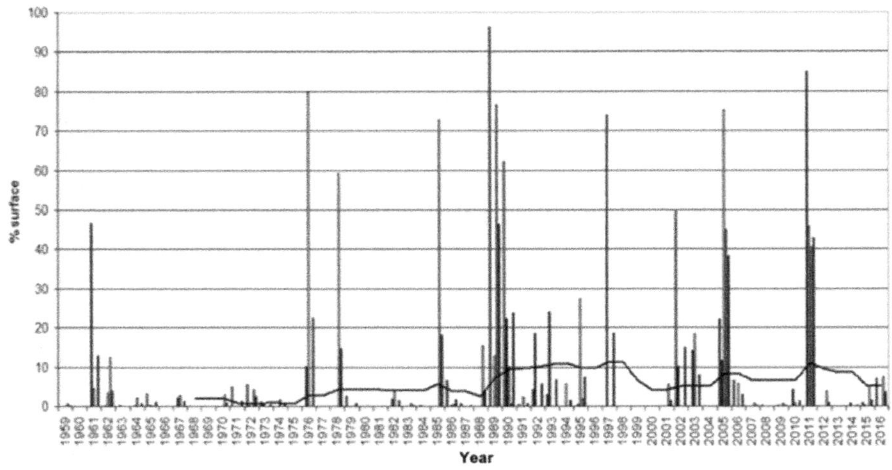

Figure 5.44. *Evolution of the surface percentage on where agricultural droughts were observed during the 1959–2016 period over the Nouvelle-Aquitaine region of France. Winter corresponds to purple color, spring to mauve color, summer to gray and autumn to blue color. The black line represents the mean decadal evolution. Source: Météo-France 2018*

The AcclimaTerra project was initiated when Hervé Le Treut was approached by the head of the Nouvelle-Aquitaine region a few years ago and asked if he was willing to do climate studies about this region (see Figure 5.45). Le Treut's idea was to mimic the work of the IPCC at the regional level and his first surprise was the number of results from senior scientists of public services on the topic of adaptation, regarding various related issues such as shorelines, mountains, wine, pine trees, etc. In a few cases, the data was also gathered from NGOs when there was no kind of research in the academic domain, for example, for specific animals' behaviors towards droughts. More than 100 researchers participated in the first AcclimaTerra report in 2013; it was printed as a book and distributed to many local citizens and regionally elected people. A second report was released in June 2018 and has been presented in different cities across the region.

This initiative is a pioneer and a model in terms of the detailed and science-based local initiative for climate adaptation and mitigation, addressing specific local topics in its regional reports (see Figure 5.46).

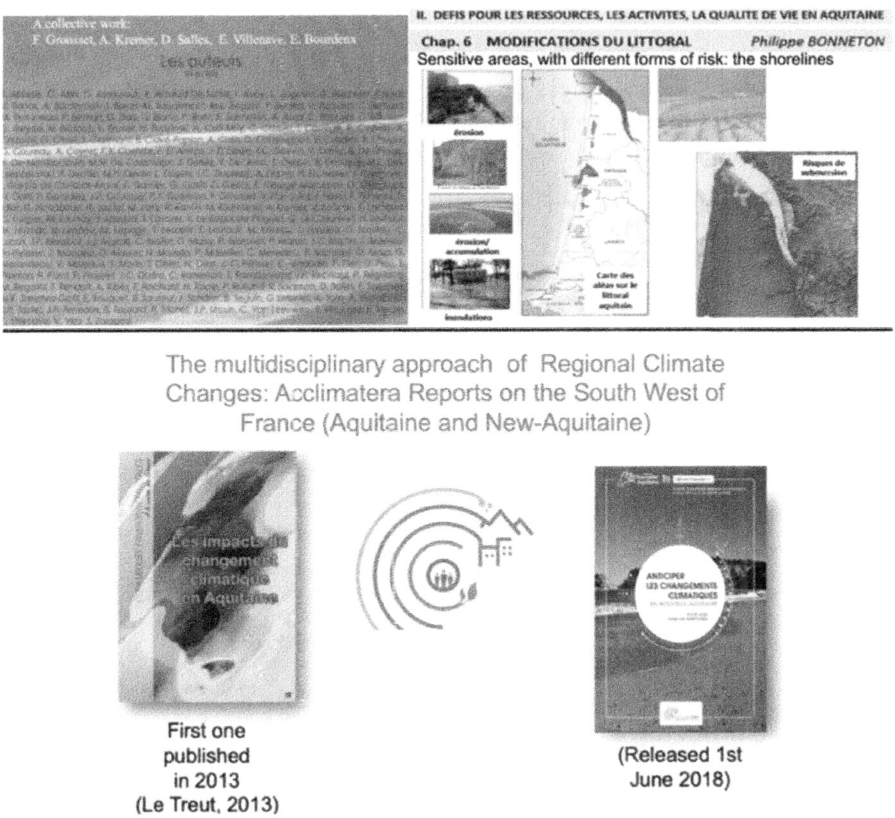

Figure 5.45. *The multidisciplinary approach of Regional Climate Changes: AcclimaTerra Reports on the South West of France (Aquitaine and Nouvelle-Aquitaine). Source: Le Treut 2013 and 2018, http://www.acclimaterra.fr/*

For example, some chapters are focused on mountains for which different topics were analyzed, including social aspects such as the strong reduction of pastoralism with a changing vegetation through global warming. Snow melting is another topic addressed linked to this issue as it is also an economical and cultural problem for winter skiing. Levels of rivers in the summer were also addressed in this study and found to be diminishing strongly, with profound consequences, notably on the types of possible agriculture. The large part of agriculture in the Nouvelle-Aquitaine region relates to corn, apart for wine, and some fruits. Corn requires water, especially in summer. A debate discussed in the AcclimaTerra report is about what decision to take in such a situation: should dams be built to get the water from winter during the summer in order to cultivate corn or not? Is it wiser to maintain the biodiversity as it is – most areas of this region have already suffered a lot from climatic conditions – protect the remaining biodiversity or shift to another kind of cereal such as sorghum for instance? Adaptation options represent a hot debate.

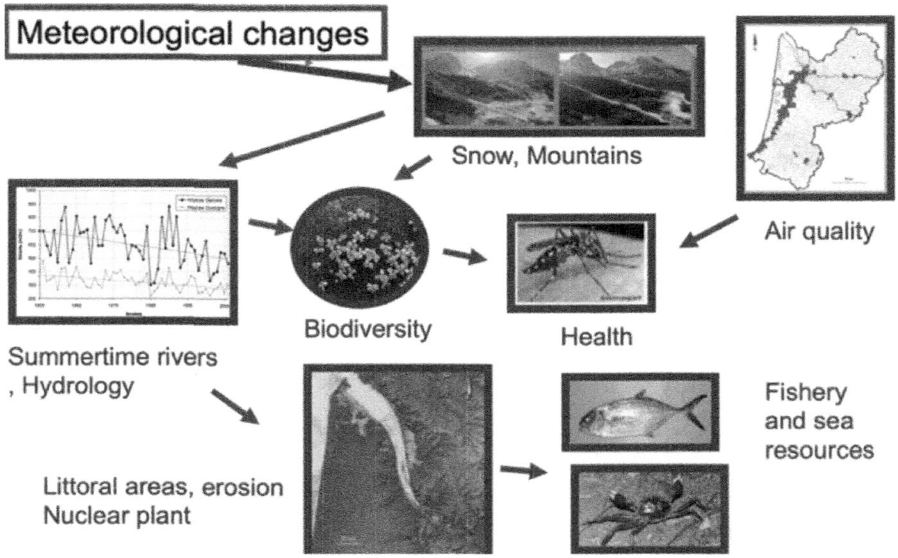

Figure 5.46. *Looking at the regional climate as a unique complex system (from a study about the Bordeaux province in France) (AcclimaTerra 2018)*

There are other changes in the local Nouvelle-Aquitaine agriculture. For instance, wine from Bordeaux is also becoming different under the effect of climate change. The current alcohol level of Bordeaux wine has become 14° on average. This average alcohol level is a lot for a Bordeaux. Not so long ago, Bordeaux was one of the rare wines in France to which there was a right to add sugar. Recently, the concern has become to avoid going up to 15°, and this increase in the average alcohol level of Bordeaux wine is related to climate change. It is not a vital concern such as those faced by African or Asian countries with drought risks and where people may die, people can survive without Chateau-Margot, but it is still an important economic and cultural problem for the region.

Another problem associated with a change in water is the change in biodiversity which can also bring different kinds of species including mosquitos. Water is also essential for the landscapes around local rivers.

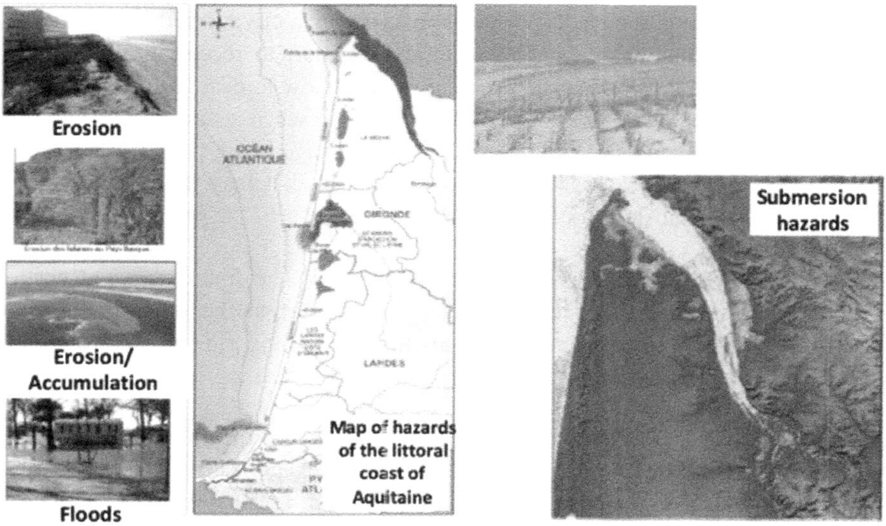

Figure 5.47. *Littoral modifications. Sensitive areas, with different forms of risk: the shorelines. Source: Bonneton and Le Treut (2018) (see: http://www.acclimaterra.fr/)*

The Bordeaux delta areas are supposed to be potentially difficult to live in as they are susceptible to submersion in the future (see Figure 5.47). The famous Chateau-Margot wine area is represented in red on the left part of Figure 5.47 around the delta. More preoccupying is the fact that a nuclear plant is located around this area, which was already at risk during the 1999 storm. This region is undoubtedly at risk. Furthermore, the water coming from mountains through rivers is also already changing its composition due to silt plugs, which cause problems for boats and for biodiversity in the delta.

The changes observed are systemic, and one lesson is that to manage even a region such as Nouvelle-Aquitaine, it is not possible to simply pass an individual law for wine, another for water, etc., as all features are related to one another.

Concerning the long sand shoreline of the Nouvelle-Aquitaine region, another specific feature is that sand dunes are about 10 meters high. The level of increment of sea level is predicted to be between 50 cm to 1 m in the next century: therefore, some dunes, if they stay in a good shape, represent a good natural protection. People have tried to manage them (they are now not allowed to walk on the sand dunes), but there are problems such as for instance in Lacanau, a city located on the coast of the ocean. A question is whether these cities will start to recede as they are not protected from erosion/floods. The problem is about how to deal with risk in such a situation: of course, we do not know for sure how and when a storm will occur. But we know that there is a risk, with potential larger damage in the future. Encouraging people to leave a place at risk is the sort of change that may take decades for a receding city, and that should start now.

The AcclimaTerra reports were widely distributed. The first report was written with Le Treut's friends, and each chapter was similar to the other in terms of organization, with topics such as wine, mountains, rivers, etc., and consensually concluding that more research should be conducted on those topics at the local level.

The second report gathered more messages to deliver to the citizens acting in the different topics addressed. With regard to fish, for example, the first report only gave a list of local species, while in the second report, the regional fish issue is discussed in terms of the size of the boats related to the northward shift in the local fish distribution: a proposed solution was that either fishermen should use larger boats and be allowed to fish in those changing areas, or that they fish other types of local fishes, not exotic species.

The first and second AcclimaTerra reports are respectively of about 300 and 500 pages, but smaller summaries of the reports were also released. A chapter of each report is dedicated to environmental law, another to the governance of the adaptation. Energy is a topic treated more globally and also nationally, but new energies can be monitored locally. The last chapter emphasizes the importance of public dialogue.

Regarding cities and territories, 10 years ago, Le Treut was very surprised not to find a single study mentioning how the city of Bordeaux could be affected by climate change: little by little, this type of very local issue is being addressed. The reports, made on a voluntary basis, have been presented to regional officials as well as in local universities and schools, and found significant success.

Following the pioneering work of AcclimaTerra under the co-leadership of Le Treut for the scientific committee and of the Nouvelle-Aquitaine region president Alain Rousset, and encouraged by the IPCC, similar local initiatives were created all over France (see Figure 5.48).

Figure 5.48. *A map of other local scientific committee initiatives similar to AcclimaTerra in France. Source: AcclimaTerra (2025)*

This network is named "GREC" in French, which can be translated into "regional panel on climate change". The main aim of these local initiatives is to deliver territorial vulnerability assessments in terms of climate change and environmental protection, considering regional singularities and related socio-economic issues. Almost every French region has its specific "GREC", which enables focuses and publications on very specific regional matter. The "GREC-SUD" of the Provence-Alpes-Côte d'Azur region has published a specific comparative analysis about soil relative humidity and set up dedicated committees about local mountains and regional natural parks. Guadeloupe has its own innovation pole with a specialty in ecological transition, named "Synergîles".

As for the IPCC at the global level, this French initiative of local network aims at delivering factual and indisputable scientific knowledge in order to inform public action in a neutral and apolitical way. The productions are not prescriptive.

These local scientific committees are usually composed of scientists working or related to the region, working on a voluntary basis only, with no funds from the state, leading to a situation with some leading and wealthy regions such as

Nouvelle-Aquitaine or PACA having more resources than other regions in France. The French territories are also quite unequal, especially between overseas territories, urban and rural regions. To bring more balance among France regions, ideas for improvements include state funding, inter-regional common works among the different "GREC" groups, and more focus on the specific maritime issues for overseas territories especially. These initiatives are all solution-oriented and include adaptation and resilience solutions additionally to attenuation actions.

5.3. From scientific diagnosis to political decision-making

We have discussed here how science can serve as a tool, not to say as a weapon, for informed decisions made either by politicians at local, national or international level, and also by company executives and within various public or private organizations.

Initiatives of mandatory training for civil servants at the graduate level under the impetus and leadership of Jean Jouzel and Luc Abbadie are a good starting point, even if their implementation by current political decisions is not sufficient (not long enough, not detailed enough, and "one shot" instead of being more continuous) and they are yet to spread the complex knowledge related to the study of the climate system, with regard to the emergency and value of the ambition underlined by Jouzel and Abbadie among other climate experts.

Scientific knowledge is key, but does not replace the need for empathy and discernment for fair public policies related to climate change. In France, the "Gilets jaunes" (Yellow Vests) crisis in 2019 that emerged after an unwisely, undemocratically imposed carbon tax affecting the most financially vulnerable peripheral population in France, and the scandal of the unconsidered implementation of "ZFE" (low emission zones, in French *Zone à Faibles Emissions*,) in territories with a high circulation of older more polluting vehicles, mainly owned by low-income populations working in suburbs poorly or not served by public transportation, or working in downtown cities, unable to afford new low-polluting vehicles, and so creating discriminatory situations, are examples of weak and counter-productive political decisions ignoring equity and socio-political fairness, done without consulting nor even considering the effective direct effect on the most vulnerable people.

Claude Riveline, fondly remembered professor of organization management at the French Ecole des Mines wrote valuable essays on this topic about the necessary balance between 'hard" and "soft" skills (Riveline 1985).

To avoid counter-productive, uselessly costly or even merely disastrous decisions going against national/international interests, high-level politicians in charge of climate policies should ideally not only have a robust scientific background in terms of competencies, but should also be sensitive and pay empathic attention to people undergoing socio-economic struggles, the latter skill being more an art than a science. The awareness and the will for a better and fairer world, protecting and delivering equal opportunities to the most vulnerable, should be the final and only guide for any good political action, including climate policies in which the final aim is the preservation of human life and civilization for the better.

The much better training and inclusiveness of people coming from low-income backgrounds within the civil servant and decision-maker community could also be an important factor for the improvement of effective decision-making, considering both science and socio-economic aspects for a wise and fair ecological and energetical transition, along with considering more advice from researchers than from consultants, the first having more long-term, expert and humanist views on topics related to climate change.

5.4. A word of conclusion

A key synthesis message of this book is that to properly tackle climate change issues, it is necessary to go into precision, learn the language of the IPCC and understand the importance of uncertainties and some key detailed processes.

Appendix

Master's Degree Level Past Exams and Elements of Discussion and Correction

This appendix presents two examples of real master's degree past exams referring to important notions related to the course delivered in the five chapters of this book.

Non-exhaustive elements of corrections and discussion are delivered for each of them, but all solutions are in the course book and are not repeated.

A critical and personal approach demonstrating the understanding of the complexity at stakes are requested for each answer, which is not subject to a "unique" way to answer it.

A.1. Past exam 1

The answers to the different questions have to be brief, especially when they repeat elements of the course. The capacity to make a synthesis from elements discussed at different moments during the course, to present arguments, will be judged in priority. There is no unique way to answer a question. The questions are fully independent and can be treated in any order. The number of points attached to each question is indicative.

The number of points by question is given as an indication only.

Duration: three hours.

For a color version of all of the figures in this appendix, see www.iste.co.uk/letreut/energetics.zip.

A.1.1. *Earth radiative budget (5.5 points)*

Figures A.1 and A.2 (from the NASA satellite ERBE) show a component of the mean monthly radiative budget of the Earth (seen from space, therefore at "the top of the atmosphere"). The color scale refers to values in W.m^{-2} (as mentioned in each diagram).

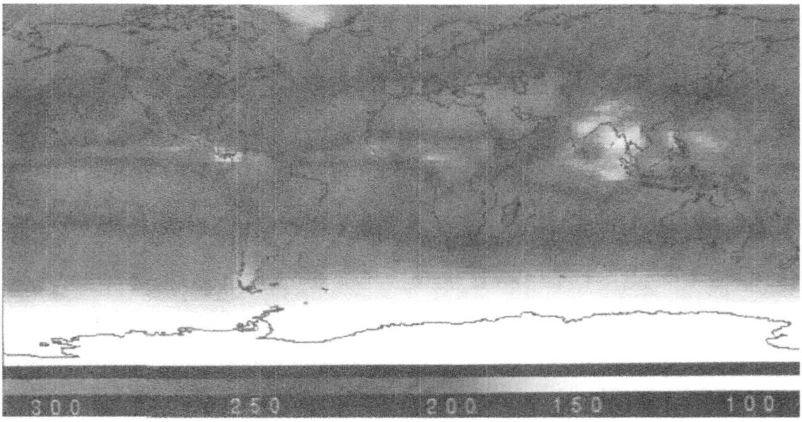

Figure A.1. *Mean monthly radiative budget of the Earth seen from space at the top of the atmosphere (TOA). Source: ERBE satellite (NASA)*

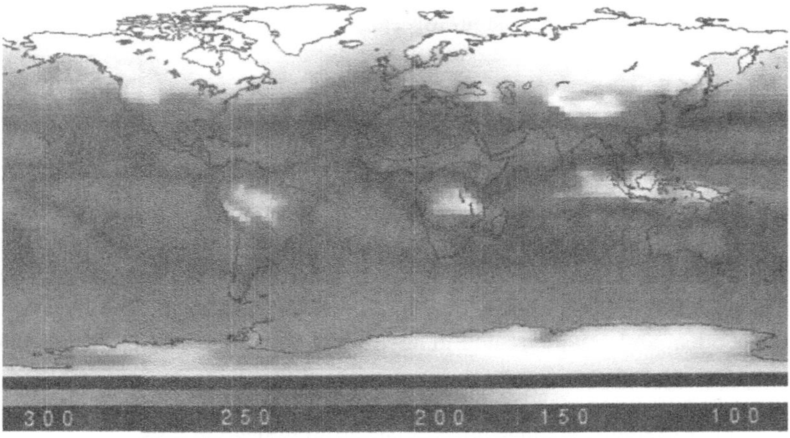

Figure A.2. *Mean monthly radiative budget of the Earth seen from space at the top of the atmosphere (TOA). Source: ERBE satellite (NASA)*

Question 1. Can you identify whether this is the outgoing terrestrial radiation or the net absorbed solar radiation (provide one or several arguments). Can you briefly explain the main features of the observed distribution?

Question 2. One of the figures corresponds to the month of January, and the other to the month of August. Can you identify which is which? Provide one or several arguments.

Question 3. What are the cloud types which can be clearly identified on these figures? And which ones are most difficult to identify? This question overlaps a little with the preceding ones.

A.1.2. *The radiative forcing of air pollutants (6 points)*

Figure A.3 (from the IPCC Summary Makers of Group I, AR5) provides the past changes of the radiative forcing (estimates in 2011 minus estimates in 1750), for various atmospheric constituents. It was shown during the courses.

The following exercise is focused on the part of the diagram which is circled.

The detailed caption with the exact quotation from AR5 SPM IPCC is the following:

Radiative forcing estimates in 2011 relative to 1750 and aggregated uncertainties for the main drivers of climate change. Values are global average radiative forcing (RF14), partitioned according to the emitted compounds or processes that result in a combination of drivers. The best estimates of the net radiative forcing are shown as black diamonds with corresponding uncertainty intervals. The numerical values are provided on the right of the figure, together with the confidence level in the net forcing (VH – very high, H – high, M – medium, L – low, VL – very low). Albedo forcing due to black carbon on snow and ice is included in the black carbon aerosol bar. Small forcings due to contrails (0.05 W m^{-2}, including contrail induced cirrus), and HFCs, PFCs and SF6 (total 0.03 W m^{-2}) are not shown. Concentration-based RFs for gases can be obtained by summing the like-colored bars. Volcanic forcing is not included as its episodic nature makes is difficult to compare to other forcing mechanisms. Total anthropogenic radiative forcing is provided for three different years relative to 1750.

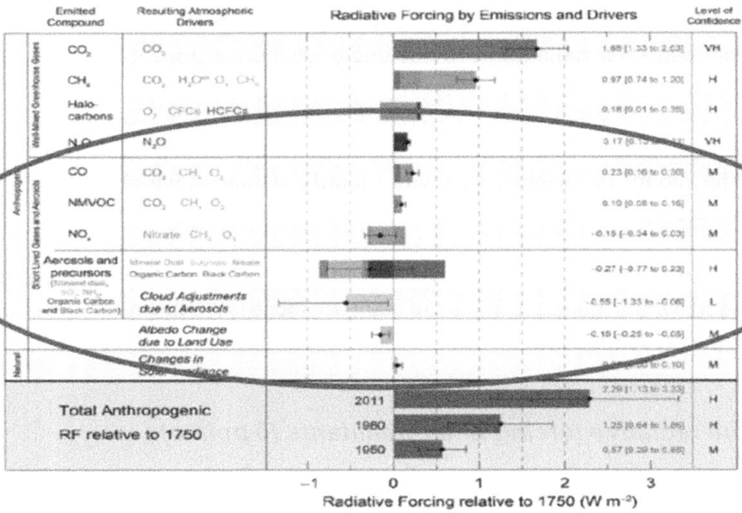

Figure A.3. *Radiative forcing estimates in 2011 relative to 1750 and aggregated uncertainties for the main drivers of climate change. Source: AR5 (2013). Reproduced with permission from Stocker et al. (2013)*

Question 1. Can you recall the definition of the radiation of constituents emitted in the atmosphere?

Question 2. For the atmospheric constituents in the circled area, can you explain why some of them have experienced a positive change in their radiative forcing, and some of them a negative change (note also that the diagram represents not only the direct effect of a given constituent, but also the associated indirect effects when the transformation of this constituent introduces new components into the atmosphere)?

Question 3. Can you comment on the uncertainty of these estimates? Why does it appear comparatively larger than the uncertainty affecting the effect of greenhouse gases?

Question 4. Can you recall the definition of CO_2 equivalent for a given emitted gas? Can you give an order of magnitude of its value for the circled constituents (treat them collectively and provide a rough estimate)? How may this definition of the "CO_2 equivalent" affect the public policies?

A.1.3. *Hadley cells and Walker cells (5.5 points)*

Figure A.4 (Météo-France, François Poulin) provides a simplified view of the "normal" atmospheric circulation above the Pacific Ocean in the equatorial region. This circulation is called the "Walker circulation" or Walker cell. The most important modifications of this circulation are called "El Niño events".

Question 1. What kind of analogy can we draw between the Hadley and the Walker cells in terms of energy transport? In which direction does the Walker cell transport energy?

Question 2. Can you recall the expected modifications of the Hadley cell circulation, in the conditions of a global warming resulting from a CO_2 increase?

Question 3. To what extent can we expect similar or different modifications in the Walker circulation in the same conditions of global warming?

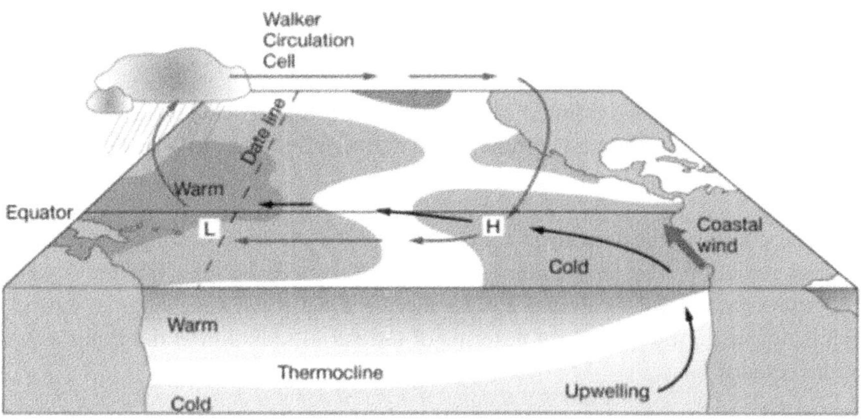

Figure A.4. *A scheme of the Walker circulation. Source: NOAA*

A.1.4. *Climate monitoring (3 points)*

During the course, we have studied the uncertainties in the amplitude of climate change associated with different feedbacks (clouds, snow, water vapor, etc.) A large part of the strategy to monitor climate change is to favor long-term measurements of key variables.

Question 1. What can we expect with such a strategy?

Question 2. Do you think it is sufficient to diminish the current quantitative uncertainties that affect our view of the future climate?

NOTE.– Real-world documents are always more complex than theoretical schemes and models, different well-argued solutions may be accepted, the quotation should be focused on the methodological approach and the use of relevant arguments.

A.2. Case study. Past Exam 1: elements of correction and discussion

A.2.1. *Earth radiative budget (5.5 points)*

The idea of this exercise is to distinguish between the terrestrial and the solar radiation for the budget at the top of the atmosphere (TOA), among the two maps.

The questions aim to get students to describe the main elements affecting the budget at the top of the atmosphere.

In the courseware, we pointed out that there are differences at the Poles where the solar radiation is extincted completely at some periods of the year during which it therefore goes to very low values.

On the contrary, the terrestrial radiation is never extincted during the year. For terrestrial radiation, the temperature may be 30°C more or less, but the absolute temperature stays stable at around 250 K: there is always, all year long, an emission which is not negligible occurring in terms of terrestrial radiation at the Poles as well as at the equator.

Therefore, when studying the two maps of the exercise, the locations with very low emissions or very low absorption are clues of a map with solar radiation, which is therefore shown in Figure A.2.

Another possible tracer would be the difference in terms of cloudiness, and this requires us to be able to look at different types of clouds. Basically, in the equatorial region, when observing clouds from above, we can draw a representative scheme as shown in Figure A.5.

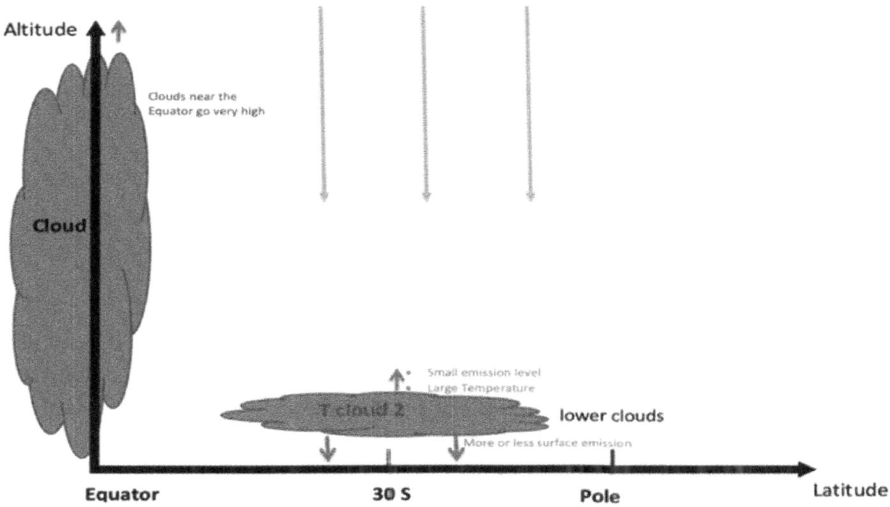

Figure A.5. *Scheme of clouds with latitude on the x-axis and altitude on the y-axis*

As represented in the scheme of Figure A.5, near 30 S latitude, there are lower clouds, which will have a different impact on solar and terrestrial radiations.

The key question in this case is: for which type of radiation – solar or terrestrial – is the contrast expected to be the strongest?

When seen from the top, the shape of the map depends on the considered case: it either shows the reflected radiation from the solar radiation, or the emitted radiation from the terrestrial component.

Between the two clouds represented in the scheme of Figure A.5, the strongest cloud contrast will be for the outgoing terrestrial radiation.

Indeed, for the terrestrial radiation, the metric that defines it at the top of the atmosphere (TOA) is the temperature of the last body that emits to space. We know that in the presence of clouds, they almost absorb all of the radiation which comes from the surface, and then reemit it. Clouds are indeed very strong greenhouse bodies with water vapor as well as particle cloud droplets. All of this plays a role in the processes of absorption and reemission.

We also have emission going into space occurring at the cloud temperature (represented as Tcloud 2, in red on the scheme of Figure A.5). The same phenomenon occurs on clouds near the tropics, but since this is higher, the radiation going into space will be emitted at a higher level.

For lower clouds as represented in Figure A.5: the emission level is small, the temperature is high, and this domain corresponds more or less to the surface emission, and therefore the terrestrial radiation in this domain would not be very visible.

There is therefore more contrast for terrestrial radiation, whereas what is emitted into space is more even in the case of solar radiation, because solar radiation is linked to the amount of droplets and to the diffusion effect studied at the beginning of this course, which does not strongly depend on temperature and which does not strongly depend on the height of clouds. It is therefore even in those two cases referring to solar radiation.

Therefore, what really makes the difference between the outgoing terrestrial radiation and the net absorbed solar radiation are the regions of low clouds.

As a synthesis:

– strong contrast between high and low clouds ⇔ tracer of a map of the outgoing terrestrial radiation;

– no contrast ⇔ tracer of a map of the net absorbed solar radiation.

A.2.2. *The radiative forcing of air pollutants (6 points)*

The aim of this exercise is to evaluate the student's ability to distinguish the differences between long-lived and short-lived gases or aerosols.

The first section of the diagram refers to long-lived gases: CO_2, CH_4, N_2O and HCFC. These long-lived gases were already included in the Kyoto Protocol. They are still the bases of the Paris Agreement to limit the greenhouse effect, and the carbon market is also based on those main greenhouse gases.

Question 2 in section A.1.2 of the exam was related to the other gases and more precisely short-lived greenhouse gases, meaning that they stay about one to a few weeks in the atmosphere, but they are constantly renewed. The idea of this exercise is to compare them. Another category circled in Figure A.3 for question 2 of section A.1.2 refers to particles.

Therefore, as a summary:

– Emitted long-lived GHGs such as CO_2, CH_4, N_2O and HCFC do have a GHG effect, but have no direct effect on air quality and human health. However, there is frequent misunderstanding about this.

– Emitted short-lived GHGs are constantly renewed in the atmosphere (to a large amount). Therefore, they have a greenhouse effect, even if it is smaller than the one of long-lived GHGs, and they have a big effect on air quality/health.

– Emitted anthropogenic particles are either:

- directly emitted to the atmosphere as particles, for instance, through deforestation;

- formed/derived from gases or liquid, for example, sulfates are generally first emitted as sulfuric acid and then it produces sulfate particles which have an important effect of reflecting back solar radiation to space;

- particles that may affect cloud reflection.

Some of these particles do reflect solar radiation and therefore, they may have a negative impact in terms of radiative forcing due to their reflection of solar radiation.

There is also a greenhouse effect from those particles because they are constantly renewed. Among those particles, black carbon (soot) is actually the only one able to also have an important GHG. All other particles have very little greenhouse gas effects.

Finally, particles, contrary to long-term GHG, do have a direct big effect on the air quality and therefore a direct effect for human health.

Therefore, pollutants, which are actually harmful for human health, usually, however, have a negative radiative forcing effect mostly because of the particles and therefore tend to have a cooling effect. This complexity is currently at the heart of many debates, and several decision-makers do not fully understand this. Emissions of vehicle pollutants for instance in country side with a few inhabitants have a smaller impact on human health than in the center of dense cities, but are considered in a differentiated way by decision-makers.

The idea of the exercise is to recall the way to reconcile gases having different aspects by using the global warming potential (GWP), defined as the integral of a gas. If we take it for CH_4, for example, by noting RF the radiative forcing, over a period of time T (we usually take T= 100 years), we can write the following equation:

$$GWP_{(CH_4)} = \frac{\int_0^T FR(CH_4)dt}{\int_0^T FR(CO_2)dt} \quad [A.1]$$

This corresponds to 1 kg of CO_2 initially (see Chapter 4), and 1 kg of CH_4 initially, the principle then consists in letting 1 kg of species CH_4 disappear. For the case of CO_2, the process accounted for is only by exchanges with the surface. In the case of CH_4, the disappearance also accounts for internal chemistry in addition to exchanges with the surface. We obtain an intermediate number, if we were just considering CH_4 at t = 0, CH_4 absorbs 100 times more than CO_2. As we take T = 100 years, there is no more CH_4 after this time. Therefore we can say it absorbs 0 times more than the CO_2 absorption at T = 100 years: there is no more (or almost no more) CH_4 left after 100 years. The half-life of CH_4 in the atmosphere is indeed about 12 years. We obtain approximately 30 (depending on the versions IPCC Assessment Reports) (see Table 4.1 in Chapter 4): this vision useful and GWP is a tool to somehow average a vision where CO_2 is very powerful in terms of absorption and a vision where it disappears from the atmosphere over time. This mathematical image where 1 kg of species disappears over time is not completely fair because, in the real world, these gases are constantly emitted and are therefore in fact renewed in the atmosphere.

Question 4 concerns the short-lived GWP. The right number is zero because if we integrate over 100 years a number that becomes 0 after one week, the sum of that would be zero. This means that all of these gases do not participate in the carbon market. Some are only considered for their capacity to change the air quality. In fact, this makes a huge difference between gases considered because of their greenhouse effect and gases considered because of their health effect.

Coming on to the 2019 debate on the carbon tax and the Yellow Vests movement, advocating for electric cars is good for air quality (electric cars for cities contribute to limit emissions of air pollutants/toxic gases), and of course the same cars emit GHGs. The question is on the overall product life cycle analysis (from production to delivery/use and supply chain). Does an electric vehicle emit less or more than a fuel vehicle? The problem is not only scientific, but also concerns of keeping a convention that has been established at some moment and has historically been in use for a while. In the 1990s, a special IPCC report mainly written by Imperial College Scientist explained how the GWP tool was constructed. This tool was created in the context of the setting up of the Kyoto Protocol and the creation of a carbon market. Economists needed one price for all fuels. This is one of the main reasons why the GWP tool was originally created. It is difficult to change this metric because it has already been used in several papers and reports and is in use in carbon markets as well. However, a 100-year GWP metric is particularly challengeable in the case of CH_4 because in order to impede the warming of the planet on a short time scale, CH_4 may have a role which is much higher than its GWP100 value with a 20-year GWP for instance (CH_4 GWP20 ~80).

Also, the idea that there must be different markets for each GHG has not been heard so far, but this is mostly a decision-making issue.

A.2.3. *Hadley cells and Walker cells (5.5 points)*

Above hot regions, air is ascending, and a cell is following a similar rule (Hadley cell) with air descending in the cold area. The gradient of temperature (cold/hot) creates a situation where the ascending air drives the air from East to West, with an effect on the ocean that is also entrained in a consistent direction with the atmosphere. Near the equator, there is no Courant effect.

The question here is how can we describe the heat transport in this region?

We had a total heat transport (energy transported E_t) simplified as

$$E_t = gz + C_pT \qquad [A.2]$$

If stable, E_t increases with height.

The directing process is the one at the top of the atmosphere where gz is larger and drives the transport direction, if stable, from the equator to higher latitudes.

The idea is to explain why it is similar in the Hadley and Walker cells.

It is actually similar because if there is stability, the heat transport will go from hot regions to cold regions, a heat mechanism which is similar for the Walker cells to the one of the Hadley cells.

Some key differences between these two mechanisms are the following:

– In Hadley cells, the cell extension between 30 S and 30 N is determined by the conservation of the angular momentum (the rotation speed of the Earth upon itself determines the limit). This is really linked to astronomical features, and to be precise, it is important to balance this rotational speed of the Earth upon itself and the gradient of temperatures between the Poles and the equator (which depends on the sphericity of the Earth/distance from the Sun).

– The main difference between these two phenomena comes from the existence of ocean basins. The fact that the combination between atmosphere and ocean depends on the limited space given to the ocean, and the amplitude of this cell really depends on a completely different mechanism.

– Therefore, we have an East/West motion which is a bit similar for Hadley and Walker cells, but the difference comes from the different way to constrain the cell due to the limited size of the ocean.

A.3. Past exam 2

The answers to the different questions have to be brief, especially when they repeat elements of the course. The capacity to make a synthesis from elements discussed at different moments during the course, to present arguments, will be judged in priority.

There is no unique way to answer a question.

The number of points by question is given as an indication only.

Duration: three hours.

A.3.1. *Temperature and cloudiness changes in a series of climate scenario (5 points)*

The following panels are taken from the last IPCC report. They show the changes estimated from a large number (32 or 42) of climate models, responding to a series of climate scenarios (RCP2.6, RCP4.5, RCP8.5), characterized respectively by an increasing amount of greenhouse gas emissions, leading to a radiative forcing in 2100 of respectively 2.6 W.m^{-2}, 4.5 W.m^{-2} and 8.5 W.m^{-2}. Colors represent the mean of the models, dots and hatches indicate whether the models agree or disagree in terms of local response.

They show:

a) The latitude/altitude distribution of temperature changes in the atmosphere, and in the ocean.

b) The geographical changes in the mean cloud distribution.

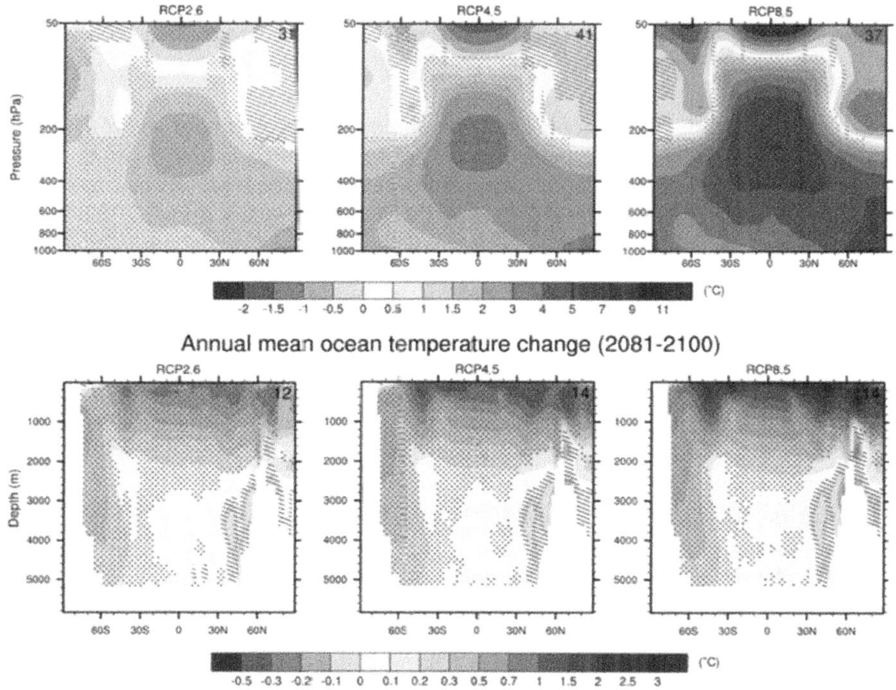

Figure A.6. *Annual mean atmospheric temperature change (2081–2100). CMIP5 multi-model changes in the annual mean zonal mean temperature in the atmosphere and ocean relative to 1986–2005 for 2081–2100 under the RCP2.6 (left), RCP4.5 (center) and RCP8.5 (right) forcing scenarios. Hatching indicates regions where the multi-model mean is less than one standard deviation of internal variability. Stippling indicates regions where the multi-model mean is greater than two standard deviations of internal variability and where 90% of the models agree on the sign of change. Source: Collins et al. (2013) and IPCC, AR5 (2013). Reproduced with permission from Stocker et al. (2023)*

Question 1. Can you use the information from the course to comment on these results (you are expected to briefly comment on the difference between the various RCP, but analyze in more detail the structure of the RCP8.5 results, and provide tentative explanations for the observed structures)?

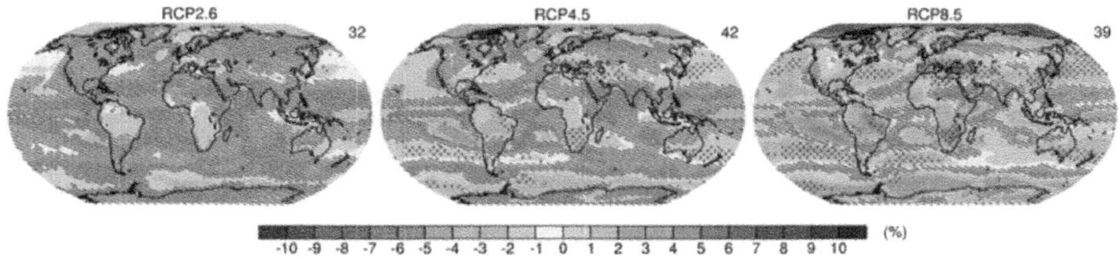

Figure A.7. *Annual mean cloud fraction change (in percentage) in 2081–2100 relative to 1986–2005 under three climate change scenarios. MIP5 multi-model changes in the annual mean total cloud fraction (in percentage) relative to 1986–2005 for 2081–2100 under the RCP2.6 (left), RCP4.5 (center) and RCP8.5 (right) forcing scenarios. Hatching indicates regions where the multi-model mean change is less than one standard deviation of internal variability. Stippling indicates regions where the multi-model mean change is greater than two standard deviations of internal variability and where 90% of the models agree on the sign of change. The number of CMIP5 models used is indicated in the upper-right corner of each panel. Source: Figure 12.17 in IPCC, AR5 (2013). Reproduced with permission from Collins et al. (2013)*

A.3.2. *Seasonal cycle of the Hadley cell (4 points)*

This scheme shows the streamlines of the atmosphere mean meridional circulation for respectively annual conditions (upper panel), Northern Hemisphere winter conditions (middle panel) and Northern Hemisphere summer conditions (lower panel).

When the circulation is stationary, we may interpret this diagram by saying that the mass circulation between two isolines is two times 10^{10} kg s^{-1}. We recall the values of C_P (1004 J kg^{-1} K^{-1}) and L (2500 kJ kg^{-1}).

The three colored maps shown in Figures A.8 and A.9 represent the radiative budget at the top of the atmosphere for the same seasons (in Wm^{-2}).

Question 1. Can you qualitatively explain how the Hadley cell may serve to re-equilibrate the energy balance of the two hemispheres in summer and in winter of the Northern Hemisphere?

Question 2. Can you show that the order of magnitude of the energy transport matches the order of magnitude of the radiative imbalance between the two hemispheres?

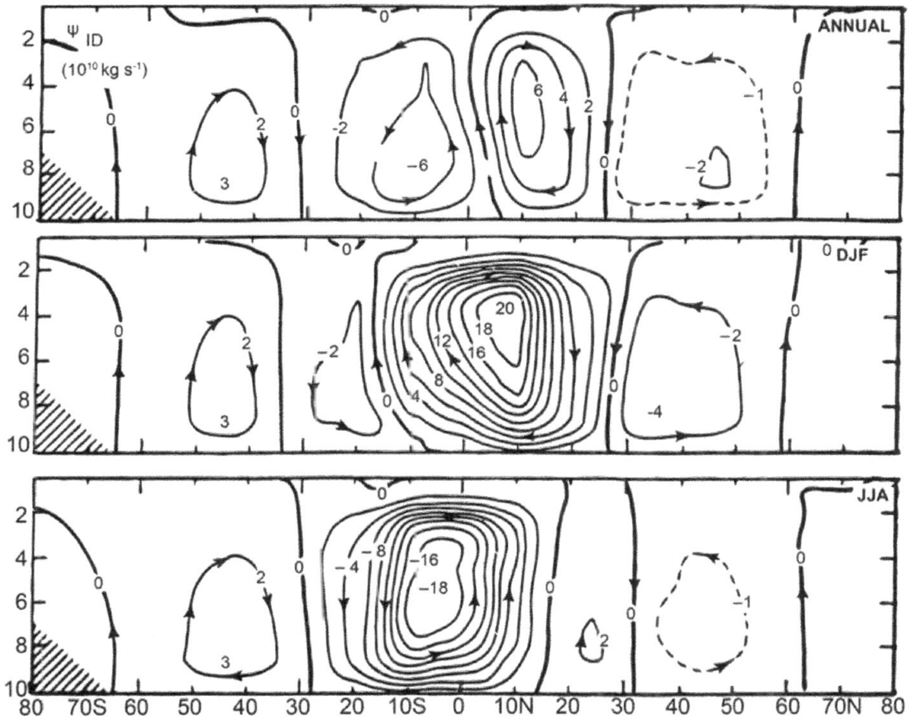

Figure A.8. *Representation of Hadley cells. Source: Peixoto et al. (1992)*

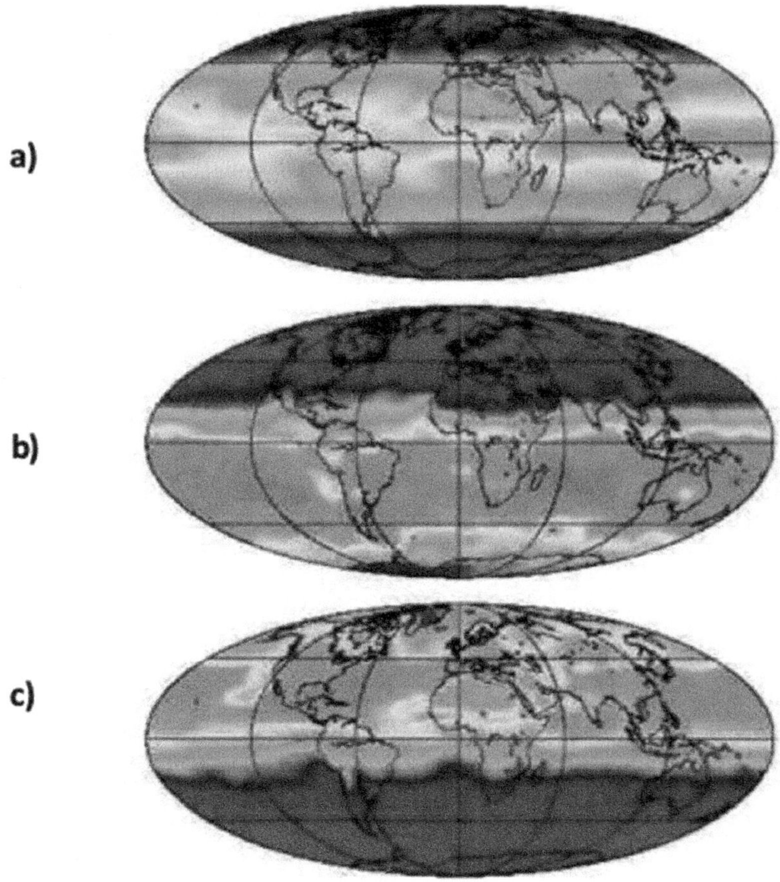

Figure A.9. *Radiative budget at the top of the atmosphere (TOA) in W m^{-2}: a) annual mean; b) under winter conditions (December, January and February); and c) under summer conditions (June, July and August). Source: the Earth Radiation Budget Experiment (ERBE)*

A.3.3. *Energy fluxes at the surface of the oceans (3 points)*

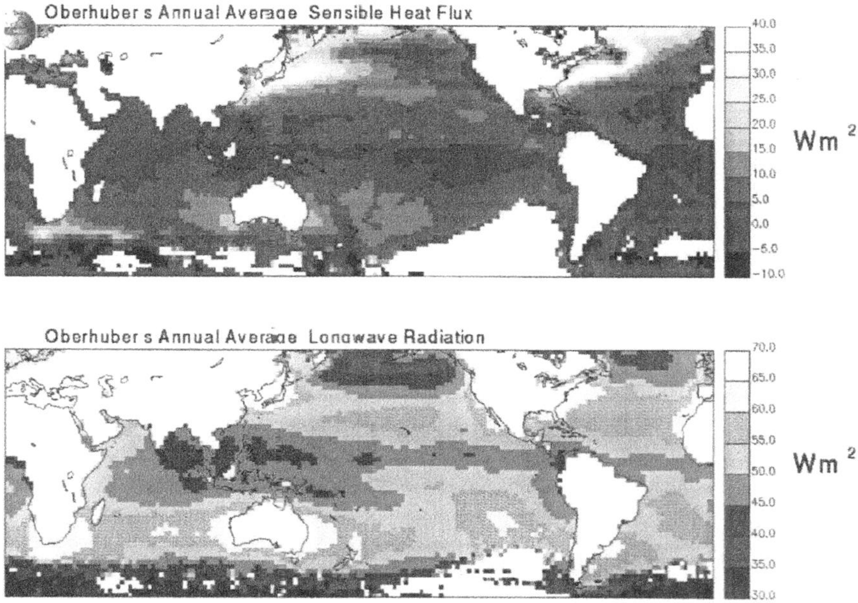

Figure A.10. *Maps of energy fluxes at the surface of the ocean for sensible heat flux (upper map) and longwave radiation (lower map). Source: Oberhuber' data*

The two maps in Figures A.9 and A.10 show two of the energy fluxes at the surface of the ocean: sensible heat flux (= flux of internal energy) and longwave radiation (= budget of the infrared terrestrial radiation).

Can you:

Question 1. Clarify what is, according to you, the sign convention?

Question 2. Comment the geographical structure of the fluxes.

A.3.4. *Brief comparison of the climate impact of aerosols and greenhouse gases (3 points)*

Question 1. Can you compare in a synthetic manner the similarities and differences of the greenhouse and aerosol radiative forcing on the climate system?

Question 2. Can you explain why the competition between the two effects may create scientific difficulties to estimate future climate changes?

A.3.5. *Vertical temperature profile (5 points)*

We have studied the following equations:

$$\frac{dF^+}{d\tau} = - F^+ + \sigma T^4 \qquad [A.3]$$

$$\frac{dF^-}{d\tau} = F^- - \sigma T^4 \qquad [A.4]$$

where F^+ is the upward flux of the terrestrial radiation, F^- is the downward flux, and σT^4 is the black body radiation and τ is the optical thickness (by convention τ is increasing with altitude).

We consider that the atmosphere is a slab of the total optical thickness τ_0, which does not absorb any solar energy, and emits to space a radiation noted R_t.

We also define the equivalent temperature of the Earth by the following equation:

$$R_t = \sigma T^4 \qquad [A.5]$$

Question 1. Can you recall (stating the main assumptions of the computation) the height at which (measured in units optical thickness from the surface) the equivalent emission level is defined as the level of the equivalent temperature T.

Question 2. Some skeptical persons say that the absorption brand of the CO_2 is saturated and that the atmosphere absorbs all of the radiation emitted by the surface in the corresponding wavelengths (this is right). They conclude that the temperature may no longer increase as the result of increasing atmospheric CO_2. Can you use the result of the previous question to show why this on the contrary is a wrong argument?

Question 3. The computation of the vertical temperature profile in question 1 neglects solar absorption. Can you imagine a simple way to account for this absorption, and derive a new shape of the vertical temperature profile (as a function of optical thickness)?

A.4. Case study. Past Exam 2: elements of correction and discussion

A.4.1. *Temperature and cloudiness changes in a series of climate scenario (5 points)*

The idea in this exercise is to comment the diagram and to show that the structure does not completely depend on amplitude effects, but that there are many aspects to detail because with increasing temperatures, the temperature change is a bit different (but not drastically).

The other important aspect is the connection with the ocean temperature. Figure A.7 is a map showing the annual mean cloud fraction, while Figure A.6 shows maps of the annual mean atmospheric temperature change (2081–2100) and annual mean ocean temperature change over the same period.

The problem is making a connection between atmospheric and ocean temperature changes and cloudiness in terms of processes.

Among the processes determining cloud change, the main process comes from the fact that when we analyze cloud change, they have to be inserted into the fact that the Hadley cell extension depends on astronomical parameters which do not change much.

The shapes of clouds in the map depend on:

– the fact that there are always ascending clouds in hot regions;

– the fact that there are low clouds over the oceans because of water;

– deserts over continents;

– mid-latitude phenomena which have their own clouds associated with c-cyclones.

The main takeaway message is that the observed shape does not depend very much on the amplitude of the warming because it is constrained by those limits which have an astronomical background, implying that they are changes, but those changes are constrained in the 10–100-km extension for the Hadley cell, which is not very much. Therefore, basically there is not much change for this reason.

A.4.2. *Seasonal cycle of the Hadley cell (4 points)*

The scheme in Figure A2.3 shows the streamlines of atmospheric meridional circulation for respectively annual conditions (upper panel), Northern Hemisphere winter conditions (middle panel) and Northern Hemisphere summer conditions (lower panel). When the circulation is stationary, we may interpret these diagrams by saying that the mass circulation between two isolines is two times 10^{10} kg s^{-1}.

There are descending cells. We recall that the stability criteria with height are that the heat energy transported E_t increases with height.

When the warming from the Sun makes the ground too hot, there is too much warming from the ground. Therefore, basically with just a direct warming of the surface, the stability criteria are not met.

The equilibrium is actually not radiative but radiative–convective: therefore, what impedes the surface from being too hot is convection. In locations where there are oceans below, there are some exchanges with them and also exchanges with the ground over continents. Basically, the most important phenomenon to maintain equilibrium is convection.

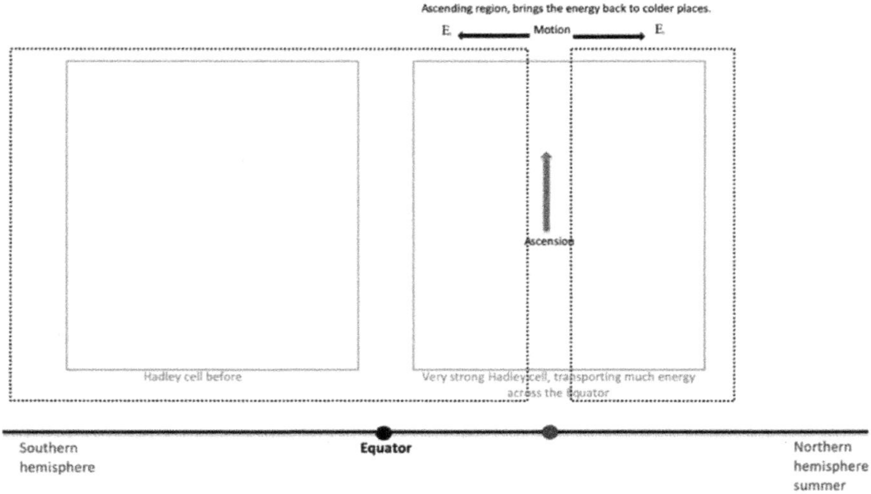

Figure A.11. *Scheme of the Hadley cell evolution during a seasonal cycle*

Similarly, for the seasonal cycle of a Hadley cell, the shape of the cell will be displaced, and what we can trust is that it is a mean over large period of time. Therefore, it corresponds to stable conditions afterwards, after the convection has played its role. If so, then we can have the kind of change represented in Figure A.11.

Let us consider the seasonal cycle, and more specifically the case of summer in the Northern Hemisphere: in that case, there is an ascending region as represented in Figure A.11. The heat transport goes from the largest heating places where air is ascending and then diverging (in the warmest places), and this brings the energy back to colder places. This means a very strong cell transports lots of energy across the equator to Southern latitudes, as represented on the cell in the bottom diagram of Figure A.8.

In Chapter 3 of this book, which is related to energy transport, we showed that Hadley cells are weak when they enter the Southern Hemisphere because of the strong fixed limits 30N–30S, which are linked to the rotational speed of the Earth upon itself. These limits of intertropical and extratropical climates do not change. The sharing of energy is primarily done inside this domain, and then it gets away.

NOTE.– The shape of the cells is not strongly modified by global warming. There is some tendency to a slight extension of the cell towards the Poles direction in a warmer climate, but it is not the first signal. The first signal is that within those cells, the heat transport may become a bit stronger because of the stronger contrast of heat occurring in a warmer climate. Therefore, there are more changes occurring on-site, at the same location: hot regions will become hotter, moist regions will be moister and hot, and dry regions will be dryer. Locally this is not completely true, but this is true globally.

Contrary to what people may generally think, the circulation in the Hadley cells tends to diminish with global warming.

Figure A.12 shows a scheme of the situation in the case of global warming/increase of temperature over Earth.

The scheme in Figure A.12 represents the heat contrast with an increased warming, which exists near the ground and above the ground. The heat contrast is reduced at low altitudes. At high altitudes, there is an increased heat contrast. The height level at which this heat contrast changes is at about 10 km. In mass, this should be a little more important. There is indeed a bit of a competition between those two effects, and models show that the gradient of temperature between high latitudes and the equator will be reduced. In lower parts of the atmosphere, the gradient of temperatures between high and low altitude regions will be reduced. This means that generally, global wind circulation is reduced in a planet facing climate change (this global wind circulation organized at the size of the Earth is visible at this link: https://svs.gsfc.nasa.gov/4240#media_group_20553). But of course, the convective wind circulation is increased in that case.

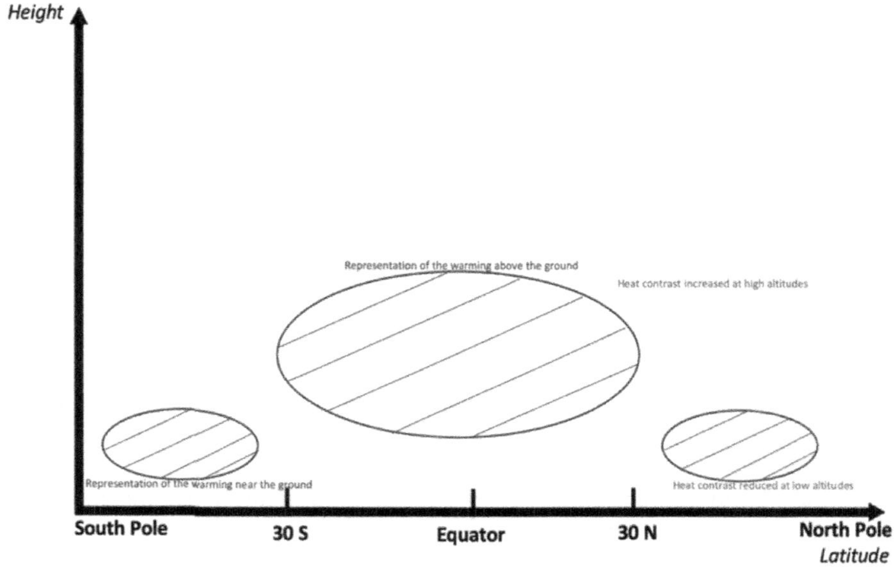

Figure A.12. *Scheme of heat contrast with an increased warming, with height represented on the y-axis, and latitude on the x-axis*

This result that may seem counterintuitive is a constant in all models: the global wind circulation is reduced, and the convective circulation is increased. This has been a problem when analyzing, for example, what happened with different storms on the western coast of France during the last century change and a few years later? There have been many discussions about whether these storms originated from climate change or not. We may imagine that there are other mechanisms that will change storm tracks and intensity in the Northern Atlantic for instance, the limit of sea ice which is changing. Therefore, many have tried to argue about the fact that climate change might anyway bring different situations with stronger storms in the Northern Atlantic. It is not proven yet but one idea is that the sea ice extent has changed. Therefore, the circulation is changed. But the answer to this problem is by no way an easy one. It was demonstrated that a change in global circulation is a phenomenon that tends to make storms weaker. So why may we have stronger storms in a warmer climate? Because there is also actually some convective way to develop those storms. We also know thanks to a few historians studying Arcanes (in Bordeaux, for example, in ancient texts in local French dialects), that several storms were reported/with no population along the storms dunes before the 20th century). This is a very important issue that must be treated also through dynamics to be satisfactory and not only by considering energetic transport aspects, on which we focused for this course book.

A.4.3. *Energy fluxes at the surface of the oceans (3 points)*

A.4.3.1. *Comparison of the two maps corresponding to heat transfer*

The maps in Figure A2.5 show heat transfers associated with longwave radiations, which are terrestrial radiation (bottom panel), and to direct heat change (top panel). One basic idea is that we know that the ocean takes more heat near the equator and we can also explain the context of heat exchange when air has been over a continent.

When there is dry air and cold air that goes to Siberia, then it is a situation with much heat given by the ocean to the atmosphere. After that, the air goes from West to East, and then reaches the Pacific Ocean. Then, there is a white color in this case, which means that in this unusual situation, the heat exchange goes from the ocean, which gives heat to the atmosphere because the atmosphere is extremely cold. This is the kind of longwave radiation that is linked to the cloud images we have seen before during this course.

References

AcclimaTerra (n.d.). [Online]. Available at: https://www.acclimaterra.fr/ [Accessed 2 December 2024].

Adler, R. (2000). Global precipitation climatology project, GPCP climatology (1988–1999). Database, University of Maryland, College Park.

Andrew, R.M. and Peters, G.P. (2024). The global carbon project's fossil CO_2 emissions dataset (2024v17). Dataset, Zenodo, Switzerland. doi: 10.5281/zenodo.13981696.

Ara Begum, R., Lempert, R., Ali, E., Benjaminsen, T.A., Bernauer, T., Cramer, W., Cui, X., Mach, K., Nagy, G., Stenseth, N.C. et al. (2022). Point of departure and key concepts. In *Climate Change 2022: Impacts, Adaptation, and Vulnerability. Contribution of Working Group II to the Sixth Assessment Report of the Intergovernmental Panel on Climate Change*, Pörtner, H.-O., Roberts, D.C., Tignor, M. et al. (eds). Cambridge University Press, Cambridge. doi: 10.1017/9781009325844.003.

Berger, A. and Loutre, M-F. (1991). Insolation values for the climate of the last 10 millions of years. *Quaternary Science Reviews*, 10(4), 297–317.

Berkeley Earth's Analysis (2024). Global temperature report for 2024 [Online]. Available at: https://berkeleyearth.org/global-temperature-report-for-2024/ [Accessed 2 December 2024].

Berthelot, M., Friedlingstein, P., Ciais, P., Monfray, P., Dufresne, J.L., Le Treut, H., Fairhead, L. (2002). Global response of the terrestrial biosphere to CO_2 and climate change using a coupled climate-carbon cycle model. *Global Biogeochemical Cycles*, 16(4), 31–1–31–15. doi: 10.1029/2001GB001827.

Bevacqua, E., Schleussner, C.F., Zscheischler, J. (2025). A year above 1.5 °C signals that Earth is most probably within the 20-year period that will reach the Paris Agreement limit. *Nat. Clim. Chang.*, 15, 262–265. doi: 10.1038/s41558-025-02246-9.

Bindoff, N.L., Willebrand, J., Artale, V., Cazenave, A., Gregory, J., Gulev, S., Hanawa, K., Le Quéré, C., Levitus, S., Nojiri, Y. et al. (2007) Observations: Oceanic climate change and sea level. In *Climate Change 2007: The Physical Science Basis. Contribution of Working Group I to the Fourth Assessment Report of the Intergovernmental Panel on Climate Change*, Solomon, S., Qin, D., Manning, M., et al. (eds). Cambridge University Press, Cambridge.

Birkmann, J., Liwenga, E., Pandey, R., Boyd, E., Djalante, R., Gemenne, F., Leal Filho, W., Pinho, P.F., Stringer, L., Wrathall, D. (2022). Poverty, livelihoods and sustainable development. In *Climate Change 2022: Impacts, Adaptation and Vulnerability. Contribution of Working Group II to the Sixth Assessment Report of the Intergovernmental Panel on Climate Change*, Pörtner, H.-O., Roberts, D.C., Tignor, M. et al. (eds). Cambridge University Press, Cambridge. doi: 10.1017/9781009325844.010.

Bony, S. and Dufresne, J.-L. (2005). Marine boundary layer clouds at the heart of tropical cloud feedback uncertainties in climate models. *Geophysical Research Letters*, 32(20). doi: 10.1029/2005GL023851.

Boucher, O. (1995). A study of aerosol-cloud-radiation interactions: Modelling and numerical simulation with a general circulation model. PhD Thesis, Sorbonne University, Paris.

Broeker, W.S. (1991). The great ocean conveyor. *Oceanography*, 4(2), 79–89. doi: 10.5670/oceanog.

Brunetti, M. and Prodi, F. (2015). The climate system. In *European Physical Journal Conferences*. EDP Sciences, Bologna. doi: 10.1051/epjconf/20159502001.

Canadell, J.G., Monteiro, P.M.S., Costa, M.H., Cotrim da Cunha, L., Cox, P.M., Eliseev, A.V., Henson, S., Ishii, M., Jaccard, S., Koven, C. et al. (2021). Global carbon and other biogeochemical cycles and feedbacks. In *Climate Change 2021: The Physical Science Basis, Contribution of Working Group I to the Sixth Assessment Report of the Intergovernmental Panel on Climate Change*, Masson-Delmotte, V., Zhai, P., Pirani, A. et al. (eds). Cambridge University Press, Cambridge. doi: 10.1017/9781009157896.007.

Capuano, T.A., Speich, S., Carton, X., Blanke, B. (2018). Mesoscale and submesoscale processes in the Southeast Atlantic and their impact on the regional thermohaline structure. *Journal of Geophysical Research: Oceans*, 123(3), 1937–1961. doi: 10.1002/2017JC013396.

Carbon Brief (n.d.). [Online]. Available at: https://www.carbonbrief.org/ [Accessed 2 December 2024].

Carissimo, B.C., Oort, A.H., Vonder Haar, T.H. (1985). Estimating the meridional energy transports in the atmosphere and ocean. *Journal of Physical Oceanography*, 15(1), 82–91.

Cassou, C. (2010). Euro-Atlantic regimes and their teleconnections. In *Seminar on Predictability in the European and Atlantic Regions from Days to Years*, Reading. [Online]. Available at: https://www.ecmwf.int/sites/default/files/elibrary/2012/73944-euro-atlantic-regimes-and-their-teleconnections_0.pdf/ [Accessed 2 December 2024].

Cess, R. (1978). Biosphere-albedo feedback and climate modeling. *Journal of Atmospheric Sciences*, 35(9), 1765–1767. doi: 10.1175/1520-0469(1978)035<1765:BAFACM>2.0.CO;2.

Charney, J. (1979). Carbon dioxide and climate: A scientific assessment. Report, National Research Council and The National Academies Press, Washington, D.C. doi: 10.17226/12181.

Ciais, P., Sabine, C., Bala, G., Bopp, L., Brovkin, V., Canadell, J., Chhabra, A., De Fries, R., Galloway, J., Heimann, M. et al. (2013). Carbon and other biogeochemical cycles. In *Climate Change 2013: The Physical Science Basis. Contribution of Working Group I to the Fifth Assessment Report of the Intergovernmental Panel on Climate Change*, Stocker, T.F., Qin, D., Plattner, G.-K. et al. (eds). Cambridge University Press, Cambridge. doi: 10.1017/CBO9781107415324.015.

CLand (n.d.). CLand Convergence Institute [Online]. Available at: https://cland.lsce.ipsl.fr/about-us/cland-convergence-institute. [Accessed 2 December 2024].

Climate Action Tracker (2024a). 2030 Emissions Gap: CAT projections and resulting emissions gap in meeting the 1.5°C Paris Agreement goal. November 2024 [Online]. Available at: https://climateactiontracker.org/global/cat-emissions-gaps/ [Accessed 2 December 2024].

Climate Action Tracker (2024b). The CAT thermometer. November 2024 [Online]. Available at: https://climateactiontracker.org/global/cat-thermometer.

CNES (2004). Illustration of the A-Train by P. Carril [Online]. Available at: https://cnes.fr/ [Accessed 2 December 2024].

CNES (2013). AVISO+ Satellite Altimetry Data [Online]. Available at: https://cnes.fr/ [Accessed 2 December 2024].

Collins, M., Knutti, R., Arblaster, J., Dufresne, J.-L., Fichefet, T., Friedlingstein, P., Gao, X., Gutowski, W.J., Johns, T., Krinner, G. (2013). Long-term climate change: Projections, commitments and irreversibility. In *Climate Change 2013: The Physical Science Basis. Contribution of Working Group I to the Fifth Assessment Report of the Intergovernmental Panel on Climate Change*, Stocker, T.F., Qin, D., Plattner, G.-K. et al. (eds). Cambridge University Press, Cambridge.

Cubasch, U., Wuebbles, D., Chen, D., Facchini, M.C., Frame, D., Mahowald, N., Winther, J.-G. (2013). Introduction. In *Climate Change 2013: The Physical Science Basis. Contribution of Working Group I to the Fifth Assessment Report of the Intergovernmental Panel on Climate Change*, Stocker, T.F., Qin, D., Plattner, G.-K. et al. (eds). Cambridge University Press, Cambridge.

Denman, K.L., Brasseur, G., Chidthaisong, A., Ciais, P., Cox, P.M., Dickinson, R.E., Hauglustaine, D., Heinze, C., Holland, E., Jacob, D. et al. (2007) Couplings between changes in the climate system and biogeochemistry. In *Climate Change 2007: The Physical Science Basis. Contribution of Working Group I to the Fourth Assessment Report of the Intergovernmental Panel on Climate Change*, Solomon, S., Qin, D., Manning, M. et al. (eds). Cambridge University Press, Cambridge.

Denton, G.H., Anderson, R.F., Toggweiler, J.R., Edwards, R.L., Schaefer, J.M., Putnam, A.E. (2010). The last glacial termination. *Science*, 328(5986), 1652–1656. doi: 10.1126/science.1184119.

Dlugokencky, E.J., Nisbet, E.G., Fisher, R., Lowry, D. (2011). Global atmospheric methane: Budget, changes and dangers. *Philos. Trans. A. Math Phys. Eng. Sci.*, 369(1943), 2058–2072. doi: 10.1098/rsta.2010.0341.

Edwards, M. and Trancik, J. (2014). Climate impacts of energy technologies depend on emissions timing. *Nature Climate Change*, 4, 347–352. doi: 10.1038/nclimate2204.

Emanuel, K.A. (1994). *Atmospheric Convection.* Oxford University Press, Oxford.

Emanuel, K.A. (2008). Hurricanes and global warming: Results from downscaling IPCC AR4 simulations. *Bulletin of the American Meteorological Society*, 89(3), 347–368.

European Space Agency (n.d.). [Online]. Available at: https://www.esa.int/ [Accessed 2 December 2024].

FAO (n.d.). [Online]. Available at: https://www.fao.org/home/fr. [Accessed 2 December 2024].

Fasullo, J.T. and Trenberth, K.E. (2008). The annual cycle of the energy budget. Part II: Meridional structures and poleward transports. *Journal of Climate, American Meteorological Society*, 2313–2325. doi: 10.1175/2007JCLI1936.

Friedlingstein, P., Jones, M.W., O'Sullivan, M., Andrew, R.M., Bakker, D.C.E., Hauck, J., Le Quéré, C., Peters, G.P., Peters, W., Pongratz, J. et al. (2021). Global carbon budget 2021. *Earth System Science Data*, 14(4), 1917–2005. doi: 10.5194/essd-14-1917-2022/.

Friedlingstein, P., O'Sullivan, M., Jones, M.W., Andrew, R.M., Bakker, D.C.E., Hauck, J., Landschützer, P., Le Quéré, C., Luijkx, I.T., Peters, G.P. et al. (2023). Global carbon budget 2023. *Earth System Science Data*, 15(12), 5301–5369. doi: 10.5194/essd-15-5301-2023.

Gastineau, G., Le Treut, H., Li, L. (2008). Hadley circulation changes under global warming conditions indicated by coupled climate models. *Tellus A*, 60(5), 863–884. doi: 10.1111/j.1600-0870.2008.00344.x.

Giorgi, F. and Lionello, P. (2008). Climate change projections for the Mediterranean region. *Global and Planetary Change*, 63(2–3). 90–104. doi: 10.1016/j.gloplacha.2007.09.005.

Global Carbon Budget (n.d.). The global carbon project [Online]. Available at: https://www.globalcarbonproject.org/ [Accessed 2 December 2024].

Gumley, L. (2005). Moderate resolution imaging spectroradiometer (MODIS) on NASA's terra satellite, image courtesy. Image, Space Science and Engineering Center, University of Wisconsin–Madison, Madison.

Gutiérrez Brizuela, N., Alford, M.H., Xie, S., Sprintall, J., Voet, G., Warner, S.J., Hughes, K., Moum, J.N. (2023). Prolonged thermocline warming by near-inertial internal waves in the wakes of tropical cyclones. *Proceedings of the National Academy of Sciences of the USA*, 120(26), e2301664120. doi: 10.1073/pnas.2301664120.

Halley, E. (1686). An account of the cause of the change of the variation of the magnetical needle with an hypothesis of the structure of the internal parts of the earth: As it was proposed to the Royal Society in one of their late meetings. *Philosophical Transactions*, 17(195). doi: 10.1098/rstl.1686.0107.

Hastenrath, S. (1982). On meridional heat transports in the world ocean. *Journal of Physical Oceanography*, 12(8), 922–927. doi: 10.1175/1520-0485(1982)012%3C0922:OMHTIT%3E2.0.CO;2.

Hays, J.D., Imbrie, J, Shackleton, N.J. (1976). Variations in the Earth's orbit: Pacemaker of the ice ages. *Science*, 194(4270), 1121–1132. doi: 10.1126/science.194.4270.1121.

Hilaire, J., Minx, J.C., Callaghan, M.W., Edmonds, J., Luderer, G., Nemet, G.F., Rogelj, J., del Mar Zamora, M. (2019). Negative emissions and international climate goals – Learning from and about mitigation scenarios. *Climatic Change*, 157, 189–219. doi: 10.1007/s10584-019-02516-4.

Hsu, P.C., Li, T., Murakami, H., Kitoh, A. (2013). Future change of the global monsoon revealed from 19 CMIP5 models. *Journal of Geophysical Research: Atmospheres*, 118(3), 1247–1260. doi: 10.1002/jgrd.50145.

Ide, K., Le Treut, H., Li, Z.-X., Ghil, M. (2001). Atmospheric radiative equilibria. Part II: Bimodal solutions for atmospheric optical properties. *Climate Dynamics*, 18, 29–49. doi: 10.1007/s003820100168.

IEA (2021). World energy balances. Study, International Energy Agency, Paris [Online]. Available at: https://iea.blob core.windows.net/assets/830fe099-5530-48f2-a7c1-11f35d510983/WorldEnergyOutlook2022.pdf [Accessed 2 December 2024].

IEA (2022). World energy outlook. Study, International Energy Agency, Paris [Online]. Available at: https://www.iea.org/data-and-statistics/data-product/world-energy-balances [Accessed 2 December 2024].

International Satellite Cloud Climatology Project (ISCCP) (n.d.). Satellite view of liquid water droplet with associated radius in microns [Online]. Available at: https://www.ncei.noaa.gov/products/climate-data-records/cloud-properties-isccp [Accessed 2 December 2024].

IPCC (2001a). *Climate Change 2001: The Scientific Basis. Contribution of Working Group I to the Third Assessment Report of the Intergovernmental Panel on Climate Change*, Houghton, J.T., Ding, Y., Griggs, D.J. et al. (eds). Cambridge University Press, Cambridge.

IPCC (2001b). *Climate Change 2001: Synthesis Report. A Contribution of Working Groups I, II, and III to the Third Assessment Report of the Integovernmental Panel on Climate Change*, Watson, R.T. and the Core Writing Team (eds). Cambridge University Press, Cambridge.

IPCC (2007a). Summary for policymakers. In *Climate Change 2007: Impacts, Adaptation and Vulnerability. Contribution of Working Group II to the Fourth Assessment Report of the Intergovernmental Panel on Climate Change*, Parry, M.L., Canziani, O.F., Palutikof, J.P. et al. (eds). Cambridge University Press, Cambridge.

IPCC (2007b). *Climate Change 2007: The Physical Science Basis. Contribution of Working Group I to the Fourth Assessment Report of the Intergovernmental Panel on Climate Change*, Solomon, S., Qin, D., Manning, M. et al. (eds). Cambridge University Press, Cambridge.

IPCC (2007c). *Climate Change 2007: Synthesis Report. Contribution of Working Groups I, II and III to the Fourth Assessment Report of the Intergovernmental Panel on Climate Change*. IPCC, Geneva.

IPCC (2012). Summary for policymakers. In *Managing the Risks of Extreme Events and Disasters to Advance Climate Change Adaptation*, Field, C.B., Barros, V., Stocker, T.F. et al. (eds). Cambridge University Press, Cambridge.

IPCC (2013). Summary for policymakers. In *Climate Change 2013: The Physical Science Basis. Contribution of Working Group I to the Fifth Assessment Report of the Intergovernmental Panel on Climate Change*, Stocker, T.F., Qin, D., Plattner, G.-K.M. et al. (eds). Cambridge University Press, Cambridge.

IPCC (2014a). Summary for policymakers. In *Climate Change 2014: Impacts, Adaptation, and Vulnerability. Part A: Global and Sectoral Aspects. Contribution of Working Group II to the Fifth Assessment Report of the Intergovernmental Panel on Climate Change*, Field, C.B., Barros, V.R., Dokken, D.J. et al. (eds). Cambridge University Press, Cambridge.

IPCC (2014b). *Climate Change 2014: Synthesis Report. Contribution of Working Groups I, II and III to the Fifth Assessment Report of the Intergovernmental Panel on Climate Change*, Core Writing Team, Pachauri, R.K., Meyer, L.A. (eds). IPCC, Geneva.

IPCC (2018). Global warming of 1.5°C. Special Report, IPCC, Geneva. doi: 10.1017/978100.

IPCC (2019a). Climate change and land: An IPCC special report on climate change, desertification, land degradation, sustainable land management, food security, and greenhouse gas fluxes in terrestrial ecosystems. Report, IPCC, Geneva.

IPCC (2019b). IPCC special report on the ocean and cryosphere in a changing climate. Report, IPCC, Geneva. doi: 10.1017/9781009157964.

IPCC (2021a). *Climate Change 2021: The Physical Science Basis. Contribution of Working Group I to the Sixth Assessment Report of the Intergovernmental Panel on Climate Change*, Masson-Delmotte, V., Zhai, P., Pirani, A. et al. (eds). Cambridge University Press, Cambridge. doi: 10.1017/9781009157896.

IPCC (2021b). Summary for policymakers. In *Climate Change 2021: The Physical Science Basis. Contribution of Working Group I to the Sixth Assessment Report of the Intergovernmental Panel on Climate Change*, Masson-Delmotte, V.P., Zhai, A., Pirani, S.L. et al. (eds). Cambridge University Press, Cambridge. doi: 10.1017/9781009157896.001.

IPCC (2021c). Technical summary. In *Climate Change 2021: The Physical Science Basis. Contribution of Working Group I to the Sixth Assessment Report of the Intergovernmental Panel on Climate Change*, Masson-Delmotte, V., Zhai, P., Pirani, A. et al. (eds). Cambridge University Press, Cambridge. doi: 10.1017/9781009157896.002.

IPCC (2022a). *Climate Change 2022: Impacts, Adaptation, and Vulnerability. Contribution of Working Group II to the Sixth Assessment Report of the Intergovernmental Panel on Climate Change*, Pörtner, H.-O., Roberts, D.C., Tignor, M. et al. (eds). Cambridge University Press, Cambridge. doi: 10.1017/9781009325844.

IPCC (2022b). *Climate Change 2022: Mitigation of Climate Change. Contribution of Working Group III to the Sixth Assessment Report of the Intergovernmental Panel on Climate Change*, Shukla, P.R. Skea, J., Slade, R. et al. (eds). Cambridge University Press, Cambridge. doi: 10.1017/9781009157926.

IPCC (2023a). *Climate Change 2023: Synthesis Report. Contribution of Working Groups I, II and III to the Sixth Assessment Report of the Intergovernmental Panel on Climate Change*, Core Writing Team, Lee, H., Romero, J. (eds). IPCC, Geneva. doi: 10.59327/IPCC/AR6-9789291691647.

IPCC (2023b). Summary for Policymakers. In *Climate Change 2023: Synthesis Report. Contribution of Working Groups I, II and III to the Sixth Assessment Report of the Intergovernmental Panel on Climate Change*, Core Writing Team, Lee, H., Romero, J. (eds). IPCC, Geneva. doi: 10.59327/IPCC/AR6-9789291691647.001.

IPSL (n.d.). [Online]. Available at: https://www.ipsl.fr/. [Accessed 2 December 2024].

Jouzel, J. (2018). Que mesure-t-on dans une carotte de glace ? Video, Canal-U. doi: 10.60527/5f9t-f318.

Jouzel, J. (2019). *Climats passés, climats futurs*. CNRS Editions, Paris.

Jouzel, J., Koster, R.D., Suozzo, R.J., Russell, G.L. (1994). Stable water isotope behavior during the last glacial maximum: A general circulation model analysis. *Journal of Geophysical Research*, 99(D12), 25791–25801. doi: 10.1029/94JD01819.

Keeling, C.D., Piper, S.C., Bacastow, R.B., Wahlen, M., Whorf, T. P., Heimann, M., Meijer, H.A. (2005). Atmospheric CO2 and 13CO2 exchange with the terrestrial biosphere and oceans from 1978 to 2000: Observations and carbon cycle implications. In *A History of Atmospheric CO2 and Its Effects on Plants, Animals, and Ecosystems*, Baldwin, I.T., Caldwell, M.M., Heldmaier, G. et al. (eds). Springer, New York.

Konsta, D., Dufresne, J.-L.D., Chepfer, H., Idelkadi, A., Cesana, G. (2016). Use of A-train satellite observations (CALIPSO-PARASOL) to evaluate tropical cloud properties in the LMDZ5 GCM. *Climate Dynamics*, 47(3), 1263–1284. doi: 10.1007/s00382-015-2900-y.

Kwiatkowski, L., Torres, O., Bopp, L., Aumont, O., Chamberlain, M., Christian, J.R., Dunne, J.P., Gehlen, M., Ilyina, T., John, J.G. et al. (2020). Twenty-first century ocean warming, acidification, deoxygenation, and upper-ocean nutrient and primary production decline from CMIP6 model projections. *Biogeosciences*, 17(13), 3439–3470. doi: 10.5194/bg-17-3439-2020.

Le Treut, H. (2013). *Les impacts du changement climatique en Aquitaine : un état des lieux scientifique*. Presses Universitaires de Bordeaux, Pessac.

Le Treut, H. (2016). Oral presentation at Institut Pierre Simon Laplace, Paris.

Le Treut, H. (2018). Anticiper les changements climatiques en Nouvelle-Aquitaine. Pour agir dans les territoires. Report, Éditions Région Nouvelle-Aquitaine, Bordeaux.

Le Treut, H., Somerville, H.R., Cubasch, U., Ding, Y., Mauritzen, C., Mokssit, A., Peterson T., Prather, M. (2007). Historical overview of climate change. In *Climate Change 2007: The Physical Science Basis. Contribution of Working Group I to the Fourth Assessment Report of the Intergovernmental Panel on Climate Change*, Solomon, S., Qin, D., Manning, M. et al. (eds). Cambridge University Press, Cambridge.

Li, Z.X., Ide, K., Le Treut, H., Ghil, M. (1997). Atmospheric radiative equilibria in a simple column model. *Climate Dynamics*, 13, 429–440. doi: 10.1007/s003820050175.

Li, J.P., Wu, Z.W., Jiang, Z.H., He, J.H. (2010). Can global warming strengthen the East Asian summer monsoon? *Journal of Climate*, 23(24), 6696–6705.

Lindsey, R. (2009a). Climate and Earth's energy budget. *NASA Earth Observatory* [Online]. Available at: https://earthobservatory.nasa.gov/features/EnergyBalance.

Lindsey, R. (2009b). The atmosphere's energy budget. *NASA Earth Observatory* [Online]. Available at: https://earthobservatory.nasa.gov/features/EnergyBalance/page6.php.

Liu, Z., Deng, Z., Davis, S.J., Giron, C., Ciais, P. (2022). Monitoring global carbon emissions in 2021. *Nature Reviews Earth & Environment*, 3, 217–219. doi: 10.1038/s43017-022-00285-w.

Lloyd, R. (2005). *Global Warming: The Physical Chemistry*. Royal Society of Chemistry Environmental Chemistry Group, London.

Lorius, C., Jouzel, J., Ritz, C., Merlivat, L., Barkov, N.I., Korotkevich, Y.S., Kotlyakov, V.M. (1985). A 150,000-year climatic record from Antarctic ice. *Nature*, 316, 591–596. doi: 10.1038/316591a0.

McInnes, L. (2007). Monthly average Sunspot numbers observed since the 18th century using data from the Royal Observatory of Belgium. Solar Influences Data Analysis Center. Dataset, Royal Observatory of Belgium, Uccle [Online]. Available at: http://sidc.oma.be/DATA/monthssn.dat [Accessed 2 December 2024].

Meinshausen, M., Smith, S.J., Calvin, K.V., Daniel, J.S., Kainuma, M.L.T., Lamarque, J.-F., Matsumoto, K., Montzka, S.A.,. Raper, S.C.B., Riahi, K. et al. (2011). The RCP greenhouse gas concentrations and their extension from 1765 to 2300. *Climatic Change*, 109(1–2), 213. doi: 10.1007/s10584-011-0156-z.

Meteo France (n.d.) Le changement climatique [Online]. Available at: https://meteofrance.com/changement-climatique. [Accessed 2 December 2024].

Mostefaoui, M. (2024). 2 – A global view of radiative processes. Video, YouTube [Online]. Available at: https://www.youtube.com/watch?v=kMwlnUJ2Z10 [Accessed 2 December 2024].

Mostefaoui, M., Ciais, P., McGrath, M.J., Peylin, P., Patra, P.K., Ernst, Y. (2024). Greenhouse gas emissions and their trends over the last 3 decades across Africa. *Earth System Science Data*, 16(1), 245–275. doi: 10.5194/essd-16-245-2024.

NASA (n.d.a). The clouds and the earth's radiant energy system (CERES) project. NASA [Online]. Available at: https://ceres.larc.nasa.gov/ [Accessed 2 December 2024].

NASA (n.d.b). The Earth radiation budget experiment (ERBE). NASA [Online]. Available at: https://science.larc.nasa.gov/erbe/ [Accessed 2 December 2024].

NASA (n.d.c). Depiction of the global circulation of Earth's atmosphere. NASA [Online]. Available at: https://www.nasa.gov/stem-content/nasa-space-science-data-coordinatedarchive/#:~:text=The%20NASA%20Space%20Science%20Data,wealth%20of%20information%20and%20images [Accessed 2 December 2025].

NASA (1998). TRMM Data Swath Close-up View (TRMM Ride). Video, NASA [Online]. Available at: https://svs.gsfc.nasa.gov/175 [Accessed 2 December 2024].

NASA (2024). GISS surface temperature analysis (v4). Report, Goddard Institute for Space Studies, New York [Online]. Available at: https://data.giss.nasa.gov/gistemp/maps [Accessed 2 December 2024].

National Climatic Data Center (NCDC) (n.d.). Vostok Ice Core Temperature and CO_2 over the past 150,000 years. *Energy Matters* [Online]. Available at: www.ncdc.gov/paleo/icecore/antartica/vostok/vostok.html [Accessed 2 December 2024].

National Snow and Ice Data Center (NSIDC) (2025). Evolution of the Arctic ice–Northern Hemisphere extent anomalies. National Snow and Ice Data Center, University of Colorado, Boulder [Online]. Available at: https://nsidc.org/home [Accessed 2 December 2024].

NOAA (n.d.) [Online]. Available at: https://www.noaa.gov/ [Accessed 2 December 2024].

Oberhuber, J.M. (1988). Oberhuber's climatological atlas of heat fluxes and radiation budget over the global oceans. Report, NCAR, Boulder [Online]. Available at: https://rda.ucar.edu/datasets/ds541.0/.

Ohmura, A. and Raschke, E. (2005). Energy budget at the Earth's surface. In *Observed Global Climate: 6*, Hantel, M. (ed.). Springer, Berlin.

O'Neill, B.C., Kriegler, E., Riahi, K., Ebi, K.L., Hallegatte, S., Carter, T.R. Mathur, R., van Vuuren, D.P. (2014). A new scenario framework for climate change research: The concept of shared socioeconomic pathways. *Climatic Change*, 122, 387–400. doi: 10.1007/s10584-013-0905-2.

Our World in Data (2023). Annual CO_2 emissions by world region. Adapted from the Global Carbon Project. Dataset, Our World in Data, Oxford [Online]. Available at: https://ourworldindata.org/grapher/annual-co-emissions-by-region?tab=chart&country=International+aviation~International+shipping.

Peixoto, J.P. and Oort, A.H. (1992). *Physics of Climate*. American Institute of Physics, Melville.

Petrescu, A.M.R., Qiu, C., Ciais, P., Thompson, R.L., Peylin, P., McGrath, M.J., Solazzo, E., Janssens-Maenhout, G., Tubiello, F.N., Bergamaschi, P. et al. (2021). The consolidated European synthesis of CH4 and N2O emissions for the European Union and United Kingdom: 1990–2017. *Earth Syst. Sci. Data*, 13, 2307–2362. doi: 10.5194/essd-13-2307-2021.

Prinn, R.G., Weiss, R.F., Fraser, P.J., Simmonds, P.G., Cunnold, D.M., Alyea, F.N., O'Doherty, S., Salameh, P., Miller, B.R., Huang, J. et al. (2000). A history of chemically and radiatively important gases in air deduced from ALE/GAGE/AGAGE. *Journal of Geophysical Research*, 115(D14), 17751–17792.

Randall, D.A., Wood, R.A., Bony, S., Colman, R., Fichefet, T., Fyfe, J., Kattsov, V., Pitman, A., Shukla, J., Srinivasan, J. et al. (2007). Climate models and their evaluation. In *Climate Change 2007: The Physical Science Basis. Contribution of Working Group I to the Fourth Assessment Report of the Intergovernmental Panel on Climate Change*, Solomon, S., Qin, D., Manning, M. et al. (eds). Cambridge University Press, Cambridge.

Rhein, M., Rintoul, S.R., Aoki, S., Campos, E., Chambers, D., Feely, R.A., Gulev, S., Johnson, G.C., Josey, S.A., Kostianoy, A. et al. (2013). Observations: Ocean. In *Climate Change 2013: The Physical Science Basis. Contribution of Working Group I to the Fifth Assessment Report of the Intergovernmental Panel on Climate Change*, Stocker, T.F., Qin, D., Plattner, G.-K. et al. (eds). Cambridge University Press, Cambridge. doi: 10.1017/CBO9781107415324.010.

Ribas, I., Porto De Mello, G.F., Ferreira, L.D., Hébrard, E., Selsis, F., Catalán, S., Garcés, A., Do Nascimento Jr., J.D., De Medeiros, J.R. (2010). Evolution of the solar activity over time and effects on planetary atmospheres. *Astrophysical Journal*, 714(1). doi: 10.1088/0004-637X/714/1/384.

Ritchie, H., Rosado, P., Roser, M. (2023). CO_2 and greenhouse gas emissions. Dataset, Our World in Data, Oxford [Online]. Available at: https://ourworldindata.org/co2-and-greenhouse-gas-emissions.

Riveline, C. (1985). Essai sur le dur et le mou. *La Jaune et la Rouge*.

Rohde, R.A. and Hausfather, Z. (2020). The Berkeley earth land/ocean temperature record. *Earth System Science Data*, 12(4), 3469–3479. doi: 10.5194/essd-12-3469-2020.

Sadourny, R. and Bougeault, P. (2001). *Dynamique de l'atmosphère et de l'océan*. Éditions de l'École Polytechnique, Palaiseau.

Schipper, E.L.F., Revi, A., Preston, B.L., Carr, E.R., Eriksen, S.H., Fernandez-Carril, L.R., Glavovic, B., Hilmi, N.J.M., Ley, D., Mukerji, R. (2022). Climate-resilient development pathways supplementary material. In *Climate Change 2022: Impacts, Adaptation and Vulnerability. Contribution of Working Group II to the Sixth Assessment Report of the Intergovernmental Panel on Climate Change*, Pörtner, H.-O., Roberts, D.C., Tignor, M. et al. (eds). Cambridge University Press, Cambridge.

von Schuckmann, K., Minière, A., Gues, F., Cuesta-Valero, F.J., Kirchengast, G., Adusumilli, S., Straneo, F., Ablain, M., Allan, R.P., Barker, P.M. et al. (2023). Heat stored in the Earth system 1960–2020: Where does the energy go? *Earth System Science Data*, 15(4), 1675–1709. doi: 10.5194/essd-15-1675-2023.

Scotese, C.R. (1997.) Paleogeographic Atlas. PALEOMAP Progress Report 90-0497, University of Texas, Arlington.

Shaun, M., Shakun, J., Clark, P.U. (2013). A reconstruction of regional and global temperature for the past 11,300 years. *Science*, 339(6124), 1198–1201. doi: 10.1126/science.1228026.

Shindell, D., Kuylenstierna, J.C.I., Vignati, E., van Dingenen, R., Amann, M., Klimont, Z. Anenberg, S.C., Muller, N., Janssens-Maenhout, G., Raes, F. et al. (2012). Simultaneously mitigating near-term climate change and improving human health and food security. *Science*, 335(6065), 183–189. doi: 10.1126/science.1210026.

Siccha, M., Morard, R., Meilland, J., Iwasaki, S., Kucera, M., Kimoto, K. (2022). Collection of X-ray micro computed tomography images of shells of planktic foraminifera with curated taxonomy. Dataset, PANGAEA, University of Bremen, Bremen. doi: 10.1594/PANGAEA.949585.

Slingo, A., Pamment, J.A., Allan, R.P., Wilson P.S. (1999). Water vapor feedbacks in the ECMWF reanalyses and hadley centre climate model. *Journal of Climate*, 13(17), 3080–3098.

Speich, S. (2017). Oral presentation at Institut Pierre Simon Laplace, Paris.

Stocker, T.F., Qin, D., Plattner, G.-K., Alexander, L.V., Allen, S.K., Bindoff, N.L., Bréon, F.-M., Church, J.A., Cubasch, U., Emori S. et al. (2013). Technical Summary. In *Climate Change 2013: The Physical Science Basis. Contribution of Working Group I to the Fifth Assessment Report of the Intergovernmental Panel on Climate Change*, Stocker, T.F., Qin, D., Plattner, G.-K. et al. (eds). Cambridge University Press, Cambridge.

Strozzi, B. (1635). Eratosthenes Teaching in Alexandria, ca., oil on canvas, 78.9 x 99.4 cm. Artwork, Montreal Museum of Fine Arts, Montreal.

Su, Z., Wang, J., Klein, P. Thompson A.F., Menemenlis, D. (2018). Ocean submesoscales as a key component of the global heat budget. *Nature Communications*, 9, 775. doi: 10.1038/s41467-018-02983-w.

Sultan, B., Roudier, P., Quirion, P., Alhassane, A. (2013). Assessing climate change impacts on sorghum and millet yields in the Sudanian and Sahelian savannas of West Africa Assessing climate change impacts on sorghum and millet yields in the Sudanian and Sahelian savannas of West Africa *Environnemental Research Letters*, 8, 014040. doi: 10.1088/1748-9326/8/1/014040.

Szopa, S., Balkanski, Y., Schulz, M., Bekki, S., Cugnet, D., Fortems-Cheiney, A., Turquety, S., Cozic, A., Déandreis, C., Hauglustaine, D. et al. (2013). Aerosol and ozone changes as forcing for climate evolution between 1850 and 2100. *Climate Dynamics*, 40, 2223–2250. doi: 10.1007/s00382-012-1408-y.

Tempfli, K., Huurneman, G.C., Bakker, W., Janssen, L.L.F., Feringa, W.F., Gieske, A.S.M., Grabmaier, K.A., Hecker, C.A., Horn, J.A., Kerle, N. et al. (2009). *Principles of Remote Sensing: An Introductory Textbook*. ITC, Enschede.

Trenberth, K.E. and Caron, J.M. (2001). Estimates of meridional atmosphere and ocean heat transports. *Journal of Climate*, 14(16), 3433–3443. doi: 10.1175/1520-0442(2001)014%3C3433:EOMAAO%3E2.0.CO;2.

Trenberth, K.E. and Cheng, L. (2022). *Environmental Research: Climate*. IOP Publishing Ltd, Bristol. doi: 10.1088/2752-5295/ac6f74.

Trenberth, K. and Fasullo, J. (2011). Tracking earth's energy: From El Niño to global warming. *Surveys in Geophysics*, 33, 413–426. doi: 10.1007/s10712-011-9150-2.

Trenberth, K., Fasullo, J., Kiehl, J. (2009). Earth's global energy budget. *Bulletin of the American Meteorological Society*, 90(3), 311–324. doi: 10.1175/2008BAMS2634.1.

Vial, J., Dufresne, J.-L., Bony, S. (2013). On the interpretation of inter-model spread in CMIP5 climate sensitivity estimates. *Climate Dynamics*, 41, 3339–3362. doi: 10.1007%2Fs00382-013-1725-9.

Warner, E., Cook-Patton, S.C., Lewis, O.T., Brown, N., Koricheva, J., Eisenhauer, N., Ferlian, O., Gravel, D., Hall, J.S., Jactel, H. et al. (2023). Young mixed planted forests store more carbon than monocultures – A meta-analysis. *Frontiers in Forests and Global Change*, 6. doi: 10.3389/ffgc.2023.1226514.

World Radiation Center (n.d.). [Online]. Available at: http://www.pmodwrc.ch/pmod.php?topic=tsi/composite/SolarConstant, International [Accessed 2 December 2024].

Yin, Q.Z. and Berger, A. (2012). Individual contribution of insolation and CO_2 to the diversity of the interglacial climates of the past 800,000 years. *Climate Dynamics*, 38(3), 709–724. doi: 10.1007/s00382-011-1013-5.

Young, H.D. and Freedman, R.A. (2012). *University Physics with Modern Physics*. Pearson, San Francisco.

Young, P.J., Archibald, A.T., Bowman, K.W., Lamarque, J.-F., Naik, V., Stevenson, D.S., Tilmes, S., Voulgarakis, A., Wild, O., Bergmann, D. et al. (2013). Pre-industrial to end 21st century projections of tropospheric ozone from the Atmospheric Chemistry and Climate Model Intercomparison Project (ACCMIP). *Atmospheric Chemistry and Physics*, 13(4), 2063–2090. doi: 10.5194/acp-13-2063-2013.

List of Authors

Hervé LE TREUT
Académie des Sciences
Paris
France

Mounia MOSTEFAOUI
École Polytechnique
Paris
France

Index

R, S

T, U, V